Advanced Electronic Technologies and Systems Based on Low-Dimensional Quantum Devices

NATO ASI Series

Advanced Science Institutes Series

A Series presenting the results of activities sponsored by the NATO Science Committee, which aims at the dissemination of advanced scientific and technological knowledge, with a view to strengthening links between scientific communities.

The Series is published by an international board of publishers in conjunction with the NATO Scientific Affairs Division

A	**Life Sciences**	Plenum Publishing Corporation
B	**Physics**	London and New York
C	**Mathematical and Physical Sciences**	Kluwer Academic Publishers
D	**Behavioural and Social Sciences**	Dordrecht, Boston and London
E	**Applied Sciences**	
F	**Computer and Systems Sciences**	Springer-Verlag
G	**Ecological Sciences**	Berlin, Heidelberg, New York, London,
H	**Cell Biology**	Paris and Tokyo
I	**Global Environmental Change**	

PARTNERSHIP SUB-SERIES

1.	**Disarmament Technologies**	Kluwer Academic Publishers
2.	**Environment**	Springer-Verlag / Kluwer Academic Publishers
3.	**High Technology**	Kluwer Academic Publishers
4.	**Science and Technology Policy**	Kluwer Academic Publishers
5.	**Computer Networking**	Kluwer Academic Publishers

The Partnership Sub-Series incorporates activities undertaken in collaboration with NATO's Cooperation Partners, the countries of the CIS and Central and Eastern Europe, in Priority Areas of concern to those countries.

NATO-PCO-DATA BASE

The electronic index to the NATO ASI Series provides full bibliographical references (with keywords and/or abstracts) to more than 50000 contributions from international scientists published in all sections of the NATO ASI Series.
Access to the NATO-PCO-DATA BASE is possible in two ways:

– via online FILE 128 (NATO-PCO-DATA BASE) hosted by ESRIN, Via Galileo Galilei, I-00044 Frascati, Italy.

– via CD-ROM "NATO-PCO-DATA BASE" with user-friendly retrieval software in English, French and German (© WTV GmbH and DATAWARE Technologies Inc. 1989).

The CD-ROM can be ordered through any member of the Board of Publishers or through NATO-PCO, Overijse, Belgium.

3. High Technology – Vol. 42

Advanced Electronic Technologies and Systems Based on Low-Dimensional Quantum Devices

edited by

Minko Balkanski
Université Pierre et Marie Curie,
Paris, France

and

Nikolai Andreev
Technical University of Sofia,
Sofia, Bulgaria

Kluwer Academic Publishers

Dordrecht / Boston / London

Published in cooperation with NATO Scientific Affairs Division

Proceedings of the NATO Advanced Study Instute on
Advanced Electronic Technologies and Systems Based on Low-Dimensional Quantum Devices
Sozopol, Bulgaria
18–28 September 1996

A C.I.P. Catalogue record for this book is available from the Library of Congress.

ISBN 0-7923-4875-3

Published by Kluwer Academic Publishers,
P.O. Box 17, 3300 AA Dordrecht, The Netherlands.

Sold and distributed in the U.S.A. and Canada
by Kluwer Academic Publishers,
101 Philip Drive, Norwell, MA 02061, U.S.A.

In all other countries, sold and distributed
by Kluwer Academic Publishers,
P.O. Box 322, 3300 AH Dordrecht, The Netherlands.

Printed on acid-free paper

Printed in the Netherlands

TABLE OF CONTENTS

2. MQW Optoelectronic Devices and Systems

NATO ASI held in Sozopol, Bulgaria, September 18-28, 1996

ADVANCED ELECTRONIC TECHNOLOGIES AND SYSTEMS BASED ON LOW-DIMENSIONAL QUANTUM DEVICES

Sponsored by

NATO Scientific and Environmental Affairs Division

Ministère de l'Enseignement Supérieur et de la Recherche, France

Université Pierre et Marie Curie, Paris, France

University St. Kliment Ohridski, Sofia, Bulgaria

Technical University, Sofia, Bulgaria

Institut des Hautes Etudes pour le development de la Culture, de la Science et de la Technologie en Bulgarie, Paris, France.

Principal School Support

NATO Program for Priority on High Technology

Bulbank, Bulgarian Bank for Foreign Trade, Sofia, Bulgaria

Université Pierre et Marie Curie, Paris, France

PREFACE

This volume on Advanced Electronic Technologies and Systems based on Low-Dimensional Quantum Devices closes a three years series of NATO - ASI's.

The first year was focused on the fundamental properties and applications. The second year was devoted to Devices Based on Low-Dimensional Semiconductor Structures. The third year is covering Systems Based on Low-Dimensional Quantum Semiconductor Devices.

The three volumes containing the lectures given at the three successive NATO - ASI's constitute a complete review on the latest advances in semiconductor Science and Technology from the methods of fabrication of the quantum structures through the fundamental physics and basic knowledge of properties and projection of performances to the technology of devices and systems.

In the *first volume*: " Fabrication, Properties and Application of Low Dimensional Semiconductors " are described the practical ways in which quantum structures are produced, the present status of the technology, difficulties encountered, and advances to be expected. The basic theory of Quantum Wells, Double Quantum Wells and Superlattices is introduced and the fundamental aspects of their optical properties are presented. The effect of reduction of dimensionality on lattice dynamics of quantum structures is also discussed.

In the *second volume*: " Devices Based on Low Dimensional Structures " the fundamentals of quantum structures and devices in the two major fields: Electro-Optical Devices and Pseudomorphic High Electron Mobility Transistors are extensively discussed.

In the *third volume*: " Advanced Electronics Technology and Systems Based on Low - Dimensional Quantum devices ", which we present now, the major developments in Quantum Structures Systems are discussed in three main chapters:

Fundamentals on Quantum Structures for Electro-Optical Devices and Systems

MQW Optoelectronic Devices and Systems

Soliton-Based Switching, Gating and Transmission Systems

Fundamentals on Quantum Structures for Electro-Optical Devices and Systems

As an example of the recent development in basic research the first chapter is devoted to Electron State Symmetries and Optical Selection Rules in the (GaAs)m(AlAs)n Superlattices Grown Along the [001], [110], and [111] Directions. Using the method of induced band representations of space groups, the full electron state symmetries and the selection rules for optical transitions in the (GaAs)m(AlAs)n superlattices (SL's) are studied.

The (GaAs)m(AlAs)n [hkl] SL's are a new class of artificially grown crystals whose structure (i. e. a space group G and an arrangement of atoms over the Wyckoff positions in a primitive cell) depends on the growth direction [hkl] and numbers of monolayers (m, n) of constituent materials.

For each direction of growth, these SL's constitute several single crystal families specified by different space groups G_1, G_2, ... G_r. By definition, within each family, the crystals have the same space group G_i but differ from each other by an arrangement of atoms over the Wyckoff positions. Thus, from the crystallographic point of view, the SL's with different numbers of monolayers m and n are distinct crystals, even those belonging to the same family.

Such a dependence of the SL crystal structure on the numbers of monoloyers influences on the phonon and electron states in these crystals. To study the optical properties of SL's one should know the complete information on their crystal structure.

Knowledge of the optical properties are essential for modeling quantum well laser diode structures. Many of interactions of quantum wells in lasers derive from the properties of the density of states function of the two-dimensional electron system. The abrupt edge of the density of states as a function of energy provides a very high differential gain above transparency leading to significant reductions in threshold current in appropriately designed devices compared with their bulk counterparts. In all quantum well lasers the threshold current follows the intrinsic linear dependence over a low temperature region but as the temperature is increased an approximately exponential increase in threshold with temperature is superimposed on the linear behavior and this becomes dominant at sufficiently high temperature. The additional current above the intrinsic linear component is often referred as the " excess current ". Experiments have shown that this current is chiefly due to non-radiative recombination via deep states in the AlGaAs barrier material forming the core of the waveguide which contains a higher density of deep states than the GaAs material comprising the well. Because the carrier density in the barrier increases exponentially with temperature compared with that in the well, this excess current component has an exponential temperature dependence.

For visible emitting lasers it is often not sufficient to consider the active region of the quantum well in isolation from the rest of the device structure. Realistic estimate of the threshold current, particularly its temperature dependence should include current paths in other parts of the device structure.

Examine of possible current paths and different assumptions on the physical parameters leads to the conclusion that the best approach to modeling the current through a laser is a self-consistent simulation of the current flow and potential by numerical solution of the current continuity equations and Poisson's equation throughout the complete structure, together with the solution of Schrödinger's equation for a non-square well. Such simulation for 670 nm GaInP lasers are presented in detail in this volume. The results give a good description of the temperature dependence of threshold current and provide tutorial illustrations of the inadequacies of the simple flat - band model.

An other important feature for the systems based on quantum structure devices are the Microcavity Semiconductor Lasers. In the chapter on Microcavity Semiconductor Lasers the theory and recent experimental developments in such lasers and their implications are reviewed. Particular attention is paid to microdisk lasers which support whispering gallery modes. InGaAsP/InP microdisks have recently been pumped optically, resulting in the fist achievement of CW room temperature lasing in these devices.

The future viability of optoelectronics as a mainstream technology in communications, computing, data storage and consumer products is contingent on developing efficient, flexible and controllable photonic emitters, detectors, filters, amplifiers, memory elements and logic devices. It is intriguing and exiting to consider utilizing the quantum nature of light itself in designing and studing photonic devices of all types. A critical part of this effort is the study of wavelength scale structures for control and selection of photon modes; structures such as microcavities and the closely related photonic bandgap materials.

A microcavity laser may be defined operationally as one whose cavity length is comparable to the emission wavelength, in at least one dimension. A simple example is the planar microcavity laser whose thin ($\approx\lambda$) active region is bounded by two parallel, highly reflecting mirrors. A related structure is the vertical cavity surface - emitting laser (VCSEL) in which a planar microcavity is modified (by implantation, oxidation, etching or optical pumping) so that the gain is of limited spatial extent: the cavity is only a few wavelengths long but much larger in lateral extent.

From a physical point of view, a microcavity may be considered as an atom or atom-like emitter inside a cavity. In a true microscopic system the dimensions are comparable to the size of the atom and are thus much smaller than the (optical) wavelength. In a macroscopic system the dimensions are much greater than the wavelength. A microcavity is therefore a mesoscopic system, in that its scale is intermediate between those of microscopic and macroscopic objects. Because of this scaling, it is often possible to adopt a semi-classical point of view, in which the active atom or atom-like species (in this case the coupled system consisting of an electron in the conduction band of the semiconductor and the corresponding hole in the valence band) is treated quantum mechanically and the electromagnetic field is considered classically. From an engineering or applied - physics perspectives, a microcavity may be considered as a filter or distributor of the radiation from the atom.

In the chapter on microcavity lasers the discussion is based on the microdisk laser, in which light undergoes total internal reflection along the perimeter of an isolated circular disk, where two-dimensional mode confinement is provided via so-called « whispering gallery modes ».

After a section which outlines some theoretical and fundamental considerations, particularly regarding the predicted effects of 1-D and 2-D microcavities on the density of photon modes. The fabrication of typical microdisk lasers, recent optical pumping experiments and the first demonstration of CW room temperature lasing semiconductor microdisk laser are described and discussed.

MQW Optoelectronic Devices and Systems

Optoelectronic devices are first discussed in a chapter on Exciton Absorption Saturation and Carrier Transport in quantum Well Semiconductors. Optical nonlinearities associated with excitonic absorption features in multiple quantum well (MQW) semiconductors offer a number of useful functions for optoelectronics devices. These include laser mode-locking elements, saturable elements for controlling the propagation of optical solutions in fibre transmission systems, all-optical directional coupler switches and self-electro-optic devices for communications, signal processing and computing. The operation and optimization of these devices rely on an understanding of the mechanisms which contribute to absorption saturation and the motion of optically generated electrons and holes in directions both parallel and perpendicular to the quantum wells. The chapter on Exciton Absorption Saturation and Carrier Transport in quantum Well Semiconductors reviews measurements of exiton absorption saturation mechanism and transport processes (in-well and cross-well) relevant to new optoelectronic electro-optic and nonlinear optical devices.

Resonant nonlinear and electro-optic interactions are particularly large in quantum wells at room temperature because of prominent excitonic features in their absorption spectra close to the band gap energy. Absorption coefficients at the peak of the exciton absorption can be in excess of $10^4\,cm^{-1}$ providing very efficient absorption in samples only a few microns thick. Optical excitation of excitons and free carriers can bleach these absorption features by a number of mechanisms including phase space filling, Coulomb screening and lifetime broadening. Pump-probe measurements using ultrashort pulses of laser light with different linear and circular polarizations can be used to identify the relative magnitudes of the various contributions.

After creation of the electron and hole pairs, a dynamical situation is produces whereby the bleaching will relax because of the motion and recombination processes for the free carriers in the sample. Drift and diffusion properties normally determine the manner in which semiconductor devices operate but many additional processes in MQW structures have to be considered if we wish to assess the ultimate performance limits of low dimensional devices. These processes include thermionic emission from, tunneling through, and trapping into the wells. The ability to control the motion of electrons and holes by designing the structures to make use of these processes gives opportunities for engineering devices with new and unique properties. Pump-probe and transient grating techniques can be used to monitor the motion and dynamics of the carriers on ultrashort time scales using excitonic saturation nonlinearities as the probe.

Soliton-Based Switching, Gating and Transmission Systems

A large chapter in this volume is devoted to the soliton - based logic gates and soliton transmission systems. This chapter is a glance into the future since solitons can be considered as candidates as "information bits" in telecommunication systems behind the year 2000.

The volume contains also a certain number of short original contributions showing the present status of the art in quantum systems.

M. BALKANSKI

N. ANDREEV

Acknowledgments

The NATO Advanced Study Institute on "Advanced Electronic Technologies and Systems based on Low-Dimensional Quantum Devices", held in Sozopol, Bulgaria, September 18-28, 1996, was made possible by an award from the Assistant Secretary General for Scientific and Environmental Affairs. We are particularly grateful to Dr. J.A. Rausell-Colom, Program Director for Priority Area and High technology, for his constant interest and helpful guidance during the preparation of the ASI.

Of great value for the present success of this school and the future development of a Center for Scientific Culture in Bulgaria is the personal involvement of Dip. Eng. Peter Kimenov, General Manager of Administration Department of Bulbank and that of Dichko Fotev, Administrator of the Sozopol Pochivna Basa of Bulbank.

We owe special thank to the rector of Sofia University, Professor Ivan Lalov, who not only supported very generously the whole process of the organization of the school, but also came to Sozopol to attest with his presence his personal interest in the development of scientific culture in Bulgaria. Many colleagues from Sofia University and from the Technical University have generously helped the organization of the school and we are grateful for their involvement.

We also wish to thanks Dr. Stoyan Tanev for giving so much of his energy and enthusiasm to the enterprise. Lucy Nedialkova has also generously helped the preparation of the book based on the lectures of the School.

ELECTRON STATE SYMMETRIES AND OPTICAL SELECTION RULES IN THE $(GaAs)_m(AlAs)_n$ SUPERLATTICES GROWN ALONG THE [001], [110], AND [111] DIRECTIONS

Yu.E.KITAEV*, A.G.PANFILOV*, and P.TRONC
ESPCI, Laboratoire d'Optique,
10 rue Vauquelin - 75005 PARIS - France
**Permanent address: A.F.Ioffe Physical-Technical Institute,*
Politekhnicheskaya 26 - 194021 St.PETERSBURG - Russia

R.A.EVARESTOV
St. Petersburg State University,
Universitetskaya nabierezhnaya 7/9 - 199034 St.PETERSBURG - Russia

Recently, using the method of induced band representations of space groups, the full electron state symmetries and the selection rules for optical transitions in the $(GaAs)_m(AlAs)_n$ superlattices (SL's) were studied. The [001], [110], and [111] growth directions have been considered [1,2]. Below, we give a review of the group-theory method used and the main results obtained.

1. Introduction

The $(GaAs)_m(AlAs)_n$ [*hkl*] SL's are a new class of artificially grown crystals whose structure (i.e. a space group G and an arrangement of atoms over the Wyckoff positions in a primitive cell) depends on the growth direction [*hkl*] and numbers of monolayers (m,n) of constituent materials.

For each direction of growth, these SL's constitute several single crystal families specified by different space groups $G_1, G_2, ..G_r$. By definition, within each family, the crystals have the same space group G_i but differ from each other by an arrangement of atoms over the Wyckoff positions. Thus, from the crystallographic point of view, the SL's with different numbers of monolayers m and n are distinct crystals, even those belonging to the same family.

Such a dependence of the SL crystal structure on the numbers of monolayers influences on the phonon and electron states in these crystals. The symmetry of phonon states and the corresponding infrared and Raman spectra selection rules have been investigated comprehensively for the $(GaAs)_m(AlAs)_n$ [001] SL's during the last years [3-7]. However, the symmetry of electron states and the corresponding optical selection rules in these SL's began to be analyzed only recently [1,2].

To study the optical properties of SL's one should know the complete information on their crystal structure. When analyzing the SL structure, we adopt an approximation that the atoms in SL's are on the sites of a zinc-blende lattice with lattice constant a being an average of the two lattice parameters of GaAs and AlAs. This can be done

M. Balkanski and N. Andreev (eds.),
Advanced Electronic Technologies and Systems Based on Low-Dimensional Quantum Devices, 1–50.

since the difference in the lattice constants of GaAs and AlAs is less than 0.2%. Taking this approximation into account the coordinates of all the atoms in the lattice are well defined. So one can determine both the space group and the atomic arrangement over the Wyckoff positions for a SL with arbitrary numbers of monolayers *m* and *n* .

It is worth to note that the $(GaAs)_1(AlAs)_1$ SL, i.e., the (1x1) SL in a short notation, grown along the [110] direction is the same as one grown along the [001] direction.

Among the $(GaAs)_m(AlAs)_n$ SL's, only the (1x1) [001/110]-grown SL can be considered as a rather equilibrium crystal structure (like GaAs and AlAs are) since the $Ga_{0.5}Al_{0.5}As$ alloy is ordered to the (1x1) SL under certain conditions of growth [8]. The other SL's can be called "non-equilibrium" crystals for they can not be grown from a solution or melting that is under equilibrium growth conditions. Note that the strong rang-order ordering and formation of superstructure have also been found in other III-V alloys (see [9] and refs. therein). This is of interest not only from the thermodynamic point of view but also in respect to many possible applications of short-period SL's such as avalanche devices and optical modulators for integrated optics [10].

Below, the electron state symmetries at the symmetry points of the BZ for the $(GaAs)_m(AlAs)_n$ SL's are analyzed. Next, the selection rules for optical transitions in these materials are derived. And at last, basing on our group-theory results and electronic structure calculations made by various authors, it is predicted which lines could be expected to be observed in polarized optical spectra.

2. Electron state symmetry

2.1. SUPERLATTICES GROWN ALONG THE [001] DIRECTION

The crystal structures of typical representatives of the two different families of $(GaAs)_m(AlAs)_n$ SL's grown along the [001] direction are presented in Figures 1 and 2. The two families correspond to even and odd values of $m+n$. The Figures 1 and 2 also show the BZ's for the corresponding space groups, D_{2d}^5 and D_{2d}^9. The formulae giving the atomic arrangements over the Wyckoff positions are presented in Table 1 (see also [1,3,4]). Here, the numbers preceding the chemical element symbols denote the number of such atoms at the Wyckoff positions shown in parentheses. Analyzing Table 1 one can see that each of the crystal families can be subdivided into several subfamilies specified by non-equivalent types of atomic arrangements over the Wyckoff positions.

For the D_{2d}^5 crystal family, there are four SL subfamilies corresponding to odd numbers of both GaAs and AlAs monolayers and two SL subfamilies with even *m* and *n*. For the four former subfamilies, high-symmetry sites (D_{2d} - 1a, 1c, and 1d) in the lattice are occupied by metal (Me = Ga, Al) atoms, whereas all As atoms are in low-symmetry Wyckoff positions (C_{2v} - 2e, 2f, and 2g). The difference between the four subfamilies is subtle and results in the variations of occupation numbers of Wyckoff positions with the same site symmetry by atoms of the same type. Thus, the period of atomic arrangements over the Wyckoff positions is equal to 4 for both *m* and *n*. For the two latter subfamilies, the Me and As atoms interchange their sites, that is high-symmetry positions are occupied by As atoms, whereas all the metal atoms arrange at low-symmetry sites. In this case, the period of atomic arrangements over the Wyckoff

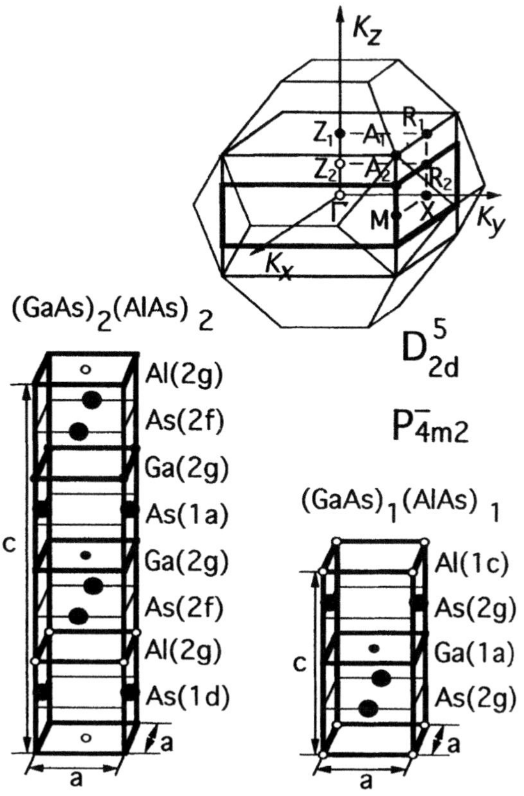

Figure 1. The crystal structures and corresponding Brillouin zones of $(GaAs)_2(AlAs)_2$ and $(GaAs)_1(AlAs)_1$ [001]-grown superlattices

positions equals 4 for *m+n*.

For the D_{2d}^9 crystal family, there are two subfamilies of SL's corresponding to either odd numbers of GaAs and even numbers of AlAs monolayers or *vice versa*. When *m* is odd and *n* is even, there is no Al atom in a D_{2d}-symmetry position, whereas none of Ga atoms is placed at a D_{2d}-symmetry site when *m* and *n* change the parities.
Knowing the space groups and atomic arrangements, one can determine the symmetry of electron states in the $(GaAs)_m(AlAs)_n$ SL's using the method of induced band representations of space groups. The main concepts of this method are given elsewhere [4,11-14]. It allows one to establish a symmetry correspondence between extended (Bloch) and localized (Wannier-type) one-electron states in crystals. The electron state symmetries for different SL's are presented in Tables 2 and 3. These Tables have the following structure. Columns 1-6 contain the atomic arrangements over the Wyckoff positions (sites in direct space) given in column 7 together with their coordinates (in units of translation vectors of the crystallographic unit cell) and site symmetry groups. Column 8 contains the Mulliken symbols of those irreducible representations (irreps)

TABLE 1. Atomic arrangements over the Wyckoff positions in $(GaAs)_m(AlAs)_n$ [001] SL's

Space group $D_{2d}^5(P\bar{4}m2)$, $m+n=2k$, $m \le n$

m=2k+1, n = 2s+1				m=2k, n=2s	
m=4i+1 n=4j+3	m=4i+3 n=4j+1	m=4i+1 n=4j+1	m=4i+3 n=4j+3	m+n=4i	m+n=4i+2
1Ga(1a) (m+n)As(2g)				mGa(2g) nAl(2g) 1As(1a)	
1Al(1d)		1Al(1c)			
$\frac{m-1}{2}$Ga(2f)	$\frac{m+1}{2}$Ga(2f)	$\frac{m-1}{2}$Ga(2f)	$\frac{m+1}{2}$Ga(2f)	1As(1d)	1As(1c)
$\frac{m-1}{2}$Ga(2e)	$\frac{m-3}{2}$Ga(2e)	$\frac{m-1}{2}$Ga(2e)	$\frac{m-3}{2}$Ga(2e)	$\frac{m+n}{2}$As(2f)	$(\frac{m+n}{2}-1)$As(2f)
$\frac{n+1}{2}$Al(2f)	$\frac{n-1}{2}$Al(2f)	$\frac{n-1}{2}$Al(2f)	$\frac{n-3}{2}$Al(2f)	$(\frac{m+n}{2}-2)$As(2e)	$(\frac{m+n}{2}-1)$As(2e)
$\frac{n-3}{2}$Al(2e)	$\frac{n-1}{2}$Al(2e)	$\frac{n-1}{2}$Al(2e)	$\frac{n+1}{2}$Al(2e)		

Space group $D_{2d}^9(I\bar{4}m2)$, $m+n=2k+1$

m=2k+1, n=2s	m=2k, n=2s+1
1Ga(1a) 1As(1c) nAl(2e)	1As(1a) 1Al(1c) mGa(2f)
(m-1)Ga(2e) (m+n-1)As(2f)	(n-1)Al(2f) (m+n-1)As(2e)

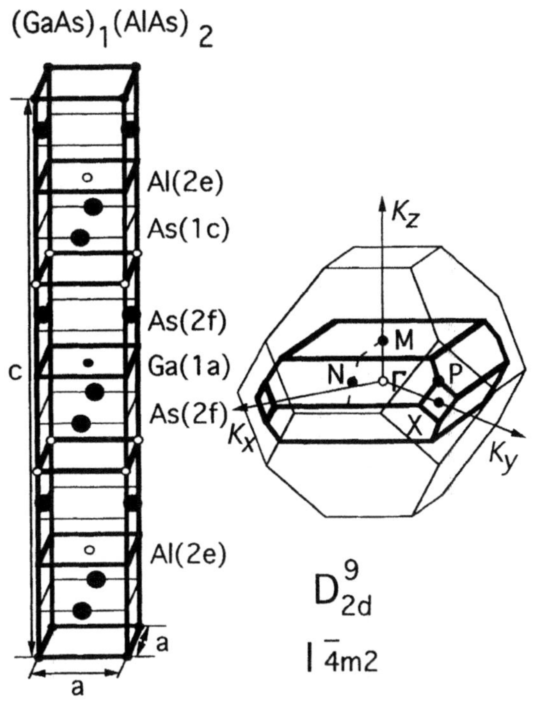

Figure 2. The crystal structure and corresponding Brillouin zones of $(GaAs)_1(AlAs)_2$ [001]-grown superlattice

of the site symmetry groups for these Wyckoff positions, according to which the localized one-electron wave functions transform, as well as the symbols of double-valued irreps (denoted by a bar over the irrep symbol) in the case where the spin-orbit interaction is taken into account.

We do not specify the explicit form of the localized functions. Their spatial distributions within the primitive cell remain undefined in terms of the group theory. In the symmetry analysis, only transformation properties of these functions are important, being described by the irreps of the site symmetry group of corresponding atoms. It should be noted that though the Wannier-type orbitals may differ from the corresponding atomic orbitals (*s, p, d* etc.) they transform according to the same irreps of a site symmetry group.

The remaining columns of Tables 2 and 3 give the labels of single- and double-valued induced representations in the **k**-basis, with the symbols of **k**-points (wave vectors), their coordinates (in units of primitive translations of the reciprocal lattice) and their point groups in rows 1 - 3 respectively, and the indices of small irreps of little groups in subsequent rows; these determine the symmetries of electron band states (Bloch states). In these Tables and in each of the following ones, the labelling of the space group irreps is that of [15], the labelling of the point group irreps is that of [16], and the site points **q** are indexed as Wyckoff positions from [17].

The expansion is limited to the *s* and *p* atomic-like localized states (one *s*-state and three *p*-states per atom) since just these functions form the uppermost valence band states and the lowest conduction band states and, therefore, determine the interband

TABLE 2. Electron state symmetries in $(GaAs)_m(AlAs)_n$ [001] SL's with the space group D^5_{2d} ($P\bar{4}m2$)

m=1 n=1	m=2 n=2	m=1 n=3	m=1 n=5	m=2 n=4	m=3 n=3	D^5_{2d} ($P\bar{4}m2$)		Γ (000) D_{2d}	M $(\frac{1}{2}\frac{1}{2}0)$ D_{2d}	A $(\frac{1}{2}\frac{1}{2}\frac{1}{2})$ D_{2d}	Z $(00\frac{1}{2})$ D_{2d}	X $(0\frac{1}{2}0)$ C_{2v}	R $(0\frac{1}{2}\frac{1}{2})$ C_{2v}
1Ga	1As	1Ga	1Ga	1As	1Ga		a_1(s)	1	1	1	1	1	1
							$b_2(p_z)$	2	2	2	2	1	1
						1a	$e(p_x,p_y)$	5	5	5	5	3,4	3,4
						(000)	$\bar{e}_1$	6	6	6	6	5	5
						D_{2d}	$\bar{e}_2$	7	7	7	7	5	5
1Al	-	-	1Al	1As	1Al		a_1(s)	1	4	3	2	3	3
							$b_2(p_z)$	2	3	4	1	3	3
						1c	$e(p_x,p_y)$	5	5	5	5	1,2	1,2
						$(\frac{1}{2}\frac{1}{2}\frac{1}{2})$	$\bar{e}_1$	6	7	6	7	5	5
						D2d	$\bar{e}_2$	7	6	7	6	5	5
-	1As	1Al	-	-	-		a_1(s)	1	1	2	2	1	1
							$b_2(p_z)$	2	2	1	1	1	1
						1d	$e(p_x,p_y)$	5	5	5	5	3,4	3,4
						$(00\frac{1}{2})$	$\bar{e}_1$	6	6	7	7	5	5
						D2d	$\bar{e}_2$	7	7	6	6	5	5
-	-	-	2Al	2As	2Al		$a_1(s;p_z)$	1,2	1,2	1,2	1,2	1,1	1,1
						2e	$b_2(p_y)$	5	5	5	5	3,4	3,4
						(00z)	$b_1(p_x)$	5	5	5	5	3,4	3,4
						C_{2v}	$\bar{e}$	6,7	6,7	6,7	6,7	5,5	5,5
-	2As	2Al	2Al	2As	2Ga		$a_1(s;p_z)$	1,2	3,4	3,4	1,2	3,3	3,3
						2f	$b_2(p_y)$	5	5	5	5	1,2	1,2
						$(\frac{1}{2}\frac{1}{2}z)$	$b_1(p_x)$	5	5	5	5	1,2	1,2
						C_{2v}	$\bar{e}$	6,7	6,7	6,7	6,7	5,5	5,5
2As	2Al	4As	6As	4Al	6As		$a_1(s;p_z)$	1,2	5	5	1,2	1,3	1,3
	2Ga			2Ga		2g	$b_2(p_y)$	5	1,2	1,2	5	1,4	1,4
						$(0\frac{1}{2}z)$	$b_1(p_x)$	5	3,4	3,4	5	2,3	2,3
						C_{2v}	$\bar{e}$	6,7	6,7	6,7	6,7	5,5	5,5

TABLE 3. Electron state symmetries in $(GaAs)_m(AlAs)_n$ [001] SL's with the space group D_{2d}^9 ($I\bar{4}m2$)#

m=1 n=2	m=2 n=3	m=1 n=4	m=1 n=6	m=2 n=5	m=3 n=6	D_{2d}^9 ($I\bar{4}m2$)		Γ (000) D_{2d}	M $(\frac{1}{2}\frac{1}{2}\text{-}\frac{1}{2})$ D_{2d}	X $(00\frac{1}{2})$ D_2	P $(\frac{1}{4}\frac{1}{4}\frac{1}{4})$ S_4	N $(0\frac{1}{2}0)$ C_s
1Ga	1As	1Ga	1Ga	1As	1Ga		$a_1(s)$	1	1	1	1	1
							$b_2(p_z)$	2	2	2	2	1
						1a	$e(p_x,p_y)$	5	5	3,4	3,4	1,2
						(000)	$\bar{e}_1$	6	6	5	5,7	3,4
						D_{2d}	$\bar{e}_2$	7	7	5	6,8	3,4
1As	1Al	1As	1As	1Al	1As		$a_1(s)$	1	2	3	4	1
							$b_2(p_z)$	2	1	4	3	1
						1c	$e(p_x,p_y)$	5	5	1.2	1.2	1.2
						$(0\frac{1}{2}\frac{1}{4})$	$\bar{e}_1$	6	7	5	6,7	3,4
						D_{2d}	$\bar{e}_2$	7	6	5	5,8	3,4
2Al	4As	4Al	6Al	6As	4Al		$a_1(s;p_z)$	1,2	1,2	1,2	1,2	1,1
					2Ga	2e	$b_2(p_y)$	5	5	3,4	3,4	1,2
						(00z)	$b_1(p_x)$	5	5	3,4	3,4	1,2
						C_{2v}	$\bar{e}$	6,7	6,7	5,5	5,6,7,8	3,3,4,4
2As	2Al	4As	6As	4Al	6As		$a_1(s;p_z)$	1,2	1,2	3,4	3,4	1,1
	2Ga			2Ga		2f	$b_2(p_y)$	5	5	1,2	1,2	1,2
						$(0\frac{1}{2}z)$	$b_1(p_x)$	5	5	1.2	1,2	1,2
						C_{2v}	$\bar{e}$	6,7	6,7	5,5	5,6,7,8	3,3,4,4

#Co-rep: N_3+N_4

optical transitions. Nevertheless, to describe the lower valence-band states and upper conduction-band ones, the *d* states should be also taken into consideration. This could be done using a similar procedure.

From Tables 2 and 3 one can easily write down the symmetry of the electron states at symmetry points of the SLBZ and determine which localized states and which atoms in a primitive cell contribute to them.

For example, in the $(GaAs)_1(AlAs)_1$ SL (in a short notation, the (1x1) SL), the *s*-atomic-like Wannier-type orbital corresponding to the Al atom in 1c Wyckoff position (shortly, *s*-orbital of Al) induces non-degenerate Bloch states Γ_1, M_4, A_3, Z_2, X_3, R_3 (see Table 2). Thus, 1 in the column Γ means small irrep Γ_1, 4 in the column M means M_4, etc.. We can also see that Γ_1 states are induced by *s*-orbitals of Ga and Al as well as by *s*- and p_z-orbitals of As, Γ_2 states are formed by p_z-orbitals of Ga and Al and by *s*- and p_z-orbitals of As and, at last, Γ_5-states are formed by p_x- and p_y-orbitals of Ga, Al, and As atoms. Thus, at the Γ point we have $4\Gamma_1$, $4\Gamma_2$, and $4\Gamma_5$ Bloch states resulting from *s* and *p* localized states (Ga, Al, and As atoms). We can conclude, for example, that the Bloch states induced by p_x- and p_y-orbitals of Ga, Al, and As atoms do not mix at the Γ point with the states induced by *s*- and p_z-orbitals since they have different symmetries. These results could significantly simplify the numerical calculations of electron-band structures of these materials since for each particular band state we could limit the number of Wannier-type orbitals that should be taken into account. This is especially important if the number of atoms per primitive cell is very large. And that is the case for most SL's. For example, for the (1x1) SL one can see that only *s*-orbitals of Al and p_x-orbitals of As contribute to the states M_4. Our group-theory results agree with the numerical calculations of contributions of different orbitals into SL band states made in [18] (see Table IV therein). For example, for the M_4-state with energy of 3.14 eV the following relative contributions of different atomic orbitals (i.e. the percentage of *s* and *p* states within the atomic spheres for each type of atom) were obtained in [18]: Ga(*s* - 0; *p* - 0), Al(*s* - 19.9; *p* - 0), As(*s* - 0; *p* -7.3). This agrees with the symmetry-analysis results given in Table 2. Note that the authors of [18] do not distinguish the contributions of different *p* states as we do.

Analyzing the results given in Tables 2 and 3, one can formulate in terms of site symmetry the definition of type-I and type-II SL's. The difference between these two types is the following: in the type I SL's, the crystalline orbitals corresponding to the top of the valence band and to the bottom of conduction band should include orbitals of atoms belonging to the layers of the same type (GaAs or AlAs). In the type II SL's, these crystalline orbitals should not. In the present description, the As atoms lying at the interfaces have obviously to be considered as belonging to both types of layers. The site symmetry method therefore allows one to establish qualitatively localizations of different crystalline orbitals.

Below, we speak about localization in the sense that the atoms from layers of only one type (GaAs or AlAs) contribute to a particular crystalline state. For example, for the (1x1) SL (see Table 2), the Bloch states Γ_1, M_1, A_1, Z_1, X_1, and R_1 induced by s-orbitals of Ga atoms are periodically localized in GaAs regions whereas the Bloch states Γ_1, M_4, A_3, Z_2 ,X_3, and R_3 induced by s-orbitals of Al atoms are periodically localized in AlAs regions. For the (2x2) SL, the Bloch states Γ_1, M_1, A_1, Z_1, X_1, and R_1, which are induced by As atoms lying between Ga planes, are also periodically localized in the GaAs regions though having another energy. Besides, the *s*- and p_z-

states of Ga and Al atoms contribute to the composite Bloch states with the same symmetries (Γ_1,Γ_2), M_5, A_5, (Z_1,Z_2), (X_1,X_3), and (R_1,R_3).

In many papers the problem of Γ - X mixing is studied both theoretically and experimentally (for references, see e.g. [19-21]). When there is no external perturbation, two states Ψ_i and Ψ_j mix if they have the same energy and if their overlap integral does not equal zero: $<\Psi_i|\Psi_j> \neq 0$. There is therefore no mixing if the symmetries of states Ψ_i and Ψ_j are different.

In the reduced-zone approximation, one uses bulk wave-functions both in GaAs and AlAs slabs. In the case of the Γ - X_z mixing, the conduction-band wave-function has, respectively, the Γ symmetry in the GaAs slabs and the X_z symmetry in the AlAs ones. When both parts of the wave-function correspond to the same energy (the confinement energy being included), the SL potential introduced with interface discontinuities lifts this degeneracy if it is allowed from parity. The perturbed wave-functions are then linear combinations of the former Γ and X_z wave-functions and have no longer any defined symmetry [22]. In our site-symmetry analysis, the exact symmetries of the wave-functions are provided, and only a further-perturbation potential could mix the exact eigenfunctions. Such a potential could arise, for example, from impurities, defects, applied fields, strains, etc.

Finally, if we take the spin-orbit interaction into consideration, the *s, p*, etc. orbitals are replaced by the $|J, m_J>$ orbitals where J and m_J are the total angular momentum and its projection. As a result, the localized states will transform according to doubled-valued irreps of a site symmetry group whereas the Bloch states will transform according to the doubled-valued irreps of the space group induced by double-valued irreps of the site symmetry group. For example, for the (1x1) SL, the s-derived $|\frac{1}{2},\pm\frac{1}{2}>$ and p-derived $|\frac{3}{2},\pm\frac{3}{2}>$ states localized at the Ga positions transform according to double-valued irrep $\bar{e}_1$ of the site symmetry group D_{2d} whereas p-derived ones $|\frac{3}{2}\pm\frac{1}{2}>$ and $|\frac{1}{2}\pm\frac{1}{2}>$ transform according to $\bar{e}_2$. The corresponding double-valued induced representations are given in Tables 2 and 3. As a result, the $4\Gamma_1$+$4\Gamma_2$+$4\Gamma_5$ states will become the $8\Gamma_6$+$8\Gamma_7$ ones.

2.2. SUPERLATTICES GROWN ALONG THE [110] AND [111] DIRECTIONS

The crystal structures of typical representatives of different crystal families are presented in Figure 3-7. It should be noted that, to keep the standard settings of space groups for the [110]-grown SL's, the y-axis is chosen to be the growth direction in contrast to [001]-grown and [111]-grown SL's where it is the z-axis. Every $(GaAs)_m(AlAs)_n$ SL grown along the [110] direction has an orthorhombic structure, except for the tetragonal (1x1) SL . In the case of even $m+n$, the SL's have simple lattices, whereas the lattices are body-centered when $m+n$ is odd. The symmetry of the SL's grown along the [111] direction is rhombohedral, the crystal lattice being hexagonal when $m+n = 3k$ and trigonal when $m+n \neq 3k$.

The space groups and formulae giving the atomic arrangements over the Wyckoff positions in the $(GaAs)_m(AlAs)_n$ SL's grown along the [110] and [111] direction are summarized in Tables 4 and 5, respectively. Analyzing Tables 4 and 5 one can see

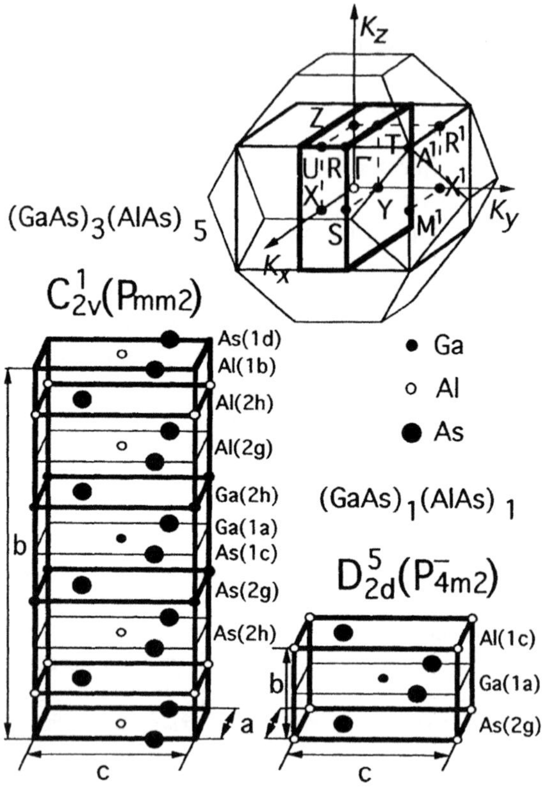

Figure 3. The crystal structures of the $(GaAs)_3(AlAs)_5$ and $(GaAs)_1(AlAs)_1$ [110]-grown superlattices and corresponding BZ's drawn by thick and thin lines, respectively. The corresponding points of the BZ of the tetragonal $(GaAs)_1(AlAs)_1$ [001] SL being isomorphic to $(GaAs)_1(AlAs)_1$ [110] are marked by superscript 1.

that there are several crystal families specified by different space groups, namely C^1_{2v}, C^7_{2v}, and C^{20}_{2v} among the SL's grown along the [110] direction as well as C^1_{3v} and C^5_{3v} among those grown along the [111] direction. In turn, some of crystal families can be subdivided into several subfamilies specified by non-equivalent types of atomic arrangements over the Wyckoff positions. For example, within the C^1_{2v} crystal family, there are four SL subfamilies corresponding to odd numbers of both GaAs and AlAs monolayers. The period of atomic arrangements over the Wyckoff positions is equal to 4 for both m and n in this case. The difference between the subfamilies is subtle and results in the variations of occupation numbers of Wyckoff positions with the same site symmetry by atoms of the same type.

The superlattice BZ's (SLBZ's) of the five space groups are shown in Figures 3-7. The SLBZ's are embedded in the parent GaAs (AlAs) BZ in order to describe zone foldings. With increasing values of m and n, the SLBZ's are reduced in the k_y ([110]-grown

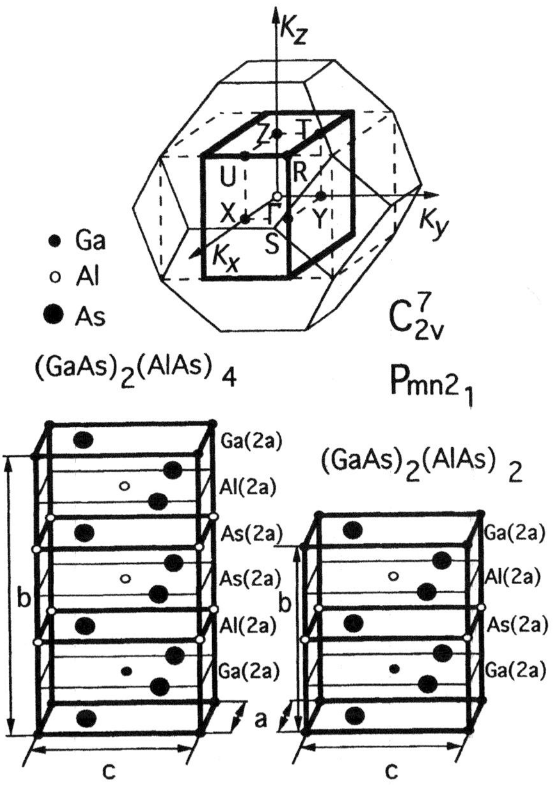

Figure 4. The crystal structures and corresponding BZ of the $(GaAs)_2(AlAs)_4$ and $(GaAs)_2(AlAs)_2$ [110]-grown superlattices.

SL's) or k_z ([111]-grown ones) direction. As a result, symmetry points of the parent BZ fold onto symmetry points of a SLBZ. Since, in both parent compounds GaAs and AlAs, the top of the valence band is located at the Γ point of the zinc-blende BZ whereas the bottom of the conduction band is located respectively at the Γ and X points, these are the most important points. Therefore, it is worth to determine which points of SLBZ they fold onto (the Γ point of the zinc-blende BZ naturally folds onto the Γ point of a SLBZ). Table 6 gives a description of the zone foldings including also the L point of the zinc-blende BZ whose energy in the conduction band is intermediate between those of the Γ and X points both in GaAs and AlAs.

According to the most of calculations (see [1] and references therein), the conduction-band minimum in the (1x1) [001]-grown SL is located at the R point of SLBZ. And it is a point which the point L of the GaAs(AlAs) BZ folds onto. In $(GaAs)_m(AlAs)_n$ [001]-grown SL's with $m=n$, one of the lowest conduction-band states is usually located at the M point, which the point X of the parent compounds' BZ folds onto.

In the case of $m+n = 2k+1$ for every SL grown in any direction under consideration, the symmetry points of AlAs(GaAs) - except for the Γ point - fold onto surface points of the SLBZ. On the contrary, when $m+n = 2k$, either X (for the [001] and [110] directions of growth) or L (for the [111] direction) fold onto the Γ point of the SLBZ

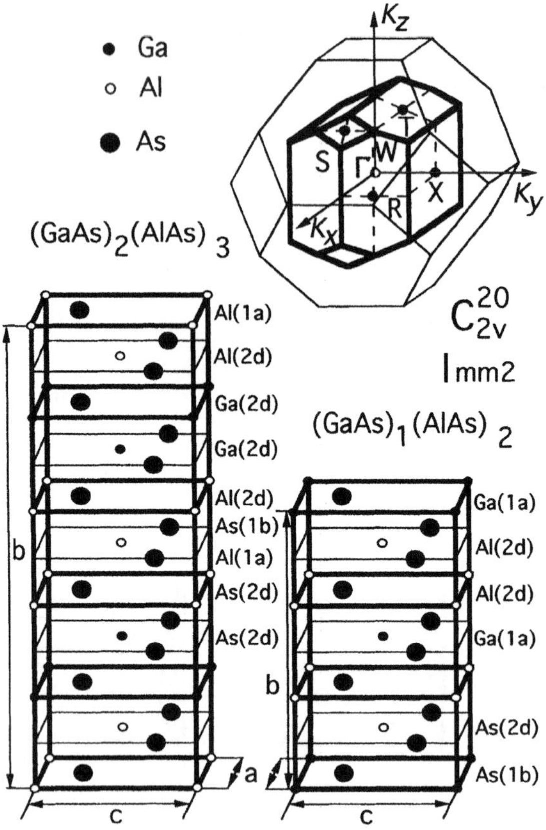

Figure 5. The crystal structures and corresponding BZ of the $(GaAs)_2(AlAs)_3$ and $(GaAs)_1(AlAs)_2$ [110]-grown superlattices

and there are chances, for some of these SL's, to exhibit a conduction-band minimum at the Γ point with a rather low energy.

Very interesting is the case of even-layered [111]-grown SL's. Here, the SLBZ points corresponding to the X point and a L-equivalent point of the GaAs BZ differ by some primitive reciprocal vectors of the SLBZ. That is to say, they fold onto the same point of the SLBZ, namely onto either the point F if $m+n = 6k \pm 2$ or onto the point M when $m+n = 6k$.

The space groups and atomic arrangements being established, we determine the symmetry of electron states. The electron state symmetries for different SL's are presented in Tables 7 to 11. The structure of Table 7 to 11 is the same as those of Tables 2,3 and differs from that of Tables 2,3 by a number of columns containing the atomic arrangements for typical representatives of each SL family (columns 1-4 in Table 7, columns 1-6 in Table 9, and columns 1-5 in Table 10, respectively). For the other SL families (Tables 8 and 11) all atoms in a primitive cell occupy the same Wyckoff position. That is why atomic arrangements are not given in the latter Tables.

TABLE 4. Atomic arrangements over the Wyckoff positions in $(GaAs)_m(AlAs)_n$ [110]-grown SL's.

Space group C^1_{2v}, (Pmm2), $m=2k+1$, $n=2l+1$; $m \leq n$			
m+n = 4i +2		m+n=4i	
m=4s+1 n=4t+1	m=4s+3 n=4t+3	m=4s+1 n=4t+3	m=4s+3 n=4t+1
1Ga(1a) 1As(1c)			
1As(1b) 1Al(1d) $(\frac{m+n}{2}-1)$As(2g) $(\frac{m+n}{2}-1)$As(2h)		1Al(1b) 1As(1d) $\frac{m+n}{2}$As(2g) $(\frac{m+n}{2}-2)$As(2h)	
$\frac{m-1}{2}$Ga(2g) $\frac{m-1}{2}$Ga(2h) $\frac{n-1}{2}$Al(2g) $\frac{n-1}{2}$Al(2h)	$\frac{m-3}{2}$Ga(2g) $\frac{m+1}{2}$Ga(2h) $\frac{n+1}{2}$Al(2g) $\frac{n-3}{2}$Al(2h)	$\frac{m-1}{2}$Ga(2g) $\frac{m-1}{2}$Ga(2h) $\frac{n-3}{2}$Al(2g) $\frac{n+1}{2}$Al(2h)	$\frac{m-3}{2}$Ga(2g) $\frac{m+1}{2}$Ga(2h) $\frac{n-1}{2}$Al(2g) $\frac{n-1}{2}$Al(2h)

Space group C^7_{2v}(Pmn2_1), $m=2k$, $n=2l$		
mGa(2a)	nAl(2a)	(m+n)As(2a)

Space group C^{20}_{2v} (Imm2), $m+n=2i+1$; $m \leq n$	
m=2s+1 n=2t	m=2s n=2t+1
1As(1b) (m+n-1)As(2d)	
1Ga(1a) (m-1)Ga(2d) nAl(2d)	1Al(1a) (n-1)Al(2d) mGa(2d)

TABLE 5. Atomic arrangements over the Wyckoff positions in $(GaAs)_m(AlAs)_n$ [111]-grown SL's

Space group C^1_{3v}(P3m1), *m+n=3i*					
m=3s, n=3t		m=3s+1, n=3t+2		m=3s+2, n=3t+1	
$\frac{m+n}{3}$As(1a), $\frac{m+n}{3}$As(1b), $\frac{m+n}{3}$As(1c)					
$\frac{m}{3}$Ga(1a)	$\frac{n}{3}$Al(1a)	$\frac{m+2}{3}$Ga(1a)	$\frac{n-2}{3}$Al(1a)	$\frac{m+1}{3}$Ga(1a)	$\frac{n-1}{3}$Al(1a)
$\frac{m}{3}$Ga(1b)	$\frac{n}{3}$Al(1b)	$\frac{m-1}{3}$Ga(1b)	$\frac{n+1}{3}$Al(1b)	$\frac{m+1}{3}$Ga(1b)	$\frac{n-1}{3}$Al(1b)
$\frac{m}{3}$Ga(1c)	$\frac{n}{3}$Al(1c)	$\frac{m-1}{3}$Ga(1c)	$\frac{n+1}{3}$Al(1c)	$\frac{m-2}{3}$Ga(1c)	$\frac{n+2}{3}$Al(1c)
Space group C^5_{3v}(R3m), *m+n≠3i*					
mGa(1a)		nAl(1a)		(m+n)As(1a)	

In Tables 7-11, the localized states of atoms which occupy low-symmetry positions which do not belong to the so-called Q-set [4] (a small set of points in the real space) induce Bloch states described by composite band representations. This means that they are direct sums of simple irreps, indices of which are presented in corresponding lines of Tables 7-11. The simple band representations [4,14] are induced by irreps of site symmetry groups of only a small set of points in the real space. The band states induced by Wannier-type orbitals corresponding to atoms occupying the Wyckoff positions which do not belong to the Q-set are composite, i.e. they are formed by a combination of simple band representations induced by orbitals of atoms occupying Wyckoff positions from the Q-set.

When the spin-orbit interaction is taken into consideration, the *s, p*, etc. localized orbitals are replaced by the $|J, m_J>$ orbitals. In the case of (1x2) [111]-grown SL taken as an example, the p-derived $|\frac{3}{2},\pm\frac{3}{2}>$ localized states of Ga will transform according to

TABLE 6. The foldings of the symmetry points Γ, X, and L in $(GaAs)_m(AlAs)_n$ SL's

compounds	m, n space group	parity	symmetry points		
GaAs (AlAs)	m+n=1 T_d^2		Γ	6X	8L
[001]-grown SL's	m+n=2i D_{2d}^5	i=2k	Γ	2Γ + 4M	4R + 4X
		i=2k+1	Γ	2Γ + 4M	8R
	m+n=2i+1 D_{2d}^9		Γ	2M + 4X	8N
[110]-grown SL's	m=n=1 D_{2d}^5		Γ	2Γ + 4M	4R + 4X
	m+n=2i C_{2v}^1, C_{2v}^7	i=2k	Γ	2Γ + 4X	4Z + 4U
		i=2k+1	Γ	2Γ + 4S	4T + 4U
	m+n=2i+1 C_{2v}^{20}		Γ	2X + 4R	4T + 4S
[111]-grown SL's	m+n=3i C_{3v}^1	m+n=2k	Γ	6M	2Γ + 6M
		m+n=2k+1	Γ	6M	2A + 6L
	m+n≠3i C_{3v}^5	m+n=2k	Γ	6F	2Γ + 6F
		m+n=2k+1	Γ	6F	2T + 6L

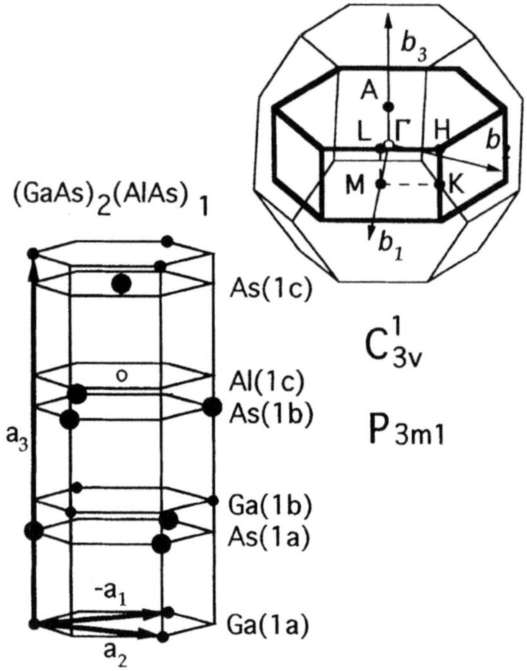

Figure 6. The crystal structure and corresponding BZ of the $(GaAs)_2(AlAs)_1$ [111]-grown superlattice

double-valued irreps $\bar{e}_1^{(1)}$, $\bar{e}_1^{(2)}$ whereas s-derived $|\frac{1}{2},\pm\frac{1}{2}>$, p-derived $|\frac{3}{2}\pm\frac{1}{2}>$, and $|\frac{1}{2}\pm\frac{1}{2}>$ ones will transform according to $\bar{e}_2$. The corresponding double-valued induced representations are also given in Tables 7 to 11. As a result, the $12\Gamma_1+6\Gamma_3$ states will become the $6\Gamma_4+6\Gamma_5+18\Gamma_6$ ones

There is a sole double-valued irrep (doubly degenerate) at every symmetry point for the group C_{2v}^1 (e.g. Γ_5 at the Γ point, cf. Table 7). There also exists only one double-valued irrep at points belonging to some symmetry lines of the SLBZ, e.g., the Λ_5 irrep for the symmetry line Γ - (Λ) - Z. On the contrary, for other symmetry lines in this SLBZ, there are two double-valued irreps (non-degenerated), which are complex-conjugated, e.g. Δ_3 and Δ_4 at points of the Γ - (Δ) - Y line. The complex-conjugated irreps forming a pair correspond to different states with the same energy. This degeneracy is connected with the inversion of time and can be lifted by applying the magnetic field which do not reduce the point symmetry of the system (that is one directed along the symmetry axis). The complex-conjugated irreps can be combined in so-called co-representations (co-reps), which are also called "physically irreducible representations" (doubly degenerate). The corresponding pairs of irreps forming co-reps are given below Tables 3, 8-11. On applying the magnetic field, the states described by

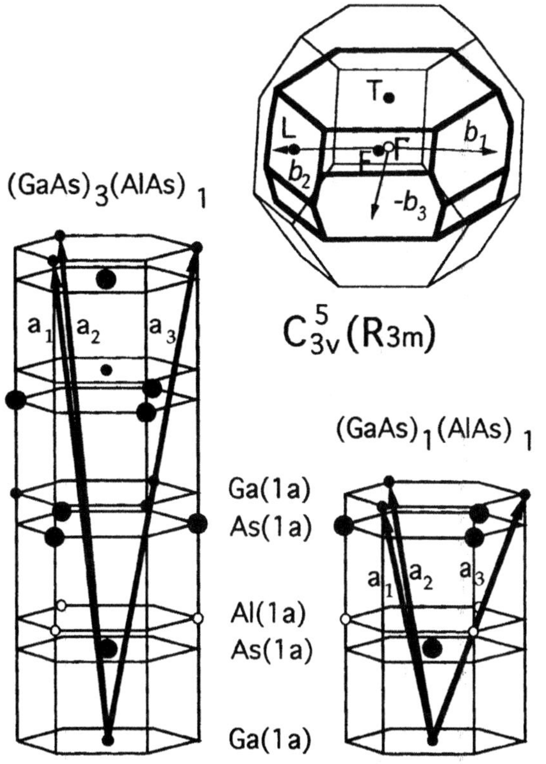

Figure 7. The crystal structures and corresponding BZ of the $(GaAs)_3(AlAs)_1$ and $(GaAs)_1(AlAs)_1$ [111]-grown superlattices

complex-conjugated irreps are split whereas the states described by doubly degenerate irreps are not.

A very similar situation takes place in the case of the groups C_{2v}^{20}, C_{2v}^{7}. There exists only a doubly degenerate irrep at some symmetry points and symmetry lines (e.g. Γ_5, Λ_5, etc.), whereas, at other symmetry points and lines, there are complex-conjugated irreps, which form corepresentations - e.g. (S_3, S_4), (Δ_3, Δ_4), etc.. So, the including of the spin-orbit interaction into consideration simplifies the band picture, reducing to one the number of representations describing electron state symmetries at a given point of the SLBZ of [110] -grown SL's.

More complicated is the case of the [111] -grown SL's. When the spin-orbit coupling is taken into account, even at the Γ point, there are both states that are described by doubly degenerate irreps, Γ_6, and states which can be described by corepresentations, (Γ_4, Γ_5). Only the latter will split on applying the magnetic field not reducing the C_{3v} symmetry. In addition, it is to be noticed that, for [111]-grown SL's, there are two doubly degenerate Γ_6 states, one being derived from the Γ_1 state and the other originated from the doubly degenerate Γ_3 one.

At lower symmetry points, the bands that are degenerated at symmetry points of SLBZ can split. For the SL's with the space groups C_{2v}^{1}, such splitting can exist -

TABLE 7. Electron state symmetries in $(GaAs)_m(AlAs)_n$ [110]-grown SL's with the space group C_{2v}^1

m=1 n=3	m=3 n=3	m=3 n=5	m=5 n=5	C_{2v}^1 (Pmm2)		Γ (000) C_{2v}	X $(\frac{1}{2}00)$ C_{2v}	Y $(0\frac{1}{2}0)$ C_{2v}	Z $(00\frac{1}{2})$ C_{2v}	S $(\frac{1}{2}\frac{1}{2}0)$ C_{2v}	T $(0\frac{1}{2}\frac{1}{2})$ C_{2v}	U $(\frac{1}{2}0\frac{1}{2})$ C_{2v}	R $(\frac{1}{2}\frac{1}{2}\frac{1}{2})$ C_{2v}
1Ga	1Ga	1Ga	1Ga		$a_1(s;p_z)$	1	1	1	1	1	1	1	1
				1a	$b_2(p_y)$	3	3	3	3	3	3	3	3
				(00z)	$b_1(p_x)$	4	4	4	4	4	4	4	4
				C_{2v}	$\bar{e}$	5	5	5	5	5	5	5	5
1Al	1As	1Al	1As		$a_1(s;p_z)$	1	1	3	1	3	3	1	3
				1b	$b_2(p_y)$	3	3	1	3	1	1	3	1
				$(0\frac{1}{2}z)$	$b_1(p_x)$	4	4	2	4	2	2	4	2
				C_{2v}	$\bar{e}$	5	5	5	5	5	5	5	5
1As	1As	1As	1As		$a_1(s;p_z)$	1	4	1	1	4	1	4	4
				1c	$b_2(p_y)$	3	2	3	3	2	3	2	2
				$(\frac{1}{2}0z)$	$b_1(p_x)$	4	1	4	4	1	4	1	1
				C_{2v}	$\bar{e}$	5	5	5	5	5	5	5	5
1As	1Al	1As	1Al		$a_1(s;p_z)$	1	4	3	1	2	3	4	2
				1d	$b_2(p_y)$	3	2	1	3	4	1	2	4
				$(\frac{1}{2}\frac{1}{2}z)$	$b_1(p_x)$	4	1	2	4	3	2	1	3
				C_{2v}	$\bar{e}$	5	5	5	5	5	5	5	5
			2Ga		$a'(s;p_y,p_z)$	1,3	1,3	1,3	1,3	1,3	1,3	1,3	1,3
	2Al	2Al	2Al	2g	$a''(p_x)$	2,4	2,4	2,4	2,4	2,4	2,4	2,4	2,4
2As	2As	4As	4As	(0yz)	$\bar{e}(1)$	5	5	5	5	5	5	5	5
				C_s	$\bar{e}(2)$	5	5	5	5	5	5	5	5
		2Ga	2Ga		$a'(s;p_y,p_z)$	1,3	2,4	1,3	1,3	2,4	1,3	2,4	2,4
	2Ga	2Al	2Al	2h	$a''(p_x)$	2,4	1,3	2,4	2,4	1,3	2,4	1,3	1,3
2Al	2As	2As	4As	$(\frac{1}{2}yz)$	$\bar{e}(1)$	5	5	5	5	5	5	5	5
				C_s	$\bar{e}(2)$	5	5	5	5	5	5	5	5

TABLE 8. Electron state symmetries in $(GaAs)_m(AlAs)_n$ [110]-grown SL's with the space group C_{2v}^7 #

		Γ	X	Y	Z	S	T	U	R
$C_{2v}^7(Pmn2_1)$		(000)	($\frac{1}{2}$00)	(0$\frac{1}{2}$0)	(00$\frac{1}{2}$)	($\frac{1}{2}\frac{1}{2}$0)	(0$\frac{1}{2}\frac{1}{2}$)	($\frac{1}{2}$0$\frac{1}{2}$)	($\frac{1}{2}\frac{1}{2}\frac{1}{2}$)
		C_{2v}	C_{2v}	C_{2v}	C_{2v}	C_{2v}	C_{2v}	C_{2v}	C_{2v}
2a	a'(s;p_y,p_z)	1,3	1	1,3	1,3	1	1,3	1	1
(0yz)	a''(p_x)	2,4	1	2,4	2,4	1	2,4	1	1
C_s	$\bar{e}$(1)	5	2,4	5	5	2,4	5	2,4	2,4
	$\bar{e}$(2)	5	3,5	5	5	3,5	5	3,5	3,5

#Co-reps: X_2+X_5; X_3+X_4; Z_1+Z_3; Z_2+Z_4; U_2+U_3; U_4+U_5
S_2+S_5; S_3+S_4; T_1+T_3; T_2+T_4; R_2+R_3; R_4+R_5

although only in the presence of the magnetic field - along the symmetry lines Γ - (Δ) - Y - (C) -S - (D) - X - (Σ) - Γ and Z - (B) -T - (E) - R - (P) -U -(A) - Z whereas it is absent along the Γ - (Λ) - Z, Y - (H) - T, X - (G) -U, and S - (Q) -R symmetry lines. Note that bands also remain degenerated along the symmetry line Γ - Z in the other [110]- and [111]-grown SL's.

One can obtain the correspondence between the irreps of different space groups. Physically, such a correspondence reflects a similarity of symmetry properties of electron states in different $(GaAs)_m(AlAs)_n$ SL's and GaAs(AlAs) crystals. It is reasonable to consider the correspondence when the alteration of the system symmetry slightly changes symmetry properties of localized states.

Alteration of such a kind can appear either under small perturbation (e.g. under pressure/mismatching) or on replacing certain atoms in a lattice (transformation of one SL into another).

The relations between the irreps of a main group and those of its subgroup can be established by subducing the irreps of the group on the subgroup. By doing this, some of the irreps remain irreducible whereas the others become reducible representations. Decomposing the latter over the irreps of the subgroup, one obtains a sum of irreps of the subgroup, which correspond to the irrep of the main group.

TABLE 9. Electron state symmetries in $(GaAs)_m(AlAs)_n$ [110]-grown SL's with the space group C_{2v}^{20} #

m=1 n=2	m=2 n=3	m=1 n=4	m=2 n=5	m=1 n=6	m=3 n=4	C_{2v}^{20} (Imm2)		Γ (000) C_{2v}	X $(\frac{1}{2}\frac{1}{2}\frac{1}{2})$ C_{2v}	S $(\frac{1}{2}00)$ C_s	R $(0\frac{1}{2}0)$ C_s	T $(00\frac{1}{2})$ C_2	W $(\frac{1}{4}\frac{1}{4}\frac{1}{4})$ C_2
1Ga	1Al	1Ga	1Al	1Ga	1Ga	1a	$a_1(s;p_z)$	1	1	1	1	1	1
						(00z)	$b_2(p_y)$	3	3	1	2	2	2
						C_{2v}	$b_1(p_x)$	4	4	2	1	2	2
							$\bar{e}$	5	5	3,4	3,4	3,4	3,4
1As	1As	1As	1As	1As	1As	1b	$a_1(s;p_z)$	1	1	1	1	2	2
						$(0\frac{1}{2}z)$	$b_2(p_y)$	3	3	1	2	1	1
						C_{2v}	$b_1(p_x)$	4	4	2	1	1	1
							$\bar{e}$	5	5	3,4	3,4	3,4	3,4
	2Ga		2Ga		2Ga	2d	$a'(s;p_y,p_z)$	1,3	1,3	1,1	1,2	1,2	1,2
2Al	2Al	4Al	4Al	6Al	4Al	(0yz)	$a''(p_x)$	2,4	2,4	2,2	1,2	1,2	1,2
2As	4As	4As	6As	6As	6As	C_s	$\bar{e}(1)$	5	5	3,4	3,4	3,4	3,4
							$\bar{e}(2)$	5	5	3,4	3,4	3,4	3,4

#Co-reps: S_3+S_4; R_3+R_4; T_3+T_4

TABLE 10. Electron state symmetries in $(GaAs)_m(AlAs)_n$ [111]-grown SL's with the space group C_{3v}^1 #

m=1 n=2	m=2 n=1	m=3 n=3	m=1 n=5	m=2 n=4		C_{3v}^1 (P3m1)	Γ (000) C_{3v}	A $(00\frac{1}{2})$ C_{3v}	K $(\frac{1}{3}\frac{1}{3}0)$ C_3	H $(\frac{1}{3}\frac{1}{3}\frac{1}{2})$ C_3	M $(\frac{1}{2}00)$ C_s	L $(\frac{1}{2}0\frac{1}{2})$ C_s
		1Ga	1Ga	1Ga		$a_1(s;p_z)$	1	1	1	1	1	1
1Ga	1Ga	1Al	1Al	1Al		$e(p_x,p_y)$	3	3	2,3	2,3	1,2	1,2
1As	1As	2As	2As	2As	1a	$\bar{e}_1^{(1)}$	4	4	5	5	4	4
					(00z)	$\bar{e}_1^{(2)}$	5	5	5	5	3	3
					C_{3v}	$\bar{e}_2$	6	6	4,6	4,6	3,4	3,4
		1Ga		1Ga		$a_1(s;p_z)$	1	1	2	2	1	1
1Al	1Ga	1Al	2Al	1As		$e(p_x,p_y)$	3	3	1,3	1,3	1,2	1,2
1As	1As	2As	2As	2As	1b	$\bar{e}_1^{(1)}$	4	4	6	6	4	4
					$(\frac{1}{3}\frac{2}{3}z)$	$\bar{e}_1^{(2)}$	5	5	6	6	3	3
					C_{3v}	$\bar{e}_2$	6	6	4,5	4,5	3,4	3,4
		1Ga				$a_1(s;p_z)$	1	1	3	3	1	1
1Al	1Al	1Al	2Al	2Al		$e(p_x,p_y)$	3	3	1,2	1,2	1,2	1,2
1As	1As	2As	2As	2As	1c	$\bar{e}_1^{(1)}$	4	4	4	4	4	4
					$(\frac{2}{3}\frac{1}{3}z)$	$\bar{e}_1^{(2)}$	5	5	4	4	3	3
					C_{3v}	$\bar{e}_2$	6	6	5,6	5,6	3,4	3,4

#Co-reps: $\Gamma_4+\Gamma_5$; A_4+A_5; M_3+M_4; L_3+L_4

TABLE 11. Electron state symmetries in $(GaAs)_m(AlAs)_n$ [111]-grown SL's with the space group C_{3v}^5 #

		Γ	T	L	F
C_{3v}^5(R3m)		(000)	$(\frac{1}{2}\frac{1}{2}\frac{1}{2})$	$(0\frac{1}{2}0)$	$(\frac{1}{2}\frac{1}{2}0)$
		C3v	C3v	Cs	Cs
1a	$a_1(s;p_z)$	1	1	1	1
(va)	$e(p_x,p_y)$	3	3	1,2	1,2
C3v	$\bar{e}_1^{(1)}$	4	4	3	3
	$\bar{e}_1^{(2)}$	5	5	4	4
	$\bar{e}_2$	6	6	3,4	3,4

#$a=a_1+a_2+a_3$;

Co-reps: $\Gamma_4+\Gamma_5$; T_4+T_5; L_3+L_4; F_3+F_4

The correspondence is presented in Table 12 where, as the main group, either the T_d group (the left part of Table 9) or the D_{2d} group (the right part) is taken. One can see that, on lowering the symmetry from a cubic T_d (GaAs or AlAs) down to a tetragonal D_{2d} (the [001]-grown SL's; pressure along a fourth-order axis), the Γ_4 states (Γ_{15}' states in usual notation) split into Γ_2 and Γ_5 ones (note different notations of irreps within different groups). When the spin-orbit interaction is taken into account, this corresponds to the splitting of Γ_8 states into Γ_6 and Γ_7 ones. On further lowering the symmetry down to an orthorhombic C_{2v} (the [110]-grown SL's; pressure along a C_2 axis), the Γ_5 states split into $\Gamma_3+\Gamma_4$ ones. Figure 8 shows the symmetry modification of the lowest conduction-band and the highest valence-band states at the center of BZ when the system symmetry changes.

TABLE 12. The correspondence between the Γ irreps of different space groups describing bulk GaAs (AlAs) and SL's grown along different directions. For the cubic T_d symmetry, the Miller-Love [15] notation is used, the commonly used labelling being given in parentheses.

T_d	D_{2d}	C_{2v}	C_{3v}	D_{2d}	C_{2v}
GaAs(AlAs)	[001]	[110]	[111]	[001]	[110]
			single-valued irreps		
$\Gamma_1(\Gamma_1)$	Γ_1	Γ_1	Γ_1	Γ_1	Γ_1
$\Gamma_2(\Gamma_2)$	Γ_4	Γ_2	Γ_2	Γ_2	Γ_1
$\Gamma_3(\Gamma_{12})$	$\Gamma_1+\Gamma_4$	$\Gamma_1+\Gamma_2$	Γ_3	Γ_3	Γ_2
$\Gamma_4(\Gamma_{15}')$	$\Gamma_2+\Gamma_5$	$\Gamma_1+\Gamma_3+\Gamma_4$	$\Gamma_1+\Gamma_3$	Γ_4	Γ_2
$\Gamma_5(\Gamma_{25}')$	$\Gamma_3+\Gamma_5$	$\Gamma_2+\Gamma_3+\Gamma_4$	$\Gamma_2+\Gamma_3$	Γ_5	$\Gamma_3+\Gamma_4$
			double-valued irreps		
$\Gamma_6(\Gamma_7)$	Γ_6	Γ_5	Γ_6	Γ_6	Γ_5
$\Gamma_7(\Gamma_6)$	Γ_7	Γ_5	Γ_6	Γ_7	Γ_5
$\Gamma_8(\Gamma_8)$	$\Gamma_6+\Gamma_7$	$\Gamma_5+\Gamma_5$	$\Gamma_4+\Gamma_5+\Gamma_6$		

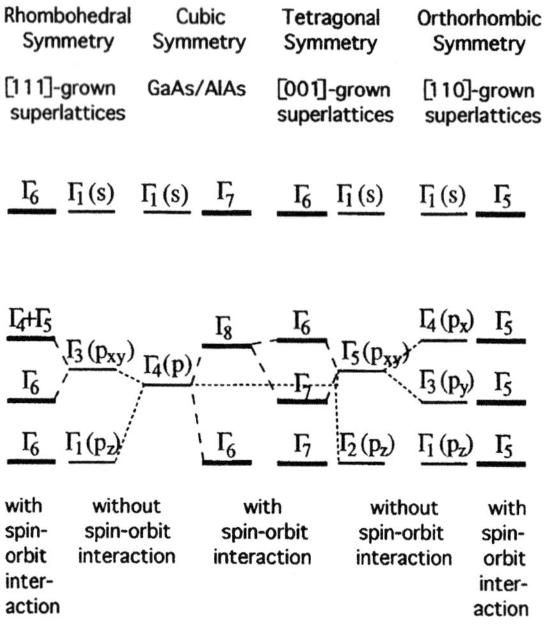

Figure 8. The correspondence between the band states at the center of the BZ of $(GaAs)_m(AlAs)_n$ SL's originating from the bottom of the conduction band and the top of the valence band of bulk GaAs (AlAs). The co-rep $\Gamma_4 + \Gamma_5$ describes a degenerate state of the $(GaAs)_m(AlAs)_n$ [111] SL.

3. Optical selection rules

The selection rules for optical transitions follow from the symmetry restrictions imposed on matrix elements of transitions from an initial (*i*) electron state to a final (*f*) one under the action of the perturbative operator *W*. In terms of group theory, for a system having the space group *G*, the optical transition matrix elements do not vanish due to symmetry if the following Kronecker products of reps of *G* contain the identity irrep

$$(D^f)^* \times D^W \times D^i \supset \Gamma_1 \qquad (1)$$

Here D^i and D^f are the irreps of the space group *G* according to which the initial and final electron states are transformed, D^W is the rep according to which the perturbative operator *W* is transformed and Γ_1 is the identity irrep of the group *G*.

In light absorption processes, the perturbative operator is $W \propto (\mathbf{ep}) \times \exp(i\mathbf{kr})$, where **e** and **k** are polarization and wave vector of light, respectively, and **p** is the momentum operator of an electron. For the light frequency range the condition $k_i a \ll 1$ (where *a* is the lattice parameter) is satisfied; that leads to an approximation called as the dipole approximation:

$$W \propto (\mathbf{ep}) \tag{2}$$

Thus, the perturbative operator W transforms according to the vector representation D^{v} of the space group G. Hence, the transitions are allowed between those pairs of electron states D^f and D^i which obey the condition

$$(D^f)^* \times D^i \cap D^v \neq 0. \tag{3}$$

Since the reps according to which the perturbative operator W is transformed correspond to the zero wave vector, only those combinations of electron states for which the Kronecker product contains the rep with $\mathbf{k} = 0$ can be allowed.

3.1. SUPERLATTICES GROWN ALONG THE [001] DIRECTION

In Tables 13 and 14, for the space groups D_{2d}^5 and D_{2d}^9 respectively, we present the $\mathbf{k}$=0 parts of Kronecker products of irreps corresponding to various combinations of initial and final electron states at the symmetry points of SLBZ. For example, $X_1 \times X_4 = \Gamma_5 + M_5$, and + in the column Γ_5 means that this irrep forms part of the direct product. The transitions are allowed between those pairs of states for which Kronecker products have irreps in common with the vector representation. For the space groups D_{2d}^5 and D_{2d}^9, the vector representation is $D^v = \Gamma_2(z)+\Gamma_5(x,y)$. Thus, the optical transition between the states X_1 and X_4 - taken above as an example - is allowed in the x and y polarizations and forbidden in the z polarization. For co-reps D_i+D_j, the corresponding Kronecker products for separate irreps $D_i \times D_i = D_i \times (D_j)^*$ and $D_i \times D_j = D_i \times (D_i)^*$ describe the transitions between the states D_i and D_j, D_i and D_i, respectively.

When the spin-orbit interaction is taken into account, the products of double-valued irreps should be considered. The symmetry correspondence between the Bloch states which transform according to single-valued irreps (without spin-orbit coupling) and those which transform according to the double-valued irreps (with spin-orbit coupling) can be obtained as follows [23]. The Bloch state D^l ($l = s, p$) corresponds to the states which transform according to the double-valued representation $\overline{D}^J = D^l \times \overline{D}^{1/2}$ where $\overline{D}^{1/2}$ is the double-valued irrep according to which the spinor function is transformed. For the space groups D_{2d}^5 and D_{2d}^9, the spinor function is transformed according to $\overline{D}^{1/2} = \overline{e}_1$. Decomposing the corresponding direct products of coordinate and spinor parts of the one-electron wave function, we obtain the set of states into which D^l transforms when the spin-orbit interaction is included. At the Γ point of BZ, the symmetry correspondence is the following: $\Gamma_1 \rightarrow \Gamma_6$, $\Gamma_2 \rightarrow \Gamma_7$, $\Gamma_5 \rightarrow \Gamma_6 + \Gamma_7$, and at the M point: $M_1 \rightarrow M_6$, $M_2 \rightarrow M_7$, $M_3 \rightarrow M_6$, $M_4 \rightarrow M_7$, $M_5 \rightarrow M_6 + M_7$.

As a result, we obtain a relation between the selection rules when the spin-orbit interaction is taken into account (lower parts of Tables 13 and 14) and those when it is not (upper parts). It can be seen that some forbidden transitions become allowed when the spin-orbit interaction is taken into account (Table 15).

Besides direct optical transitions, the phonon-assisted ones may be of importance. The phonon-assisted optical transitions are allowed between those initial and final electron states belonging to different points of the BZ which obey the following conditions

$$(D^{virt})^* \times D^i \cap D^v \neq 0, \quad (D^{virt})^* \times D^{ph} \times D^f \supset \Gamma_1, \tag{4a}$$

TABLE 13. Selection rules for direct optical transitions in $(GaAs)_m(AlAs)_n$ [001] SL's with the space group D_{2d}^5

	Γ_1	Γ_2	Γ_3	Γ_4	Γ_5	allowed polarizations
			single-valued irreps			
D=Γ, M, A, Z						
D_ixD_i (i=1-4)	+					forbidden
D_1xD_2, D_3xD_4		+				z
D_1xD_3, D_2xD_4			+			forbidden
D_1xD_4, D_2xD_3				+		forbidden
D_ixD_5 (i=1-4)					+	x,y
D_5xD_5	+	+	+	+		z
D=X, R						
D_ixD_i (i=1-4)	+	+				z
D_1xD_2, D_3xD_4			+	+		forbidden
D_ixD_j (i=1,2; j=3,4)					+	x,y
			double-valued irreps (with spin-orbit interaction included)			
D=Γ, M, A, Z						
D_ixD_i (i=6,7)	+		+		+	x,y
D_6xD_7		+		+	+	x,y; z
D=X, R						
D_5xD_5	+	+	+	+	+	x,y; z

TABLE 14. Selection rules for direct optical transitions in $(GaAs)_m(AlAs)_n$ [001] SL's with the space group D_{2d}^9

	Γ_1	Γ_2	Γ_3	Γ_4	Γ_5	allowed polarizations
			single-valued irreps			
D=Γ, M						
D_ixD_i (i=1-4)	+					forbidden
D_1xD_2, D_3xD_4		+				z
D_1xD_3, D_2xD_4			+			forbidden
D_1xD_4, D_2xD_3				+		forbidden
D_ixD_5 (i=1-4)					+	x,y
D_5xD_5	+	+	+	+		z
X_ixX_i (i=1-4)	+			+		forbidden
X_1xX_2, X_3xX_4		+	+			z
X_ixX_j (i=1,2; j=3,4)					+	x,y
P_1xP_1, P_2xP_2	+			+		forbidden
P_3xP_3, P_4xP_4	+		+			forbidden
P_1xP_2, P_3xP_4		+		+		z
P_ixP_j (i=1,2; j=3,4)					+	x,y
N_ixN_i (i=1,2)	+	+			+	x,y; z
N_1xN_2			+	+	+	x,y
		double-valued irreps (with spin-orbit interaction included)				
Γ_ixΓ_i (i=6,7)	+		+		+	x,y
Γ_6xΓ_7		+		+	+	x,y; z
M_ixM_i (i=6,7)	+		+		+	x,y
M_6xM_7		+		+	+	x,y; z
X_5xX_5	+	+	+	+	+	x,y; z
P_ixP_i (i=5-8)	+		+			forbidden
P_5xP_6, P_7xP_8		+		+		z
P_ixP_j (i=5,6; j=7,8)					+	x,y
N_ixN_i (i=3,4)			+	+	+	x,y
N_3xN_4	+	+			+	x,y; z

TABLE 15. The modification of selection rules for direct optical transitions when including spin-orbit interaction (the space groups D_{2d}^5, D_{2d}^9).

The labels of both the irreps and additional light polarizations (x, y, or z) taken in parentheses refer to the case where spin-orbit coupling is taken into account.

		conduction-band states			
		$\Gamma_1(\Gamma_6)$	$\Gamma_2(\Gamma_7)$	$\Gamma_5(\Gamma_6)$	$\Gamma_5(\Gamma_7)$
valence-band states	$\Gamma_1(\Gamma_6)$	(x,y)	(x,y)z	x,y	x,y(z)
	$\Gamma_2(\Gamma_7)$	(x,y)z	(x,y)	x,y(z)	x,y
	$\Gamma_5(\Gamma_6)$	x,y	x,y(z)	(x,y)	(x,y)z
	$\Gamma_5(\Gamma_7)$	x,y(z)	x,y	(x,y)z	(x,y)

where D^{virt} and D^{ph} are the irreps according to which the virtual electron and phonon state are transformed. In the case (4a), the virtual electron state belongs to the same wave vector as an initial state. Other transitions are also possible, with the virtual electron state belonging to the same wave vector as a final state. In the latter case, the states obey the following conditions

$$(D^i)^* \times D^{ph} \times D^{virt} \supset \Gamma_1, \quad (D^f)^* \times D^{virt} \cap D^v \neq 0. \qquad (4b)$$

For the short-period SL's under consideration, there is a consensus among all the published calculations that the uppermost valence-band states at the Γ point are about 1eV higher in energy than the states at the other symmetry points [18,24-28]. Therefore, due to such a large difference in energy, one can neglect the transitions via the virtual valence-band states at symmetry points with $\mathbf{k} \neq 0$. By doing this, one has to consider only the (4a) transitions for the optical absorption spectra and the (4b) ones for the luminescence spectra. Transitions involving optical phonons with $\mathbf{k} = 0$ (the Γ_1, Γ_2, and Γ_5 optical phonons) are also to be considered in the case where two conduction bands and/or two valence bands have energies close one to the other at the Γ point .

Using Tables 13 and 14, one can obtain the selection rules for phonon-assisted optical transitions. Such rules for the D_{2d}^5 and D_{2d}^9 space groups are presented in Tables 16 and 17. If one takes into consideration the spin-orbit interaction, the rules are modified (Tables 18 and 19 for the D_{2d}^5 and D_{2d}^9 space groups, respectively).

TABLE 16. Selection rules for phonon-assisted optical transitions in $(GaAs)_m(AlAs)_n$ [001] SL's with the space group D_{2d}^5 (neglecting the spin-orbit interaction)

initial state (val.band)	virtual state (cond.band)	final state (cond.band)	assisting phonons*	allowed polarizations
Γ_1	Γ_1	any	any	forbidden
	Γ_2	D_i	D_j $D=\Gamma,M,A,Z$ $(i,j)=(1,2);(2,1);(3,4);(4,3);(5,5)$	z
		D_i	D_i $D=R,X$	
	Γ_5	D_5	$D_1+D_2+D_3+D_4$ $D=\Gamma,M,A,Z$	x,y
		D_i $(i\neq5)$	D_5 $D=\Gamma,M,A,Z$	
		D_i $(i=1,2)$	D_3+D_4 $D=X,R$	
		D_i $(i=3,4)$	D_1+D_2 $D=X,R$	
Γ_2	Γ_1	D_i	D_i $D=\Gamma,M,R,X,A,Z$	z
	Γ_2	any	any	forbidden
	Γ_5	D_5	$D_1+D_2+D_3+D_4$ $D=\Gamma,M,A,Z$	x,y
		D_i $(i\neq5)$	D_5 $D=\Gamma,M,A,Z$	
		D_i $(i=1,2)$	D_3+D_4 $D=X,R$	
		D_i $(i=3,4)$	D_1+D_2 $D=X,R$	
Γ_5	Γ_1	D_i	D_i $D=\Gamma,M,R,X,A,Z$	x,y
	Γ_2	D_i	D_j $D=\Gamma,M,A,Z$ $(i,j)=(1,2);(2,1);(3,4);(4,3);(5,5)$	x,y
		D_i	D_i $D=R,X$	
	Γ_5	D_5	$D_1+D_2+D_3+D_4$ $D=\Gamma,M,A,Z$	z
		D_i $(i\neq5)$	D_5 $D=\Gamma,M,A,Z$	
		D_i $(i=1,2)$	D_3+D_4 $D=X,R$	
		D_i $(i=3,4)$	D_1+D_2 $D=X,R$	

* The Γ_3, Γ_4, Z_3, and Z_4 phonons are absent (see e.g. [4]).

TABLE 17. Selection rules for phonon-assisted optical transitions in $(GaAs)_m(AlAs)_n$ [001] SL's with the space group D_{2d}^9 (neglecting the spin-orbit interaction)

initial state (val.band)	virtual state (cond.band)	final state (cond.band)	assisting phonons	allowed polarizations
$\Gamma_i(i=1,2)$	$\Gamma_i(i=1,2)$	any	any	forbidden
Γ_1	Γ_2	D_i	D_j $D=\Gamma, M$ $(i,j)=(1,2);(2,1);(5,5)$	z
		L_i	L_j $L=X,P$ $(i,j)=(1,2);(2,1);(3,4);(4,3)$	
		$N_i(i=1,2)$	N_i	
Γ_2	Γ_1	D_i	D_i $D=\Gamma,M^*,X,P,N$	z
$\Gamma_i(i=1,2)$	Γ_5	D_5	D_1+D_2	x,y
		$D_i(i=1,2)$	D_5 $D=\Gamma,M$	
		$L_i(i=1,2)$	L_3+L_4 $L=X,P$	
		$L_i(i=3,4)$	L_1+L_2	
		$N_i(i=1,2)$	N_1+N_2	
Γ_5	Γ_1	D_i	D_i $D=\Gamma,M^*,X,P,N$	x,y
	Γ_2	D_i	D_j $D=\Gamma, M$ $(i,j)=(1,2);(2.1);(5,5)$	x,y
		L_i	L_j $L=X,P$ $(i,j)=(1,2);(2,1);(3,4);(4,3)$	
		$N_i(i=1,2)$	N_i	
	Γ_5	D_5	D_1+D_2	z
		$D_i(i=1,2)$	D_5 $D=\Gamma,M$	
		L_i (i=1,2)	L_3+L_4 $L=X,P$	
		L_i (i=3,4)	L_1+L_2	
		$N_i(i=1,2)$	N_1+N_2	

* The Γ_3, Γ_4, M_3, and M_4 phonons are absent (see e.g. [4]).

TABLE 18. Selection rules for the light absorption in $(GaAs)_m(AlAs)_n$ [001] SL's with the space group D^5_{2d} (with the account of spin-orbit interaction). The labels v and c denote the valence-band and conduction-band virtual states, respectively. $i = 6, 7; j = 13 - i$.

initial state (val.band)	virtual state	final state (cond.band)	assisting phonons		allowed polarizations
Γ_i	direct transitions	Γ_i			x,y
		Γ_j			x,y;z
		D_5	$D_1+D_2+D_3+D_4$	D=R,X	
		D_i	D_1+D_5	D=Γ,Z	
Γ_i	$\Gamma_{i,c}$	D_j	D_2+D_5	D=Γ,Z	x,y
		D_i	$D_1+D_3+D_5$	D=M,A	
		D_j	$D_2+D_4+D_5$	D=M,A	
		D_5	$D_1+D_2+D_3+D_4$	D=R,X	
		D_i	D_2+D_5	D=Γ,Z	
Γ_i	$\Gamma_{j,c}$	D_j	D_1+D_5	D=Γ,Z	x,y;z
		D_i	$D_2+D_4+D_5$	D=M,A	
		D_j	$D_1+D_3+D_5$	D=M,A	
Γ_i	$\Gamma_{i,v}$	Γ_i	$\Gamma_1+\Gamma_5$		x,y
Γ_i	$\Gamma_{i,v}$	Γ_j	$\Gamma_1+\Gamma_5$		x,y;z
Γ_i	$\Gamma_{j,v}$	Γ_i	$\Gamma_2+\Gamma_5$		x,y;z
Γ_i	$\Gamma_{j,v}$	Γ_j	$\Gamma_2+\Gamma_5$		x,y

TABLE 19. Selection rules for the light absorption in $(GaAs)_m(AlAs)_n$ [001] SL's with the space group D_{2d}^9 (with the account of spin-orbit interaction). The labels v and c denote the valence-band and conduction-band virtual states, respectively. i = 6, 7; j = 13 - i.

initial state (val.band)	virtual state	final state (cond.band)	assisting phonons		allowed polarizations
Γ_i	direct transitions	Γ_i			x,y
		Γ_j			x,y;z
		D_i	D_1+D_5	D=Γ,M	
		D_j	D_2+D_5	D=Γ,M	
Γ_i	$\Gamma_{i,c}$	N_3,N_4	N_1+N_2		x,y
		X_5	$X_1+X_2+X_3+X_4$		
		P_{i-1}	P_1+P_3		
		P_{j-1}	P_2+P_4		
		P_{i+1}	P_1+P_4		
		P_{j+1}	P_2+P_3		
		D_j	D_1+D_5	D=Γ,M	
		D_i	D_2+D_5	D=Γ,M	
Γ_i	$\Gamma_{j,c}$	N_3,N_4	N_1+N_2		x,y;z
		X_5	$X_1+X_2+X_3+X_4$		
		P_{i-1}	P_2+P_4		
		P_{j-1}	P_1+P_3		
		P_{i+1}	P_2+P_3		
		P_{j+1}	P_1+P_4		
Γ_i	$\Gamma_{i,v}$	Γ_i	$\Gamma_1+\Gamma_5$		x,y
Γ_i	$\Gamma_{i,v}$	Γ_j	$\Gamma_1+\Gamma_5$		x,y;z
Γ_i	$\Gamma_{j,v}$	Γ_i	$\Gamma_2+\Gamma_5$		x,y;z
Γ_i	$\Gamma_{j,v}$	Γ_j	$\Gamma_2+\Gamma_5$		x,y

Besides, in Tables 18 and 19 those transitions involving the optical Γ phonons are included, which obey the (4b) conditions. Note that one can practically always find appropriate phonons for the indirect transitions. These transitions are allowed between those initial and virtual states (in the case (4a)) or virtual and final ones (in the case (4b)) for which Kronecker products have irreps in common with the vector representation, that is to say, for which the zero-phonon transitions are allowed.

Thus, a particular interband transition can be allowed either completely or in certain polarizations due to the appropriate symmetries of involved states, and/or due to the spin-orbit interaction, and/or due to the electron-phonon interaction. In the two latter cases, its oscillator strength will be weak. It is worth to note that the oscillator strength is also determined by atomic arrangement in the Wyckoff positions. The higher the occupation number of a position, the more is the strength of a transition involving band states induced by the atoms in this position. It can be considered in terms of the "bulk genesis" of some bands when the number of GaAs and/or AlAs layers is rather high, that is to say when the occupation numbers of some Wyckoff positions are much larger than those of others (by an order of magnitude or more).

Of course, the most important ones are the transitions between the uppermost states in the valence band and the lowest states in the conduction band. Group theory analysis alone does not permit to determine even the relative energy positions of different states. It is possible only to assume that relative energy positions of different band states could change when *m* and *n* are varied. This is due to the fact that distinct localized functions contribute to the band states with the same symmetry in different SL's. For example, one can see for the (2x2) and (1x3) SL's that As and Me(Ga and Al) atoms interchange their positions, i.e. in the (2x2) SL the a, d, and f Wyckoff positions are occupied by As atoms and g positions by Ga and Al atoms whereas in the (1x1) SL the As atoms replace Me atoms and, in turn, Me atoms replace As. It is clear that the number of band states with a particular symmetry depends completely on which Wyckoff positions are either occupied or not whereas the energies of these states depend on chemical nature of elements that occupy these positions. Such rearrangement of atoms over the Wyckoff positions as for (1x1) and (2x2) SL's will ultimately result in the relative shifts of band states on energy scale. This is a qualitative picture of this effect. However, some predictions could be made only by comparison with the band-structure calculations [18,24-28](see below).

3.2. SUPERLATTICES GROWN ALONG THE [110] AND [111] DIRECTIONS

For the space groups C_{2v}^1, C_{2v}^7, and C_{2v}^{20}, the vector representation is $D^v = \Gamma_1(z)+\Gamma_3(y)+\Gamma_4(x)$; for the space groups C_{3v}^1 and C_{3v}^5, it is $D^v = \Gamma_1(z)+\Gamma_3(x,y)$. The upper parts of Tables 20 to 23 give the selection rules without the account of spin-orbit coupling. The latter mixes states and smoothes the difference between them, especially for the [110]-grown SL's. When the spin-orbit interaction is taken into consideration, all direct optical transitions are completely allowed between the Γ states in the [110]-grown SL's (cf. the last line of Table 20). This is another manifestation of the fact that these band states do not differ from each other in symmetry (see Sect.2). However, in actual SL's, the spin-orbit interaction is not extremely strong. That is why, when analyzing optical spectra, it is worth to take into account the genesis of states, that is the symmetry correspondence between Bloch states with and without including this interaction, and modification of selection rules.

TABLE 20. Selection rules for direct optical transitions in $(GaAs)_m(AlAs)_n$ [110] SL's with the space group C^1_{2v} for the symmetry points D=Γ, X, Y, Z, S, T, U, R

	Γ_1	Γ_2	Γ_3	Γ_4	allowed polarizations
			single-valued irreps		
$D_i x D_i$ (i=1-4)	+				z
$D_1 x D_2$, $D_3 x D_4$		+			forbidden
$D_1 x D_3$, $D_2 x D_4$			+		y
$D_1 x D_4$, $D_2 x D_3$				+	x
	double-valued irreps (with spin-orbit interaction included)				
$D_5 x D_5$	+	+	+	+	x,y,z

As we described above (Section 3.1), the Bloch state D^l ($l = s, p$) corresponds to the states which transform according to the double-valued representation $\overline{D}^J = D^l \times \overline{D}^{1/2}$ where $\overline{D}^{1/2}$ is the double-valued irrep according to which the spinor function is transformed. For the space groups C^1_{2v}, C^7_{2v}, and C^{20}_{2v} the spinor function is transformed according to $\overline{D}^{1/2} = \overline{e}$, whereas for the space groups C^1_{3v} and C^5_{3v} according to $\overline{e}_2$. Decomposing the corresponding direct products of coordinate and spinor parts of one-electron wave function, we obtain the set of states into which D^l transforms when spin-orbit interaction is included. At the center of BZ, the symmetry correspondence is the following: $\Gamma_1 \rightarrow \Gamma_6$, $\Gamma_3 \rightarrow \Gamma_4 + \Gamma_5 + \Gamma_6$ (the C_{3v} groups in question), Γ_i (i=1-4) $\rightarrow \Gamma_5$ (the C_{2v} groups).

Thus, for the [110]-grown SL's, a transition between two Γ_1 states is allowed only in the z polarization, and the transition between Γ_5 states originating from them is allowed in the x and y polarizations as well. So, in actual $(GaAs)_m(AlAs)_n$ [110]-grown SL's, all the direct Γ - Γ optical transitions should be observed in the polarized spectra, though some of them being allowed in certain polarizations only due to the spin-orbit interaction.

As to other points of BZ of [110]-grown $(GaAs)_m(AlAs)_n$ SL's and every BZ point of [111]-grown ones, the picture is not so simple. If both the spin-orbit interaction is taken into account and the magnetic field is absent, some of band states can be described by co-reps, having the same energies. The selection rules for the transitions between

TABLE 21. Selection rules for direct optical transitions in $(GaAs)_m(AlAs)_n$ [110] SL's with the space group C_{2v}^7

	Γ_1	Γ_2	Γ_3	Γ_4	allowed polarizations
			single-valued irreps		
$D=\Gamma$, Y					
$D_i x D_i$ (i=1-4)	+				z
$D_1 x D_2$, $D_3 x D_4$		+			forbidden
$D_1 x D_3$, $D_2 x D_4$			+		y
$D_1 x D_4$, $D_2 x D_3$				+	x
D=Z, T					
$D_i x D_i$ (i=1-4)			+		y
$D_1 x D_2$, $D_3 x D_4$				+	x
$D_1 x D_3$, $D_2 x D_4$	+				z
$D_1 x D_4$, $D_2 x D_3$		+			forbidden
D=X, S, U, R					
$D_1 x D_1$	+	+	+	+	x,y,z
	double-valued irreps (with spin-orbit interaction included)				
$D=\Gamma$, Y, Z, T					
$D_5 x D_5$	+	+	+	+	x,y,z
D=X, S					
$D_i x D_i$ (i=2-5)				+	x
$D_2 x D_3$, $D_4 x D_5$			+		y
$D_2 x D_4$, $D_3 x D_5$		+			forbidden
$D_2 x D_5$, $D_3 x D_4$	+				z
D=U, R					
$D_i x D_i$ (i=2-5)		+			forbidden
$D_2 x D_3$, $D_4 x D_5$	+				z
$D_2 x D_4$, $D_3 x D_5$				+	x
$D_2 x D_5$, $D_3 x D_4$			+		y

TABLE 22. Selection rules for direct optical transitions in $(GaAs)_m(AlAs)_n$ [110] SL's with the space group C_{2v}^{20}

	Γ_1	Γ_2	Γ_3	Γ_4	allowed polarizations
		single-valued irreps			
$D=\Gamma, X$					
$D_i x D_i$ (i=1-4)	+				z
D_1xD_2, D_3xD_4		+			forbidden
D_1xD_3, D_2xD_4			+		y
D_1xD_4, D_2xD_3				+	x
S_1xS_1, S_2xS_2	+		+		y,z
S_1xS_2		+		+	x
R_1xR_1, R_2xR_2	+			+	x,z
R_1xR_2		+	+		y
$D=T, W$					
D_1xD_1, D_2xD_2	+	+			z
D_1xD_2			+	+	x,y
		double-valued irreps (with spin-orbit interaction included)			
$\Gamma_5x\Gamma_5$, X_5xX_5	+	+	+	+	x,y,z
S_3xS_3, S_4xS_4		+		+	x
S_3xS_4	+		+		y,z
R_3xR_3, R_4xR_4		+	+		y
R_3xR_4	+			+	x,z
T_3xT_3, T_4xT_4, W_3xW_4			+	+	x,y
W_3xW_3, W_4xW_4, T_3xT_4	+	+			z

TABLE 23. Selection rules for direct optical transitions in $(GaAs)_m(AlAs)_n$ [111] SL's with the space groups C_{3v}^1 (the Γ, A, H, K, L, and M symmetry points) and C_{3v}^5 (the Γ, T, L, and F symmetry points of the SLBZ)

	Γ_1	Γ_2	Γ_3	allowed polarizations
		single-valued irreps		
D=Γ, T, A				
D_1xD_1, D_2xD_2	+			z
D_1xD_2		+		forbidden
D_1xD_3, D_2xD_3			+	x,y
D_3xD_3	+	+	+	x,y,z
D=H, K				
D_ixD_i (i=1-3)	+	+		z
D_1xD_2, D_1xD_3, D_2xD_3			+	x,y
D=L, F, M				
D_1xD_1, D_2xD_2	+		+	x,y,z
D_1xD_2		+	+	x,y
	double-valued irreps (with spin-orbit interaction included)			
D=Γ, T, A				
D_4xD_4, D_5xD_5		+		forbidden
D_4xD_5	+			z
D_4xD_6, D_5xD_6			+	x,y
D_6xD_6	+	+	+	x,y,z
D=H, K				
D_ixD_i (i=4-6)	+	+		z
D_4xD_5, D_4xD_6, D_5xD_6			+	x,y
D=L, M, F				
D_3xD_4	+		+	x,y,z
D_3xD_3, D_4xD_4		+	+	x,y

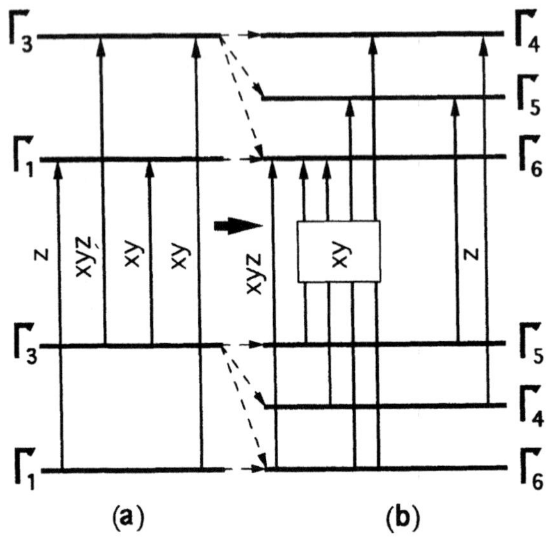

Figure 9. The modification of allowed direct optical transitions on including spin-orbit interaction (the space groups C^1_{3v} and C^5_{3v}).
(a) - without spin-orbit interaction, (b) - including spin-orbit interaction

such combined states can be easily obtained since the co-reps are direct sums of complex-conjugated counterparts. The magnetic field lifts the degeneracy related to the time-inversion. As a result, states which are described by complex-conjugated irreps may have different energies and be involved in transitions with different selection rules (Tables 21 to 23).

For the $(GaAs)_m(AlAs)_n$ [111]-grown SL's, taking into account the symmetry correspondence between single- and double-valued irreps, we can obtain a relation between the selection rules when the spin-orbit interaction is taken into consideration (lower parts of Table 23) and those when it is not (upper part). This correspondence is schematically presented in Figure 9. From Table 23, it follows that, whereas the transition between two Γ_6 states is completely allowed when both of them are derived from the Γ_3 states, the oscillator strength of a transition between two Γ_6 states originated either from the Γ_1 and Γ_3 states or from two Γ_1 states depends on the intermixing of Γ_1 and Γ_3 states by the spin-orbit interaction, that is on the strength of the latter. When the interaction is taken into account, a Γ_1-Γ_1-originated transition (allowed in the z polarization) becomes weakly allowed in the x and y polarizations, whereas the Γ_1-Γ_3-originated one (allowed in the x and y polarizations) does in the z polarization. These peculiarities of different Γ_6 states can help to interpret the optical spectra. Additional information can be obtained in experiments with the magnetic field not reducing the system symmetry. The latter does not split doubly degenerate Γ_6

TABLE 24. Selection rules and assisting phonons for indirect optical transitions in $(GaAs)_m(AlAs)_n$ [110] SL's with the space groups C_{2v}^1, C_{2v}^7, and C_{2v}^{20}

initial state (val. band)	virtual state (cond. band)	allowed polarizations	final states: D=Γ,Y,Z,T (C_{2v}^1,C_{2v}^7) D=X, S, U, R (C_{2v}^1) D= Γ, X (C_{2v}^{20})				D=X,S,U,R (C_{2v}^7)	D=S, R (C_{2v}^{20})		D=T, W (C_{2v}^{20})	
			without the spin-orbit interaction								
			D_1	D_2	D_3	D_4	D_1	D_1	D_2	D_1	D_2
			assisting phonons								
Γ1	Γ1	z	D_1	D_2	D_3	D_4	D_1	D_1	D_2	D_1	D_2
	Γ2	forbidden	-	-	-	-	-	-	-	-	-
	Γ3	y	D_3	D_4	D_1	D_2	D_1	D_2	D_1	D_2	D_1
	Γ4	x	D_4	D_3	D_2	D_1	D_1	D_1	D_2	D_2	D_1
Γ2	Γ1	forbidden	-	-	-	-	-	-	-	-	-
	Γ2	z	D_2	D_1	D_4	D_3	D_1	D_2	D_1	D_1	D_2
	Γ3	x	D_3	D_4	D_1	D_2	D_1	D_2	D_1	D_2	D_1
	Γ4	y	D_4	D_3	D_2	D_1	D_1	D_1	D_2	D_2	D_1
Γ3	Γ1	y	D_1	D_2	D_3	D_4	D_1	D_1	D_2	D_1	D_2
	Γ2	x	D_2	D_1	D_4	D_3	D_1	D_2	D_1	D_1	D_2
	Γ3	z	D_3	D_4	D_1	D_2	D_1	D_2	D_1	D_2	D_1
	Γ4	forbidden	-	-	-	-	-	-	-	-	-
Γ4	Γ1	x	D_1	D_2	D_3	D_4	D_1	D_1	D_2	D_1	D_2
	Γ2	y	D_2	D_1	D_4	D_3	D_1	D_2	D_1	D_1	D_2
	Γ3	forbidden	-	-	-	-	-	-	-	-	-
	Γ4	z	D_4	D_3	D_2	D_1	D_1	D_1	D_2	D_2	D_1
			with the spin-orbit interaction								
			D_5				D_i(i=2-5)	D_i(i=3,4)		D_i(i=3,4)	
Γ5	Γ5	x,y,z	$D_1+D_2+D_3+D_4$				D_1	D_1+D_2		D_1+D_2	

TABLE 25. Selection rules and assisting phonons for indirect optical transitions in $(GaAs)_m(AlAs)_n$ [111] SL's with the space groups C^1_{3v} and C^5_{3v}

Without the spin-orbit interaction

initial state (val. band)	virtual state (cond. band)	allowed polarizations	$D=\Gamma,A$ (C^1_{3v}), $D=\Gamma,T$ (C^5_{3v}): D_1	D_3	$D=H,K$ (C^1_{3v}): D_1	D_2	D_3	$D=L,M$ (C^1_{3v}), $D=L,F$ (C^5_{3v}): D_1	D_2
			assisting phonons						
Γ_1	Γ_1	z	D_1	D_3	D_1	D_2	D_3	D_1	D_2
Γ_1	Γ_3	x,y	D_3	D_1+D_3	D_2+D_3	D_1+D_3	D_1+D_2	D_1+D_2	D_1+D_2
Γ_3	Γ_1	x,y	D_1	D_3	D_1	D_2	D_3	D_1	D_2
Γ_3	Γ_3	x,y,z	D_3	D_1+D_3	D_2+D_3	D_1+D_3	D_1+D_2	D_1+D_2	D_1+D_2

With the spin-orbit interaction

initial state (val. band)	virtual state (cond. band)	allowed polarizations	$D=\Gamma,A$ / $D=\Gamma,T$: D_4	D_5	D_6	$D=H,K$: D_4	D_5	D_6	$D=L,M$ / $D=L,F$: D_3	D_4
			assisting phonons							
Γ_4	Γ_5	forbid.	-	-	-	-	-	-	-	-
	Γ_4	z	D_3	none	D_3	D_3	D_1	D_2	D_2	D_1
	Γ_6	x,y	D_1	D_3	D_1+D_3	D_1+D_2	D_2+D_3	D_1+D_3	D_1+D_2	D_1+D_2
Γ_5	Γ_5	z	none	D_1	D_3	D_3	D_1	D_2	D_1	D_2
	Γ_4	forbid.	-	-	-	-	-	-	-	-
	Γ_6	x,y	D_1	D_3	D_1+D_3	D_1+D_2	D_2+D_3	D_1+D_3	D_1+D_2	D_1+D_2
Γ_6	Γ_5	x,y	none	D_1	D_3	D_3	D_1	D_2	D_1	D_2
	Γ_4	x,y	D_3	none	D_3	D_3	D_1	D_2	D_2	D_1
	Γ_6	x,y,z	D_1	D_3	D_1+D_3	D_1+D_2	D_2+D_3	D_1+D_3	D_1+D_2	D_1+D_2

states, whereas "physically degenerated" (Γ_4, Γ_5) states can split, multiplying the number of direct optical transitions.

For the [001] short-period SL's as well as for both parent compounds, the uppermost valence-band states at the Γ point are about 1eV higher in energy than the states at the other symmetry points [18,24-28]. One can expect that it will be the case for the $(GaAs)_m(AlAs)_n$ [110]- and [111]-grown SL's. Therefore, due to such a large difference in energy, for the phonon-assisted transitions, one can neglect the transitions via the virtual valence-band states at symmetry points with wave vector $\mathbf{k} \neq 0$. That is why we consider both initial and virtual electron states belonging to the same $\mathbf{k} = 0$ (Γ states).

The results are presented in Tables 24 - 25. For [110]-grown SL's when spin-orbit interaction is taken into account, not only all transitions between initial and virtual Γ states are allowed (since all direct Γ - Γ optical transitions are completely allowed) but, moreover, any phonon (with an appropriate wave vector) can participate in transitions between virtual and final states. When neglecting the spin-orbit interaction, a phonon-assisted process is allowed if the transition between initial and virtual states is allowed. The only question is, therefore, which of phonons participates in the transition. Table 24 gives these assisting phonons. Table 25 presents the selection rules for the $(GaAs)_m(AlAs)_n$ [111]-grown SL's.

The transitions involving optical phonons with $\mathbf{k} = 0$, which are of special importance when two conduction bands have energies close one to the other at the Γ point, are also included in Tables 24 -25. When two valence bands are close to each other at the center of SLBZ, there may also exist Γ-phonon-assisted optical transitions via virtual valence-band states, the selection rules being similar to the ones given in the Tables.

4. Discussion

In this section, we will focus on [001] SL's since very little work has been performed, to our knowledge, on band structures of [110] and [111] SL's.

There are many contradictions in the results of electronic structure calculations given by different authors [18, 24-28] for [001] SL's. Therefore, it is worth to give a comparative analysis of different methods presently used because several review papers devoted to this problem (see, e.g. [20,21,29]) did not treat short-period SL's in a completely satisfactory way.

We may divide all the approaches used into the ones based on the properties of constituent compounds and those dealing with the SL as a new crystal.

i) The approaches of first type are called "engineering methods" [21] or "boundary-condition approaches" [29]. The most frequently used envelope-function method [30] within the Kane-Luttinger model [31] considers that each GaAs and AlAs slab has a band structure close to that of the corresponding bulk crystals and that the band structures of constituent crystals conjugate at interface regions. These approaches can satisfactorily describe only SL's with not too short period since in the interface region the band description is not valid at all [32].

ii) The approaches of second type are called "first-principles" or "supercell" ones and can be applied for both SL's and usual bulk GaAs and AlAs crystals [33]. Only these approaches could be applied to the very-short-period SL's. All the first-principles methods use the band approximation. These methods can be classified according to the choice of a crystal potential, which is the same for all the electrons of the crystal, (method of pseudopotential; local density approximation, LDA; linear muffin-tin-orbital method, LMTO etc.) and according to a choice of basis functions (LCAO or tight-binding; orthogonalized plane waves, OPW; Wannier functions; linearized augmented plane waves, LAPW etc.). The most of them are self-consistent, the form of potential being corrected by iteration procedure. By analyzing these methods one should bear in mind that, if the choice of potential corresponds to different physical models, the difference between LCAO, OPW or LAPW methods lies in different choice of basis functions for the expansion of the one-electron eigenfunction. These are atomic orbitals in LCAO, plane waves in OPW etc. The Wannier function method involves the orbitals constructed from the atomic orbitals of all atoms in the primitive cell.

The calculations of electron-band structures of short-period SL's were performed within both engineering methods (for references see e.g. [18,21,29]) and first-principles approaches [18,24-28]. We shall try to apply our group-theory results to the recent first-principles band-structure calculations presented in [18,27,28] (LDA), [24,26] (pseudopotential), [25] (LMTO). Though some trends of all the calculations coincide, the comparison of our results with the data of calculations is hindered by a lack of the complete symmetry assignment of energy bands in the latter, except for [18,27], where the symmetries of uppermost valence-band states and lowest conduction-band ones are given for every SL's. The spin-orbit interaction was taken into account only in [24,25].

For every [001]-grown short-period SL, the results approximately coincide when the upper valence-band states are considered (the results are summarized in Figure 10), the uppermost valence-band state being Γ_6. Next one has the Γ_7 symmetry. The separations between these states are Δ_1=21 meV [25] for the (1x1) SL, and Δ_2=12 meV [25] for the (2x2) SL. According to the evaluations of [24,25],

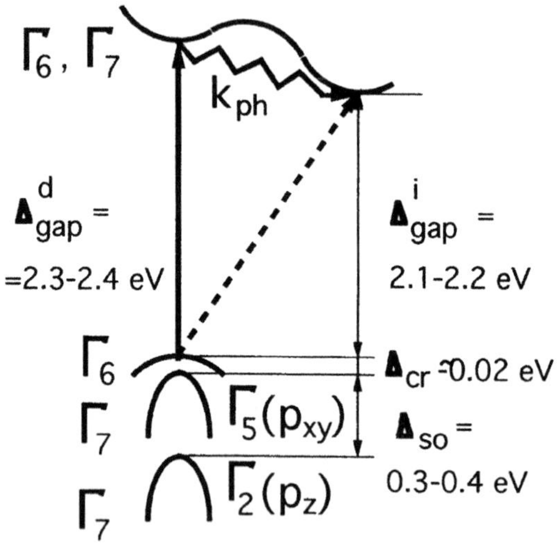

Figure 10. A schematic band-structure diagram of short-period $(GaAs)_m(AlAs)_n$ [001] superlattices

Δ is approximately the same for all the short-period SL's, varying in the range of 0.1-0.2 eV. The spin-orbit splitting between the two (Γ_6 and Γ_7) band states and the lower Γ_7 one is approximately the same as in the case of bulk GaAs (0.34$\pm$0.02) eV[25]. When neglecting the spin-orbit interaction, the uppermost states are Γ_5 and Γ_2 (or Γ_{5v} and Γ_{4v} in the notation of [18,24,27]) with the gap between them being Δ_1=50 meV, Δ_2=20 meV [18] or even less [24]. Therefore, upon including the spin-orbit interaction, the valence-band state Γ_6 can be considered as derived from the Γ_5 one (Sect.3.1), whereas the Γ_7 state as derived from both Γ_2 and Γ_5 states, provided that the contributions of the remote upper states can be neglected.

The (1x1) SL has an indirect gap according to the most of the calculations, the conduction-band minimum being at the R point of the SLBZ (R_1-state that corresponds to the R_5 state when spin-orbit interaction is included). The calculated band gap is equal to 1.88 eV [18], 1.69 eV [25], 1.85 eV [27], and 1.93 eV [28]. Another minimum is at the M point (M_5-state). It lies at 2.10 eV [18], 1.92 eV [26], 2.13 eV [27], 2.12 eV [28]. At the Γ point, the lowest conduction-band state lies higher than the R_1-state by 0.29 eV [18], 0.24 eV [25], 0.26 eV [24], or 0.09 eV [28].

There are essential contradictions between the results of the calculations concerning the ordering in energy of the conduction bands at the Γ point. It results in different optical spectra to be expected. According to [18,24], in the (1x1) SL, the Γ_1 state is higher than the Γ_2 state by 10 meV [18] or 300 meV [24]. In [27], a different order was

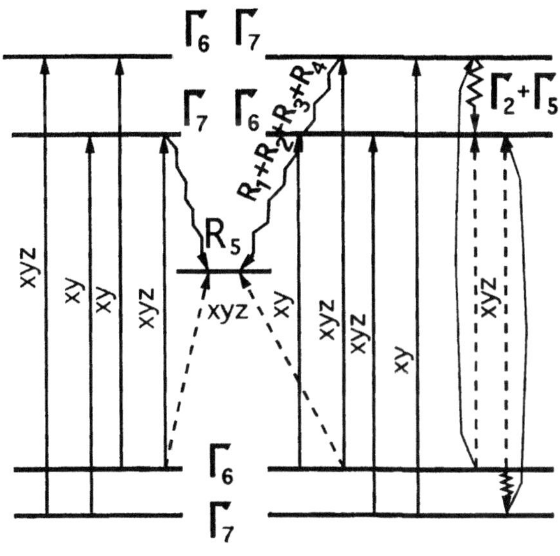

Figure 11. Possible energy-level diagram and dipole-allowed (solid lines) and phonon-assisted (dashed lines) optical transitions in the $(GaAs)_1(AlAs)_1$ [001] SL

obtained, the lowest conduction-band state being Γ_1 rather than Γ_2 (the latter is 120 meV higher). Unfortunately, none of the calculations takes into consideration the spin-orbit interaction. Since the single-valued representations Γ_1 and Γ_2 correspond, respectively, to the double-valued ones Γ_6 and Γ_7, it means that the different calculations lead to opposite conclusions concerning the symmetry of the lowest conduction-band Γ state (Γ_6 or Γ_7). Besides, the spin-orbit interaction-induced mixing with lower-lying states may induce energy shifts of the states. In [25], the lowest state at the Γ point in (1x1) SL is supposed to be Γ_6 as in GaAs and AlAs bulk crystals but without proofs of this statement.

That is why we consider two possible energy-level diagrams in the case of (1x1) SL (Figure 11) depending on either the Γ_6 or the Γ_7 conduction-band state is lower. If the lowest conduction-band state has the Γ_7 symmetry, the direct transition with the lowest energy is allowed in every polarization. Two next transitions are forbidden in the z polarization. And, finally, the fourth direct transition with the highest energy is allowed in every polarization. On the contrary, if that band state has the Γ_6 symmetry, the first and fourth direct transitions are forbidden in the z polarization, the two others being allowed. In both cases, the transition between the Γ_6 valence-band state (derived from the Γ_5 one) and the Γ_7 conduction-band state (derived from Γ_2 one) is allowed in the z polarization only in the strength of the spin-orbit interaction, whereas the transition between the Γ_7 valence-band state (originated from both the Γ_2 and Γ_5 ones) and the Γ_6 conduction-band state (derived from the Γ_1 one) is completely allowed (cf. Table 15). Polarized-absorption experiments could therefore allow to determine the order of the states.

The phonon-assisted transitions could also provide additional information about the band structure of the (1x1) SL. As a matter of fact, the indirect optical absorption edge corresponds to the transition into the R conduction-band state and is 0.2-0.3 eV lower than the direct one. This should be separated from the so-called "indirect" transitions that may appear due to the breaking of wave-vector-conservation rule by defects, disorder, impurities, interface roughness, etc. in actual SL's. Since, at the center of the BZ, the gap between two uppermost valence-band states is small, it may lead to the appearance of rather weak Γ-phonon-assisted transitions in polarized spectra at the energy positions of forbidden direct transitions. For example, in the case of uppermost valence-band state and the lowest conduction-band one having the same symmetry Γ_6 (the right part of Figure 11), the direct transition is forbidden in the z polarization whereas the (Γ_2+Γ_5)-phonon-assisted transition with the same energy is allowed. Moreover, there exist two possible transitions involving either the virtual Γ_7 valence-band state or the virtual Γ_7 conduction-band one (cf. Table 18). It can manifest in both absorption and luminescence polarized spectra. One should note that the fine structure of the phonon-assisted transitions is determined not only by the electronic structure but also by the peculiarities of the phonon density of states at the corresponding point of the BZ (the maximal phonon energy is about 50 meV [34]).

The contradictions between the calculations are important in the case of the (2x2) SL. The authors of [18] have obtained that the (2x2) SL has a direct gap with the lowest conduction-band state Γ_1 (Γ_{1c} in their notation) at 2.02 eV, M_4 state lying 40 meV higher, and the next Γ_1 state lying 210 meV higher. In [25], the lowest state is Γ_6 (spin-orbit interaction was taken into account) at 2.03 eV. In [26], a M state is also 40 meV higher than the lowest Γ state lying at 1.85 eV. In [27], a different order of the conduction-band states was obtained. According to the calculations of [27], the conduction-band minimum is at the M point of the SLBZ (M_4 state), so this SL has an indirect gap as well as the (1x1) SL. At the Γ point, the lowest conduction-band state is Γ_1 (that corresponds to the results of [18]) but the next Γ_1 state is only 50 meV higher. According to [28], the energy positions of conduction-band minima at the Γ and M points coincide (2.10 eV). Thus, all the calculations give a small difference in energy positions of the lowest Γ_1 (Γ_6) and M_4 (M_7 with spin-orbit interaction included) states in the case of the (2x2) SL.

On analyzing a number of experimental data, the authors of [19,20] conclude that, for this SL as well as for every $(GaAs)_m(AlAs)_n$ [001]-grown SL with $1 < m,n < 4$, the conduction-band minimum lies at the M point of the SLBZ (folded X_{xy} state in their terminology), a Γ-state (folded X_z state) being some tens meV higher. The authors of [20] note a possible role of the interchange of Ga and Al atoms across the interface, which may result in the essential lowering of the M-state energy.

The calculations give that, for the (2x2) SL, the lowest conduction-band states at the center of the BZ are the Γ_1 states, the nearest non-Γ_1 state lying as far as 0.5 eV higher (a Γ_2 state). Provided that the spin-orbit interaction-induced mixing with the remote states is negligible, one can conclude that the lowest conduction-band Γ-states have the Γ_6 symmetries. The direct optical transition with the lowest energy is then forbidden in the z polarization, as well as the next one involving the higher-lying Γ_6 state. At the same time, the transitions between the Γ_7 valence-band state and Γ_6 conduction-band ones are completely allowed. The energy-level diagram will be similar to that shown in Figure 12.

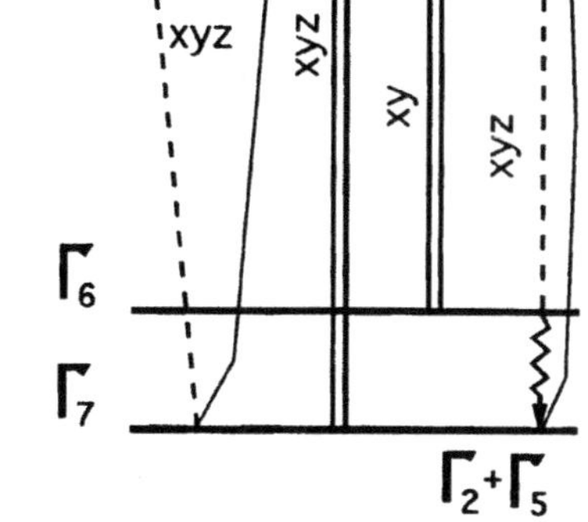

Figure 12. Possible energy-level diagram and dipole-allowed (solid lines) and phonon-assisted (dashed lines) optical transitions in the $(GaAs)_2(AlAs)_2$ [001] SL

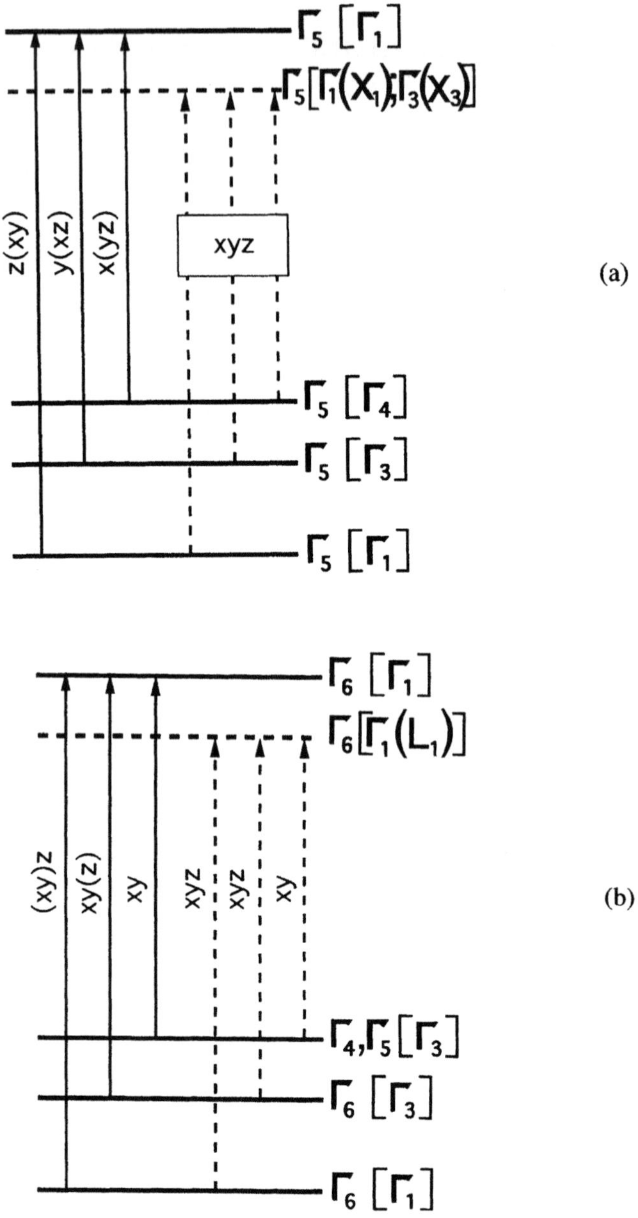

Figure 13. Possible energy-level diagrams and dipole-allowed direct optical transitions in $(GaAs)_m(AlAs)_n$ superlattices:
a) [110]-grown; b) [111]-grown.
The Γ–states of SL's not originating from Γ–states of the parent materials and corresponding optical transitions involving these states are shown by dashed lines. The states without spin-orbit interaction are given in brackets, the corresponding folded bulk states are shown in parentheses.

As in the case of the (1x1) SL, there exist Γ-phonon-assisted transitions what makes it possible to observe the lowest energy Γ-Γ optical transition in the z polarization as well. There may be also observed phonon-assisted indirect optical Γ-M transitions, which are weakly allowed in every polarization.

We can roughly outline the picture of interband transitions in $(GaAs)_m(AlAs)_n$ [110]- and [111]-grown SL's. Figure 13 schematically presents energy-level diagrams for these SL's. For not to overload the diagrams, only states at the center of SLBZ and direct Γ - Γ transitions are included in the Figure. Energy levels corresponding to possible (in the case $m + n = 2k$) folded band states and corresponding transitions involving these states are shown by dashed lines. Allowed polarizations of the transitions are pointed out at corresponding arrows, polarizations allowed only due to the spin-orbit interaction being taken in parentheses. In square brackets, the genesis of the band states is indicated, i.e. irreps of original states determining the distinct features of the band states are given. Since a group theory analysis alone does not permit to determine the energy positions of different band states, these diagrams could be specified on obtaining additional information from both band-structure calculations and experiments.

The analysis of the selection rules, results of band calculations, and the data of polarized-light optical experiments is to create the picture of electron-band structures of these superlattices. Studying the fine structure of the light-absorption spectra of high-quality SL's and comparing the oscillator strengths of lines in different polarizations one would get information of the nature of the lines (direct transitions, phonon-assisted ones) and evaluate the strengths of both spin-orbit and electron-phonon interactions.

Similar analysis of polarized-light optical spectra could be made in the case of the $(GaAs)_m(AlAs)_n$ SL's with greater m and n.

Summing up, we can say that only experiments could decide in favour of a particular model of band-structure calculations. Nevertheless, when comparing the obtained results with the experimental data one should take into account not only the possible uncertainties arising from the theoretical approximations but also the imperfections of actual SL's. Both the group-theory results and band-structure calculations clearly show that the optical spectra are very sensitive to any variation of SL geometry. The actual structures can differ from the adopted model because of different reasons. For example, the lattice-parameter difference between GaAs and AlAs bulk crystals can induce electron energy shifts. This is particularly important if one takes into account the role of the substrate [19]. Besides, the interface roughness can locally vary m and n by one or several units and thereby change the SL symmetry.

5. Conclusion

Summing up, the main results can be formulated as follows.

5.1. SUPERLATTICES GROWN ALONG THE [001] DIRECTION

i) The [001] SL's belong to two crystal families specified by space groups D^5_{2d} and D^9_{2d} depending on $m+n$ is even or odd. In both cases, there are several subfamilies differing by odd or even numbers of monolayers both in barriers and wells. The main difference between these subfamilies is the interchange of occupation of the same Wyckoff positions by As and Me (Ga, Al) atoms.

ii) When one goes from one SL to another, the variations of unit cell geometry resulting in rearrangement of atoms over the Wyckoff positions lead to drastic changes in symmetry of electron and phonon states and in contributions of a particular atomic orbital in the electron states. The latter results in the localization of particular electron bands within either the barriers or the wells. All these changes manifest themselves in optical spectra mainly due to a change of the symmetry of the conduction-band bottom state.

iii) The selection rules have been derived both for direct and phonon-assisted transitions, the spin-orbit interaction having been taken into account. The latter may be of great importance in short-period SL's. The analysis of the selection rules, results of more reliable band calculations, and the data of polarized-light optical experiments will allow to create more precise picture of electron-band structures of these materials.

5.2. SUPERLATTICES GROWN ALONG THE [110] AND [111] DIRECTIONS

i) The $(GaAs)_m(AlAs)_n$ [110] SL's belong to 3 families specified by space groups C_{2v}^1, C_{2v}^7, and C_{2v}^{20} whereas the $(GaAs)_m(AlAs)_n$ [111] ones constitute two families C_{3v}^1 and C_{3v}^5. In turn, depending on specific numbers of monolayers of materials constituting the unit cell, the crystal families described by C_{2v}^1, C_{2v}^{20}, and C_{3v}^1 space groups are divided into subfamilies corresponding to non-equivalent atomic arrangements over the Wyckoff positions in the unit cell.

ii) In contrast to [001] SL's, for [110] and [111] SL's, the rearrangement of atoms over the Wyckoff positions within each family is more subtle and results in variation of occupation of Wyckoff positions with the same site symmetry by atoms of the same type.

iii) It was found, that for the [111] SL's there is a splitting of Γ-states due to the spin-orbit interaction in contrast to the [110] SL's where such splitting is absent. Moreover, with spin-orbit interaction being taken into account, all Γ-states have the same symmetry in the [110] SL's.

iv) The selection rules for direct and phonon-assisted optical transitions including the case when spin-orbit interaction is taken into account were established. An hierarchy of optical transitions, i.e. allowed, allowed due to spin-orbit interaction and forbidden ones has been obtained.

Acknowledgement

We acknowledge the HTECH.CRG 960664 NATO grant.

6. References

1. Kitaev Yu.E., Panfilov A.G., Tronc P., and R.A.Evarestov (1997) *J.Phys. Condensed Matter* **9**, 257.
2. Kitaev Yu.E., Panfilov A.G., Tronc P., and R.A.Evarestov (1997)

J.Phys. Condensed Matter **9**, 277.
3. Kitaev Yu. E. and Evarestov R. A. (1988) *Fiz.Tverd.Tela* **30,** 2970 (Engl. Transl. (1988) *Sov.Phys.-Solid State* **30,** 1712).
4. Evarestov R.A. and Smirnov V.P. (1993) *Site Symmetry in Crystals: Theory and Applications* (*Springer Series in Solid State Sciences*) vol 108, ed. M. Cardona, Springer, Heidelberg.
5. Bairamov B.H., Gant T.A., Delaney M., Kitaev Yu.E., Klein M.V., Levi D., Morkoç H., and Evarestov R.A. (1989) *Zh.Eksp.Teor.Fiz.* **95,** 2200 (Engl. Transl. (1989) *Sov.Phys.-JETP* **68,** 1271).
6. Bairamov B.H., Evarestov R.A., Kitaev Yu.E., Jahne E., Delaney M., Gant T.A., Klein M.V., Levi D., Klem J., and Morkoç H. (1991) *J.Phys.Chem.* **95,** 10772.
7. Cardona M., Suemoto T., Christensen N.E., Isu T., and Ploog K. (1987) *Phys.Rev. B* **36,** 5906.
8. Kuan T.S., Kuech T.F., Wang W.I., and Wilkie E.L. (1985) *Phys.Rev.Letters* **54,** 201.
9. Gomyo A., Suzuki T., and Iijima S. (1988) *Phys.Rev.Letters* **60,** 2645.
10. Allen L.T.P., Weber R.E., Wasburn J., and Pao Y.C. (1987) *Appl.Phys.Lett.* **51** 670
11. Cloizeaux J. (1963) *Phys.Rev.* **129,** 554.
12. Kovalev O.V. (1975) *Fiz.Tverd.Tela* **17,** 1700 (Engl. Transl. (1975) *Sov.Phys.-Solid State* **17,** 1106).
13. Zak J. (1981) *Phys.Rev.B* **23,** 2824.
14. Evarestov R.A. and Smirnov V.P. (1984) *Phys. Stat. Sol. b* **122,** 231, 559.
15. Miller S.C. and Love W.F. (1967) *Tables of Irreducible Representations of Space Groups and Co-Representations of Magnetic Space Groups* , Pruett, Boulder.
16. Bradley C.J. and Cracknell A.P. (1972) *The Mathematical Theory of Symmetry in Solids* , Clarendon, Oxford.
17. *International Tables for Crystallography* (1983), *vol A, Space Group Symmetry,* ed.T.Hahn, Reidel, Dordrecht.
18. Wei S.-H. and Zunger A. (1988) *J.Appl.Phys.* **63,** 5794.
19. Li G.-H. (1992) *Semiconductor Interfaces and Microstructures* , ed Z.C. Feng, World Scientific, Singapore, p 120.
20. Ge W., Schmidt W.D., Sturge M.D., Pfeiffer L.N., and West K.W. (1994) *J.Lumin.* **59,** 163.
21. Sham L.J. and Lu Y.-T. (1989) *J.Lumin.* **44,** 207.
22. Fu Y., Willander M., Ivchenko E.L., and Kiselev A.A. (1993) *Phys.Rev B* **47,** 13498.
23. Parmenter P.G. (1955) *Phys.Rev.* **100,** 573.
24. Andreoni W. and Car R. (1980) *Phys.Rev.B* **21,** 3334.
25. Gopalan S., Christensen N.E., and Cardona M. (1989) *Phys.Rev B* **39,** 5165.
26. Yan-Ten Lu and Sham L.J. (1989) *Phys.Rev.B* **40,** 5567.
27. Zhang S.B., Hybertsen M.S., Cohen M.L., Louie S.G., and Tomanek D. (1989) *Phys. Rev. Letters* **63,** 1495.
Zhang S.B., Cohen M.L., Louie S.G., Tomanek D., and Hybertsen M.S. (1990) *Phys.Rev.B* **41,** 10058.
28. Mäder K.A. and Zunger A. (1994) *Phys.Rev.B* **50,** 17393.
29. Smith D.L. and Mailhot C. (1990) *Rev.Mod.Phys.* **62,** 173.

30. Bastard G. (1992) *Wave mechanics applied to semiconductor heterostructures* Les Editions de Physiques, Les Ulis.
31. Kane E.O. (1957) *J.Phys.Chem.Sol.* **1** , 249.
32. Volkov V.A. and Pinsker T.N. (1979) *Surf.Sci.* **81,** 181.
33. Bassani F. (1966) *Semiconductors and Semimetals* , vol. 1, ed. R.K.Willardson and A.C.Beer, Academic Press, N.Y.-London, p. 21.
34. Menendez J. (1989) *J.Lumin.* **44,** 285.

ELECTRONIC STRUCTURE OF ALAS/GAAS SUPERLATTICES WITH AN EMBEDDED CENTERED GAAS QUANTUM WELL

V.DONCHEV, TZV.IVANOV AND K.GERMANOVA
Faculty of Physics, Sofia University "St.Kl.Ohridski"
5, blvd. J.Bourchier, Sofia-1164, Bulgaria

Quantum wells (QWs) of GaAs embedded in short-period AlAs/GaAs superlattices (SLs) present both interesting physical properties and practical advantages in the fabrication of optoelectronic devices [1-6].

In this work we calculate the electronic structure of GaAs QWs embedded in AlAs/GaAs SLs in order to study the nature and the conditions of existence of the interaction between the SL and the embedded well (EW).

The structures considered are formed by a GaAs QW embedded in an AlAs/GaAs SL (4.5 periods on both sides) which is surrounded by semi-infinite uniform regions of $Al_xGa_{1-x}As$ with x equal to the mean Al content in the SL. The growth direction z is [100]. The conduction E_c and valence E_v band profiles of the structures considered can be seen in Fig.1. An unperturbed SL with 9 wells (and two boundary wells as shown in Fig.1) and a single QW (SQW) with AlAs barriers have been also studied for comparison.

The calculations are made in the framework of the envelope-function approximation [7]. The solutions of the Schrödinger equation ψ in each layer are matched at each interface in order to obtain continuity of ψ and $\frac{1}{m^*}\frac{d\psi}{dz}$ [4,5,7]. One band model is used with effective masses m^* 0.0665, 0.327 and 0.09 for electrons, heavy holes (hh) and light holes (lh), respectively in GaAs and 0.15, 0.478 and 0.208 for electrons, hh and lh, respectively in AlAs. Taking into account the mirror symmetry and putting the z-axis zero in the middle of the EW it is enough to consider only the half of the structure assuming consecutively even (cos) and odd (sin) solutions ψ in the EW.

The results for a GaAs QW of 42 monolayers (11.86nm) embedded in an $(AlAs)_4/(GaAs)_8$ SL are listed in Table 1 together with the confinement energies for a SQW. The electron states with numbers n=0,1,13 and 14 correspond in energies and wave function localizations to the states with

M. Balkanski and N. Andreev (eds.),
Advanced Electronic Technologies and Systems Based on Low-Dimensional Quantum Devices, 51–54.

TABLE 1. Confinement energies (in meV) of a GaAs SQW with AlAs barriers and a GaAs QW of 42 monolayers embedded in an $(AlAs)_4/(GaAs)_8$ SL. n is the level number in the case of the EW, where the localization of each state is also given (EW-embedded well; SL-superlattice; BW-boundary well)

n	Electrons			Heavy holes			Light holes		
	SQW	EW		SQW	EW		SQW	EW	
0	28.6	28.4	(EW)	6.8	6.8	(EW)	19.9	19.6	(EW)
1	115.2	114.0	(EW)	27.3	27.1	(EW)	80.3	78.5	(EW)
2		215.7	(BW)	61.5	61.0	(EW)		124.7	(BW)
3		215.7	(BW)		78.6	(BW)		124.7	(BW)
4	261.2	248.6	(EW;SL)		78.6	(BW)		148.2	(SL;EW)
5		258.3	(SL)		94.3	(SL)		150.1	(SL)
6		262.7	(SL;EW)		94.4	(SL)		163.3	(SL;EW)
7		283.2	(SL)		97.6	(SL)		170.4	(SL)
8		288.1	(SL;EW)		98.1	(SL)	182.7	183.5	(EW;SL)
9		314.7	(SL)		102.1	(SL)		200.7	(SL)
10		318.5	(SL;EW)		102.6	(SL)		208.2	(SL;EW)
11		347.2	(SL)		106.2	(SL)		231.6	(SL)
12		348.7	(SL;EW)		106.5	(SL)		234.2	(SL;EW)
13	467.6	471.5	(EW)	109.2	109.7	(EW)	327.3	334.2	(EW)
14	732.3	735.1	(EW)	170.5	169.7	(EW)	505.5	518.0	(EW)
15		1039.1	(EW)	245.1	243.6	(EW)			

n=0,1,3 and 4 for a SQW with AlAs barriers. The SL states are grouped in 4 couples of one odd and one even state. This is a result from the degeneracy of the levels of the two SL subsets with 4 wells each [4]. The odd states are unperturbed in comparison with a SL with 9 equal wells because their wave functions are zero inside the central well. This is not the case for the even states (see below).

The energy of the third state of the SQW (261.2meV) is within the first mini-band of the unperturbed SL. In the case of the EW the energy of the corresponding state (248.6meV) is shifted outside the SL mini-band, but is very close to it. His wave function shows tree maxima in the EW similarly to the case of a SQW. However a strong delocalization in the SL is also observed (Fig.1a).

On the other hand the SL even states (n=6,8,10,12) are shifted in energy in comparison with the case of the unperturbed SL. Besides, they penetrate in the EW and their wave functions inside show also the tree maxima characteristic for the third state of the SQW (Fig.1b). This penetration is more pronounced for states which are closer in energy to the state (n=4)

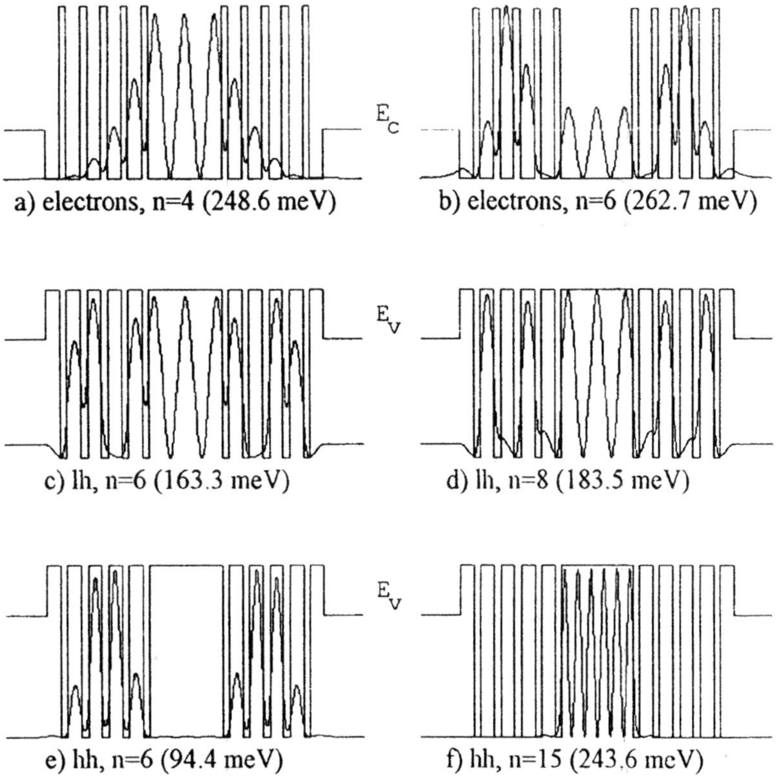

Figure 1. **Wave functions squared and energies for some electron and holes states of a GaAs QW of 42 monolayers embedded in an $(AlAs)_4/(GaAs)_8$ SL**

coming from the EW.

Thus, an interaction between the SL and the EW is observed. It is connected to the fact that the third electron state of the EW is close in energy to the SL mini-band.

This interaction exist also for light holes (see Table 1). In this case it is stronger (Fig.1c,d) because the level coming from the EW (n=8) is inside the SL mini-band. In the case of heavy holes the levels of the SQW are far enough from the SL mini-bands and therefore the interaction is very week. The SL states practically do not penetrate in the EW (Fig.1e). All the states coming from the EW are localized inside (Fig.1f) and correspond in energy to the levels of the SQW.

In order to study more completely the interaction between the SL and the EW and the conditions where it is manifested we have considered also a GaAs QW with variable width embedded in an $(AlAs)_8/(GaAs)_{16}$ SL. We have concentrate to the behaviour of the first SL mini-band and the first level originating from the EW. When the width of the EW is very different from that of the wells in the SL the EW works as a single QW. The SL breaks down to two identical subsets of 4 wells each, the EW acting as an effective barrier between them. The degeneracy of their levels give 4 pairs of closely spaced levels which wave functions practically do not penetrate in the EW.

When the width of the EW approaches that of the wells in the SL the first level of the EW enter the SL mini-band. Its wave function is delocalized and it becomes undistinguishable from the SL states, which are now 9. Their wave functions are similar to those of an unperturbed SL with 9 wells. The even states shift in energy away from the odd ones and now penetrate in the central well. These results are in good agreement with those reported in [4] for a SL of InGaAsP/InP with a narrowed or widened central well. Thus, in the case considered the interaction between the SL and the EW is observed when the first state of the EW is close in energy to or inside the SL mini-band. The nature of the interaction is similar to that in the previous case (Table 1 and Fig.1).

In conclusion a centered EW breaks down the SL in two subsets and acts as an effective barrier between them. An interaction between the SL and the EW occurs when a state of the EW is close in energy to or inside the SL mini-band. It is manifested by i) a delocalization of the corresponding EW state in the SL and ii) an energy shift and a penetration of the SL even states in the EW.

This work was supported by the Bulgarian National Science Fund and the Research Fund of the Sofia University "St.Kl.Ohridski".

References

1. Sakaki, H., Tsuchiya, M. and Yoshino, J. (1985) Energy levels and electron wave functions in semiconductor quantum wells having superlattice alloylike material (0.9nm GaAs/0.9nm AlGaAs) as barrier layers, *Appl.Phys.Lett.* **47**, 295-297.
2. Arriaga, J. and Velasko, V. R. (1995) Electronic states of a semi-infinite superlattice with an embedded quantum well, *J.Phys.:Condens.Matter* **7**, 3493-3500.
3. Donchev, V., Ivanov, I. and Germanova, K. (1996) Optical and Theoretical Assessment of GaAs Quantum Wells Having Superlattices as Barrier Layers, in M.Balkanski (ed.), *Devices Based on Low-Dimensional Semiconductor Structures*, Kluwer Academic Publishers, Dordrecht, pp.175-178.
4. Waki, M. and Watari, K. (1995) Localization of Electronic States in Semiconductor Superlattices, *Superlattices and Microstructures* **17**, 111-115.
5. Kucharczyk, R. and Steslicka, M. (1992) A finite superlattice with the embedded quantum well, *Solid State Commun.* **81**, 557-561.
6. Blood, P., Fletcher, E.D., Foxon, C.T. and Griffiths, K. (1989) Recombination processes in quantum well lasers with superlattice barriers, *Appl.Phys.Lett.* **55**, 2380-2382.
7. Altarelli, M. (1886) in G.Allan, G.Bastard, N.Boccara, M.Lannoo and M.Voos (eds.) *Heterojunctions and Semiconductor Superlattices*, Spinger, Berlin, p.12.

ELECTRONIC STATES IN GRADED COMPOSITION QUANTUM WELLS UNDER A CONSTANT ELECTRIC FIELD

S. VLAEV, A. MITEVA (*) and V. DONCHEV (**)
Bulgarian Academy of Sciences - Institute of General and Inorganic Chemistry, Bul. Acad. G. Bonchev bl.11, Sofia 1113, Bulgaria,
(*) *Bulgarian Academy of Sciences - Space Research Institute, P.O.Box 799, Sofia 1000, Bulgaria, and*
(**) *Sofia University, Department of Condensed Matter Physics, Bul. J. Boucher 5, Sofia 1164, Bulgaria*

1. Model and Method

In this work we present a numerical calculation of electron and hole states in single Al(x)Ga(1-x)As quantum wells (QWs) with rectangular and linear concentration profile under application of a constant electric field applied perpendicular to the QW layers.

We have used for numerical calculations the tight-binding (TB) approximation and the algorithm described in [1]. It makes possible the application of surface Green function matching method for matching a final inhomogeneous region with semi-infinite homogeneous regions. Besides it allows realistic TB calculations for electronic states in rectangular and graded composition QWs in the presence of a constant electric field (F). We describe the presence of an external electric field perpendicular to the interfaces with shifting of the diagonal terms of the empirical TB Hamiltonian matrix by the corresponding potential drop Δ across one monolayer. An electric field of F = 70.8 kV/cm corresponds to Δ = 1 meV. The electric field is applied between the first (n = 1) and the last (n = N) monolayer (ML) of the QWs, where the layer index n grows along the growth direction. The cation is in n = 1. In the barrier regions the field is F = 0 for both QWs. The width of both QWs is 12.4 nm or N = 44 ML. The growth direction is [100]. The Al concentration x is a constant in every layer, but is a function of n. In the barriers for both QWs x = 0.36. In the rectangular QW (RQW) x = 0. In the linear QW (LQW) x varies linearly with n from 0.02 at the left edge to 0.36 at the right one. The calculations are made for the temperature T = 0 K.

M. Balkanski and N. Andreev (eds.),
Advanced Electronic Technologies and Systems Based on Low-Dimensional Quantum Devices, 55–58.

2. Results and Discussion

Figure 1 shows the calculated main optical transition energies E(e1-hh1), E(e1-lh1), E(e2-hh2) without and in the presence of a constant electric field for linear and rectangular QWs. The transition energies are larger in the linear than in the rectangular QWs without and under application of the same electric field. For both QWs the transition energies decrease with the increasing of the applied electric field. This is more pronounced for the RQW.

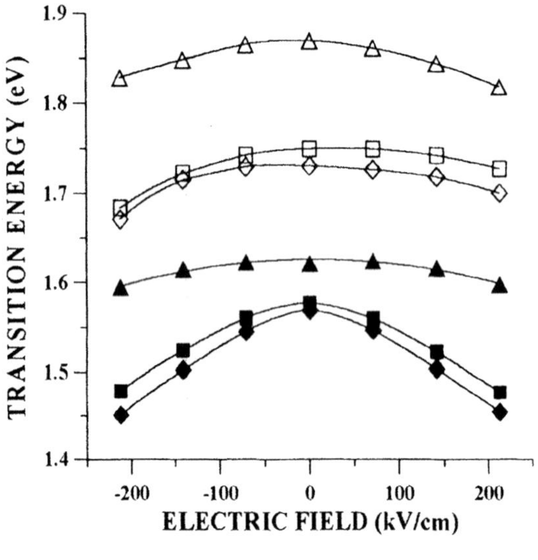

Figure 1. Transition energy as a function of applied electric field for LQW (empty symbols) and RQW (full symbols): E(e1-hh1) - diamonds; E(e1-lh1) - squares; E(e2-hh2) - triangles.

The comparison with the experimental data for QWs with similar concentration profiles [2-5] is shown in Table 1. The temperature dependence of the GaAs band gap [6] is taken into account by making the corresponding corrections in the experimental data. It is seen that there is a good correspondence between our calculated values and the experimental ones. Some discrepancies can be due to the differences in the concentration profiles and excitonic effects (which have not been taken into account).

We have calculated the spatial distributions of the spectral strength for each well state and its orbital components in the QWs under study. Figure 2 and Figure 3 show the spatial distributions of the total spectral strength for the conduction e1 and the valence band hh1 bound states without (a) and in the presence (b) of the constant electric field F = 70.8 kV/cm. In both QWs the corresponding distributions for e1 and hh1 have the amplitudes displaced in the same direction in the presence of the electric field. The displacement for the RQW is larger than for the LQW, which is a result of the concentration profile difference.

TABLE 1.Calculated and experimental values for the Stark shift (in eV) of E(e1-hh1) for LQW and RQW.

Field	LQW			RQW		
(kV/cm)	This work	[4]	[5]	This work	[2]	[3]
70.8	0.004	0.011	0.012	0.022	0.006	0.015
141.6	0.012			0.065	0.030	
QW width (nm)	12.4	10.1	10.0	12.4	10.0	10.8

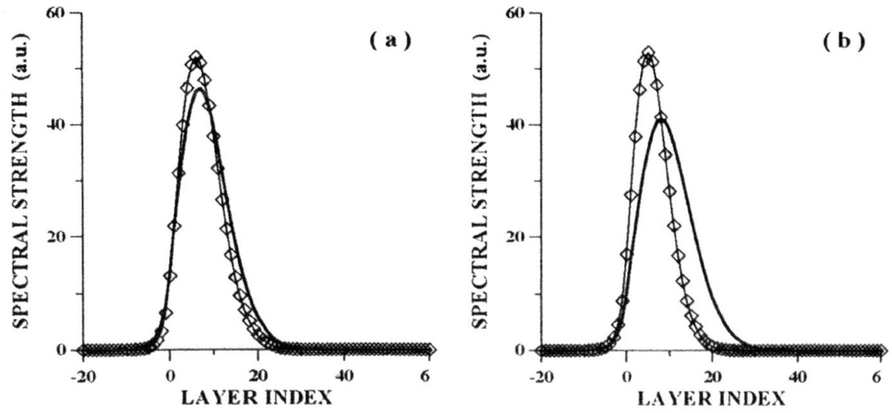

Figure 2: Spectral strength for LQW, (a) with F = 0, (b) with F = 70.8 kV/cm, e1 (solid line), hh1 (solid line with diamonds).

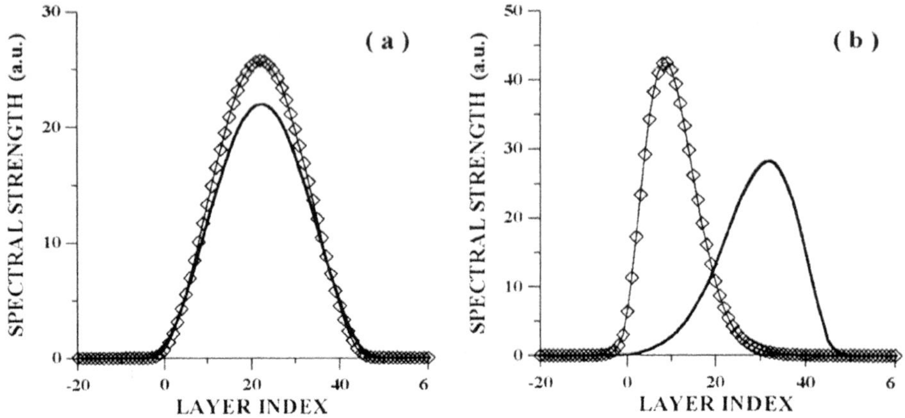

Figure 3: Spectral strength for RQW, (a) with F = 0, (b) with F = 70.8 kV/cm, e1 (solid line), hh1 (solid line with diamonds).

3. Conclusion

The RQW shows a larger Stark shift of the corresponding transition energies than the LQW which is better for the fabrication of field-controlled optical devices. However, the LQW under application of the same electric field will have less decrease of the intensity of the main optical transition E (e1-hh1) than the RQW which is also desired for achieving high device performance. So, the graded gap quantum well structures can exhibit some new and useful optical properties for use in optoelectronic devices.

This work was financially supported by Grant X-646 from the Bulgarian National Fund for Scientific Investigations.

References

1. Vlaev, S., Velasco, V.R., and Garcia-Moliner, F. (1994) Electronic states in graded-composition heterostructures, *Phys. Rev.*B **49**, 11222-11229.
2. Polland, H.-J., Schultheis, L., Kuhl, J., Gobel, E.O., and Tu, C.W. (1985) Lifetime enhancement of two dimensional excitons by the quantum-confined Stark effect, *Phys. Rev. Lett.* **23**, 2610-2613.
3. Ishikawa, T., Nishimura, S., and Tada, K. (1990) Quantum-confined Stark effect in a parabolic-potential quantum well, *Jpn. J. Appl. Phys.* **29**, 1466-1473.
4. Ishikawa, T. and Tada, K. (1989) Observation of quantum-confined Stark effect in a graded-gap quantum well, *Jpn. J. Appl. Phys.* **28**, L1982-L1984.
5. Hiroshima, T. and Nishi, K. (1987) Quantum-confined Stark effect in graded-gap quantum wells, *J. Appl. Phys.* **62**, 3360-3365.
6. Madelung, O.(1991) *Semiconductors Group IV Elements and III-V Compounds*, Springer Verlag, Berlin.

DIMENSION RELATED EFFECTS ON THE STRUCTURE PERFECTION IN Si/SiC MULTILAYER STRUCTURES

E.VALCHEVA, T.PASKOVA, O.KORDINA*,R.YAKIMOVA*,E.JANZEN*
Faculty of Physics, Sofia University, 5, J.Bourchier Blvd., 1126 Sofia, Bulgaria
**Department of Physics and Measurements Technology, Linkoping University, 58183 Linkoping, Sweden*

1. Introduction

The development of modern electronics needs production of multiple semiconductor structures with a thick substrate and multiple thin layers on top. Recently, a great effort has been made on heterostructural growth between materials with very different lattice constants such as CVD growth of SiC and Si [1]. There is about 20% mismatch between the lattice constants of cubic SiC and Si that leads to stresses and strains inside the layers. These stresses and strains exert important influence on the physical and technical performance of the layers.

The aim of this study is to explore experimentally the structural characteristics of multilayer structures based on Si/SiC in connection with dimensions of separate layers. Two configurations of five (Si/SiC) bilayers with different thickness were grown by CVD method. The samples were characterised by X-ray diffraction (XRD) and infrared transmission (IR) spectroscopy.

2. Experimental

The epitaxial cubic SiC/Si multilayer structures were grown at Linkoping University using CVD system with a hot-wall reactor. The temperature used for SiC growth was 1200°C and for Si growth was 900°C. The growth rates for both the SiC and Si layers were 0.6±0.1 μm/h. All details of growth procedure were given elsewhere [2]. Figure 1 shows a schematic drawing of the investigated multilayer structures (sample I and II) and their specific thickness. Sample III is a single SiC layer used as a reference.

X-ray diffraction and infrared transmission measurements were performed at Sofia University. The x-ray diffraction patterns were taken using a standard x-ray diffractometer URD-6 with Cu Kα radiation, with a graphite monohromator and with a

M. Balkanski and N. Andreev (eds.),
Advanced Electronic Technologies and Systems Based on Low-Dimensional Quantum Devices, 59–62.

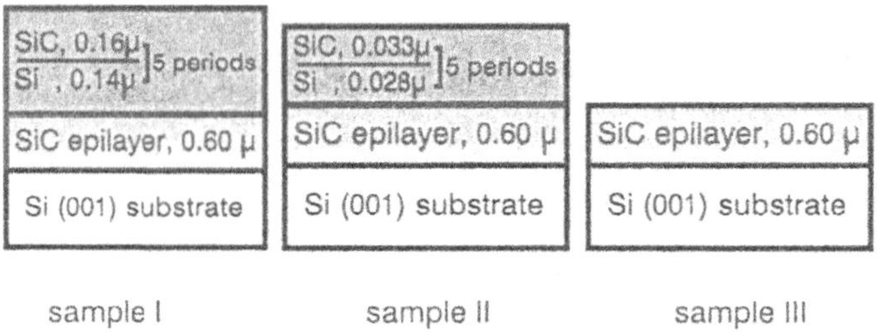

Figure 1. Schematic drawing of the samples.

step size of 0.05°. The angular range varied from 10° to 100°. Infrared transmission measurements were made using a infrared spectrometer in the range from 450 to 1700 cm^{-1} at room temperature.

3. Results and discussion

Figure 2 shows the Θ-2Θ diffractograms obtained from the samples. The diffraction pattern of sample I (fig.2a) consists of the (002) and (004) reflections of the 3C-SiC epilayers and the (002) and (004) reflections of the Si substrate and the epilayer. The high intensities of this peaks indicate a predominantly (001) oriented growth of the layers structure. The diffractogram reveals additional (022) and (111) peaks due to the appearance of misoriented (011)-Si and (111)-Si phases and (022) and (111) reflexes of SiC due to the misoriented (011)-SiC and (111)-SiC domains.

The x-ray spectrum of the multilayer structure with thinner layers (fig.2b) reveals only the reflexes of (001) orientation of both Si and SiC similar to the reference sample spectrum (fig.2c).

All three IR spectra show a deep band situated between 790 and 960 cm^{-1} known to be due to the lattice absorption of SiC. It represents the range of the phonon dispersion curve with the edge frequencies characteristic for TO and LO phonons. Other features common for all spectra are due to the silicon: the dominant in the Si IR absorption spectrum phonon mode TO(X)+TA(X) at 610 cm^{-1} and LO(L)+LA(L) at 820 cm^{-1} appearing as a shoulder on the right hand side of the SiC absorption band. Moreover, in the range 950-1500 cm^{-1} of the spectrum of sample I (the multilayer structure with thicker layers) strong and sharp features appear:a twin dip at 990 and 1010 cm^{-1}, a sharp dip at 1175 cm^{-1} and double structure at 1370 and 1450 cm^{-1}. According to the IR active multiphonon absorption mode of Si they can be ascribed to:

964 cm^{-1}	-	2TO(X,Σ,L)
1040 cm^{-1}	-	2O(Γ)
1175 cm^{-1}	-	CH_3 rocking absorption
1363 cm^{-1}	-	TO(L) + LO(L) + TO(X)
1448 cm^{-1}	-	3TO(Σ) or TO(L) + TO(L) + TO(X)

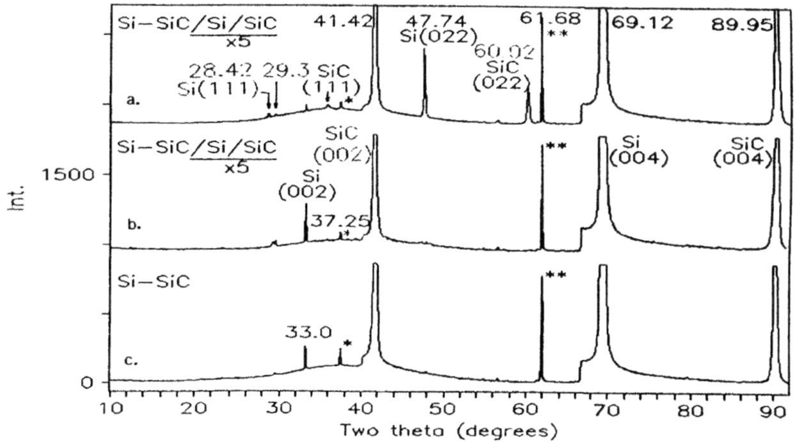

Figure 2. X-ray diffraction spectra of the samples. The peaks labeled by (*) and (**) are due to the β-lines of the Cu radiation.

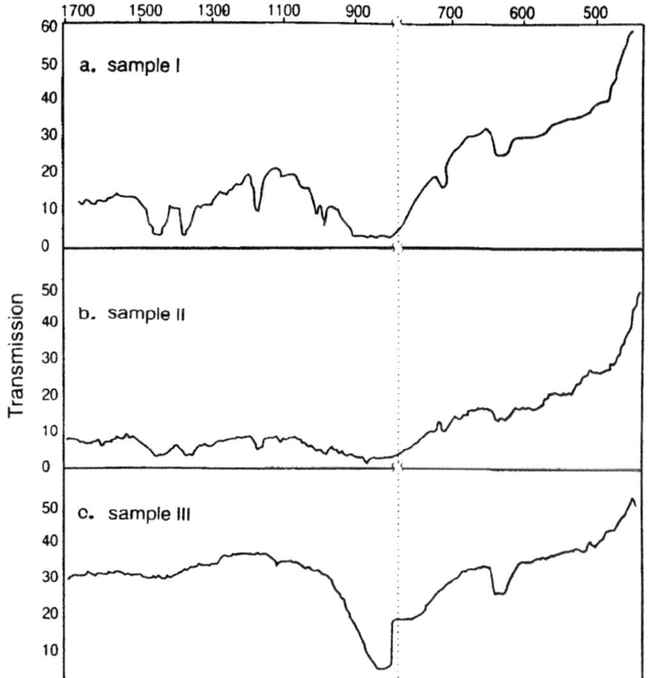

Figure 3. IR spectra of the samples

The spectrum of the thinner layer structure (sample II) weakened multiphonon absorption is observed very similar to the behavior of the reference IR spectrum (fig.3c).

Our previous investigations [2,3] of sequentially growth of SiC and Si layers (SiC single layer on Si substrate, SiC/Si bilayers and SiC/Si/SiC trilayers) show that the first SiC layer is epitaxial to the Si substrate in (001) orientation, while the next grown layers reveal misoriented Si and SiC domains with (011) and (111) orientation. It is worth to note that every new layer replicates the misorientation of the previous and reveals an additional misoriented domains. The misorientation growth causes the appearance of the additional x-ray reflections and IR multiphonon absorption. These specific features are signifficantly reduced and eliminated with the decrease of the layer thickness.

As it is known critical thickness exists for the appearance of misfit dislocations in heterostructures with a big difference in lattice constants. The critical thickness depends on the lattice parameter mismatch between the layers [4]. For SiC and Si the lattice mismatch is 20% and the critical thickness is calculated to be about 50 A. Since the thickness of the layers in sample I is much greater than the critical thickness the presence of misorientation domains is relevant. In the sample in which the layer thickness is lower than the critical thickness, the strains due to the lattice mismatch and difference in thermal expansion coefficients, are expected and they could be responsible for the absence of misoriented domains and improvement of the crystal growth process.

In summary, the results show a strong influence of the dimensions of the layers on the structural characteristics in Si/SiC multiple structures.

Acknowledgements

This work was financially supported by the Swedish Board for Industrial and Technical Development (NUTEK), Sweden, the Research Foundation of Sofia University, under contract 235/1996 and by the National Research Foundation, Bulgaria, under contract F-521.

References

1. Feng, Z.L., Choyke, W.J., and Poweell, J.A. (1988) Raman determination of layer stresses and strains for heterostructures and its application to the cubic SiC/Si system, *J.Appl.Phys.* **64**, 6827-6835
2. Bjorketun, L.-O., Hultman, L., Kordina, O., and Sundgren, J.-O. (1996) Texture evolution in Si/SiC layered structures deposited on Si (001) by chemical vapor deposition, submitted
3. Paskova, T., Valcheva, E., Kordina, O., Surtchev, M., Yakimova, R., and Janzen, E. (1996) CVD grown Si/SiC based multilayer structures, 9th IS on Condensed Matter Physics, 9-13 Sept., submitted
4. Matthews, J.W., and Blakeslee, A.E. (1977) Almost perfect epitaxial multilayers, *J.Vac.Sci.Tecnol.* **14**, 989-991

EFFECT OF THE NON-PARABOLICITY AND DYNAMICAL SCREENING ON THE SECOND-HARMONIC GENERATION IN DOUBLY RESONANT ASYMMETRIC QUANTUM WELL SYSTEMS

M. ZALUZNY* and V. BONDARENKO**

Institute of Physics, M. Curie-Sclodowska University, 20-031,Lublin, pl. M. Curie-Sclodowskiej 1, Poland

***Institute of Physics, National Academy of Sciences of Ukraine,pr. Nauky 46, Kyiv-28, 252650, Ukraine*

The second-harmonic generation due to the intersubband transitions in the near doubly resonant three level non-parabolic quantum well systems is discussed theoretically employing the self-consistent field method and the density matrix formalism.

It is well known that the dynamical screening (the depolarization effect, DE) in the systems with parallel subbands leads to a depolarization shift between the intersubband spacing and the intersubband infrared absorption resonance (the responce). However, what is less known is that the DE modifies also the absorption line due to the non-parabolicity of the constituent materials, with the contribution to the line broadening resulting from the non-parabolicity is, to a large extent, compensated by the DE (see, e.g. [1]).

It is established that the DE modifies the second-harmonic generation (SHG) spectrum in two level asymmetric quantum well (QW) systems. When the subbands are parabolic the resonance in the SHG spectrum occurs when 2hv (or hv) coincides with the depolarization shifted intersubband energy E_{21}' and not the bar intersubband space E_{21} (see, e.g. [2]). In the case of non-parabolic subbands the DE leads to a depolarization shift and line-narrowing of the SHG spectrum near the resonance [3]. Situation is more complex in the doubly resonant three level system, where

M. Balkanski and N. Andreev (eds.),
Advanced Electronic Technologies and Systems Based on Low-Dimensional Quantum Devices, 63–64.

even in the case of the parabolic subbands the DE leads not only to the shift of the peak but also can enhance its maximal value [2].

In this report, extending our previous calculations [4], the SHG spectrum is numerically calculated for the double-resonant non-parabolic $Al_{0.4}Ga_{0.6}As/GaAs/Al_xGa_{1-x}As/Al_{0.4}Ga_{0.6}As$ step QW structure, with the GaAs well thickness to be of 45 Å, the total well thickness to be of 110 Å, $N_s=10^{12}$ cm^{-2}. We use the usually quoted value of 5 meV for the broadening parameter Γ. The calculations show that in our system the condition $E_{21}'=E_{32}'=h\nu$ is achieved when the interface energy barrier is $V_{step}\cong V_{res}=110$ meV for $h\nu\cong h\nu_{res}=100$ meV.

It is found that in the absence of the DE the non-parabolic dispersion of the subbands leads to the broadening of the peak and shifts its maximum towards lower energy. Situation is different in the presence of the DE: (i) the DE shifts the peak towards the higher energy; (ii) the DE leads to substantial enhancement of the maximum peak value. For example, at T=300 K (4 K) the depolarization induced enhancement factor f is of about 6 (3.5) (in the parabolic case it is of about 3). Note, that the enhancement strongly depends not only on temperature but also on the broadening parameter. At T=300 K the increase of Γ from 5 meV to 7.5 meV reduces f nearly two times.

From the presented results we can conclude that correct description of the influence of the non-parabolicity of the subbands on the SHG spectrum in the high quality near doubly resonant systems have to take into account the DE.

REFERENCES

1. Zaluzny M. (1994) Influence of nonparabolicity on collective intersabband spin- and charge-density excitation spectra, *Physical Review* **B49**, 2923-2926.
2. Zaluzny M. (1995) Influence of the depolarization effect on second harmonic generation in asymmetric quantum wells, *Physical Review* **B51**, 9757-9763.
3. Zaluzny M. (in press) On the second harmonic generation spectrum in asymmetric quantum wells, *Acta Physica Polonica A*.
4. Zaluzny M. and Bondarenko V. (1996) Influence of the depolarization effect on third harmonic generation in quantum wells, *Journal of Applied Physics* **79**.

OPTICAL DIAGNOSTICS OF QUANTUM DOTS IN $GaAs/In_xGa_{1-x}As$ HETEROSTRUCTURES

V.Ya.ALESHKIN, V.M.DANIL'TSEV, O.I.KHRYKIN,
Z.F.KRASIL'NIK, D.G.REVIN, V.I.SHASHKIN
Institute for Physics of Microstructures of Russian Academy of Sciences
GSP-105, Nizhny Novgorod, 603600, Russia;
e-mail:revin@ipm.sci-nnov.ru

Diagnostics of quantum dots (QDs) realised on the base of pseudomorphous strained heterostructures is complicated by the problems resulted from the formation of QDs by self-organising in growth process. They include in particular the dispersions of QDs dimensions and solid solution content, the variety of zero-dimensional (0-D) object forms (such as, pyramids and discs), the presence of two-dimensional (2-D) wetting layers etc. In this paper we report on the study of 0-D electron states in $GaAs/In_xGa_{1-x}As$ heterostructures grown by the metalorganic chemical vapour deposition. High growth rates: v > 100 nm/min facilitate pyramid-like nanometric scale islands nucleation from the $In_xGa_{1-x}As$ solid solution. For x > 0.4 such islands arise even at low growth rates due to a marked lattice mismatch of the layers forming the heterostructure, which favours QD rather than 2-D growth.

QDs electron states were investigated by high excitation photoluminescence (PL), photoconductivity (PC) and interband absorption techniques in broad temperature range. The common experiment idea was the QDs observation via the peculiarities of combine density of states corresponded to 0-D character of QDs which do not exist in quantum wells (QWs). For the comparison the same measurements were carried out in the perfect single QW (PL exciton line width of 3.7 meV at 77 K).

The changes arising in the PL spectrum at the increase of the excitation power reflect filling of both the upper excited levels in QDs and the levels in QDs of smaller dimensions. For $In_xGa_{1-x}As$ (x ~ 0.6) based QD structures there was the notch between the regions of GaAs and QDs emission in the PL spectrum at high excitation power (Fig.1) whereas such notch was not observed in PL spectra of QWs (Fig.2). The large value of PL peak width (~ 52 meV) seems to result from both the QDs size dispersion and the nonuniformity of solid solution in QDs. At all excitation levels the contour of PL line was a superposition of the δ-like lines of the radiative recombination of electrons and holes occupied the ground states in separate QDs (under high excitation levels also between excitation states). Combine density of states of the QD ensemble has a maximum as a function of the energy while QW one is characterised by the stepped increase with the energy. The corresponding combine density of states was observed also in PC spectra. The similar investigations were carried out with the series

M. Balkanski and N. Andreev (eds.),
Advanced Electronic Technologies and Systems Based on Low-Dimensional Quantum Devices, 65–66.

of InAs based QD structures. The QD nucleation moment and the change of their dimensions versus the amount of the deposited InAs were revealed (Fig.3).

The research described in this publication was made possible in part by Grant # 2-013/3 from Russian Scientific Program "Physics of Solid State Nanostructures".

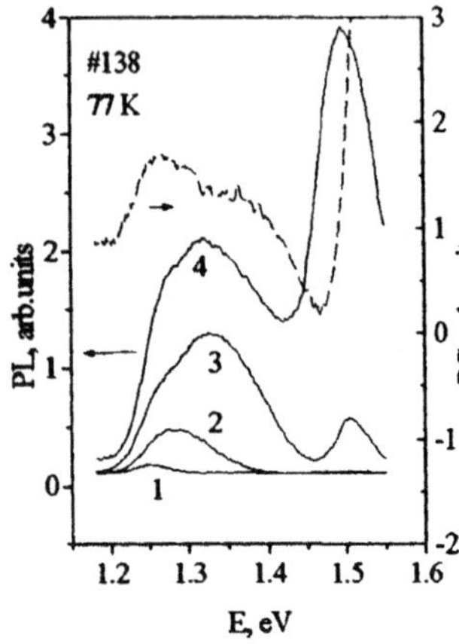

Figure 1. PL spectra for various pump levels and PC one of quantum dots. Pump power, W/cm^2 : 1 - 10^3, 2 - 10^4, 3 - 10^5, 4 - 10^6, λ_{exc}=532 nm.

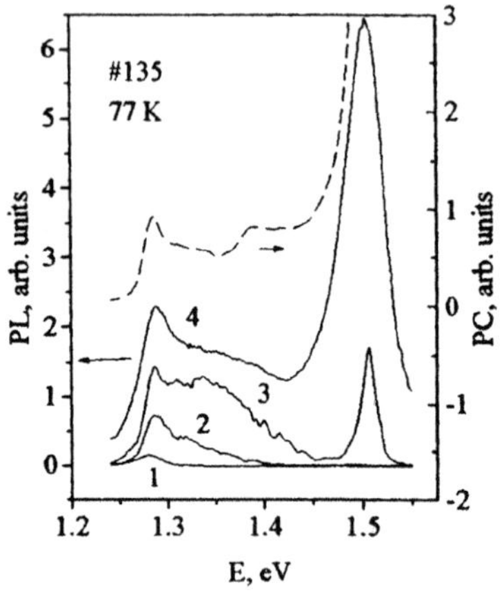

Figure 2. PL spectra for various pump levels and PC one of 4 nm GaAs/$In_{0.3}Ga_{0.7}As$/GaAs quantum well. Pump power, W/cm^2 : 1 -10^3, 2 - 10^4, 3 - 10^5, 4 - 10^6, λ_{exc}=532 nm.

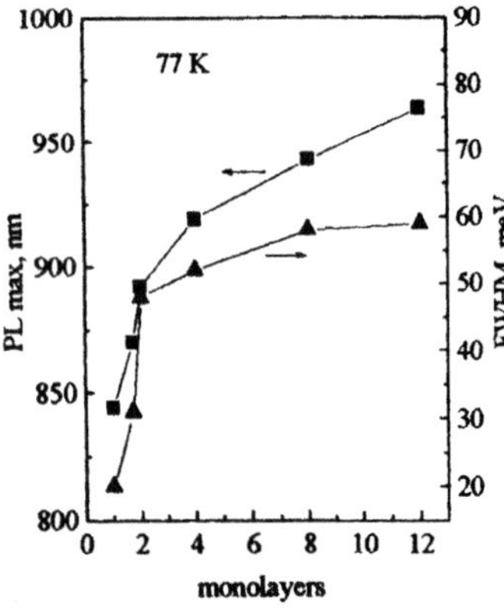

Figure 3. PL peak maximum position and FWHM of InAs/GaAs heterostructures as function of deposited (in relative monolayers) InAs.

GENERATION OF HIGH-FREQUENCY OSCILLATIONS BY ELECTROMAGNETIC SHOCK WAVE (EMSW) IN NONLINEAR TRANSMISSION LINES ON THE BASIS OF MULTILAYER HETEROSTRUCTURES

A.B.KOZYREV and A.M.BELYANTSEV
Institute for Physics of Microstructures of Russian Academy of Sciences
603600, Nizhny Novgorod, GSP-105, Russia

Recently the monolithic GaAs nonlinear transmission lines (NLTL) were proposed to use for generation of waveforms with picosecond fall time through generation of EMSW [1] and for generation of microwave oscillations [2]. In particular, [2] deals with producing microwave voltage bursts (solitons). However this way of generation is ineffective since the spectrum of oscillations is broad and their amplitude drops rapidly.

In this paper we propose the method of direct transformation of input pulse into narrow spectrum radiopulse with central frequency in the range 100-150 GHz. The method is based on the generation of the high-frequency wave propagating synchronousely with EMSW front on NLTL with nonlinear capacitance and dispersion [3]. The requirements on the dispersion of NLTL and nonlinear capacitance of multilayer heterostructure (MLHS) which are necessary for attainment of the radiopulse with narrow spectrum width are determined. For this purpose the computer simulation of nonlinear wave evolution on NLTL with quantum barrier varactor (with functional *C(V)* characteristic) [4] and on NLTL with asymmetric MLHS $AlAs - n\,GaAs - i\,Al_xGa_{1-x}As - n^+\,GaAs - AlAs$ (with hysteretic *C(V)* characteristic) [3] has been carried out. It was found that the general requirement on the dispersion of NLTL is existence of minimum of group velocity at synchronism. In particular, such dispersion exhibits NLTL of the type of *LC*-chain with capacitance cross links (Fig.1).

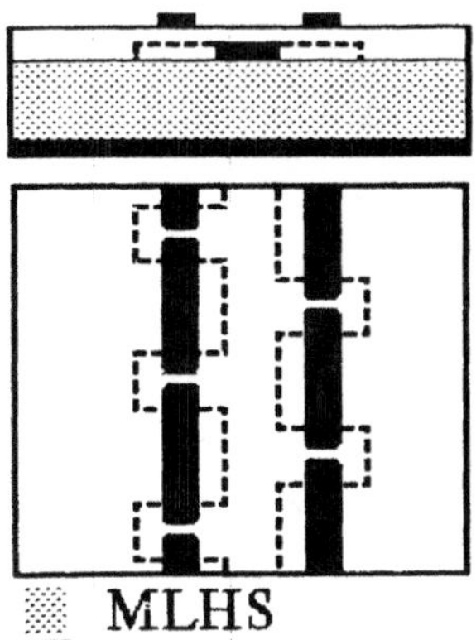

Figure 1. Layout of NLTL.

In the case of NLTL with quantum barrier varactors the higher is capacitance ratio C_{max}/C_{min} the narrower is spectrum width and at $C_{max}/C_{min} \sim 20 \div 25$ it is close to monochromatic one. Results of simulation are shown in Fig.2. Frequency of generated oscillations slightly depends on the amplitude of the input pulse and

M. Balkanski and N. Andreev (eds.),
Advanced Electronic Technologies and Systems Based on Low-Dimensional Quantum Devices, 67–68.

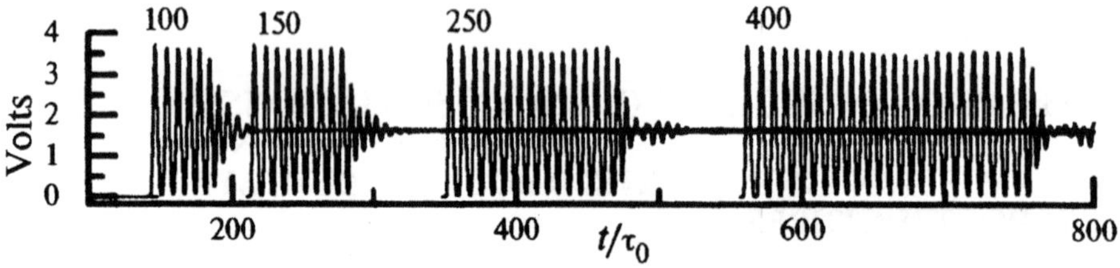

Figure 2. Voltage waveforms at 100, 150, 250, 400 sections of NLTL ($\tau_0 = \sqrt{L_0 C_0}$, L_0 and C_0 are basic inductance and capacitance of NLTL).

completely determined by dispersive and nonlinear properties of NLTL.

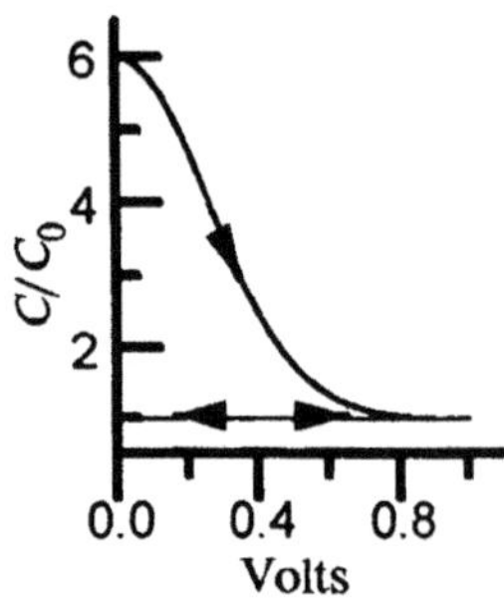

Figure 3. Hysteretic *C(V)* characteristic.

In the case of NLTL with asymmetric MLHS the effective generation of quasimonochromatic oscillations can be obtained with more smooth *C(V)* characteristics ($C_{max}/C_{min} \sim 3 \div 5$). However, nonlinear capacitance in this case should exhibit hysteretic properties, i.e., its value should saturate behind the EMSW front and stay in saturated state for a long time (Fig.3). Results of computer simulation for this case are shown in Fig.4. In this case there is a great advantage in controlling the generated frequency by variation of the amplitude of the input pulse (in compare with NLTL with quantum barrier varactor). The upper limit of generated frequency for considered structures is determined by persistence of capacitance variation and for MLHS is above 1 THz. The amplitude of voltage oscillations is limited by the breakdown voltage and is of the order of 1-5 V.

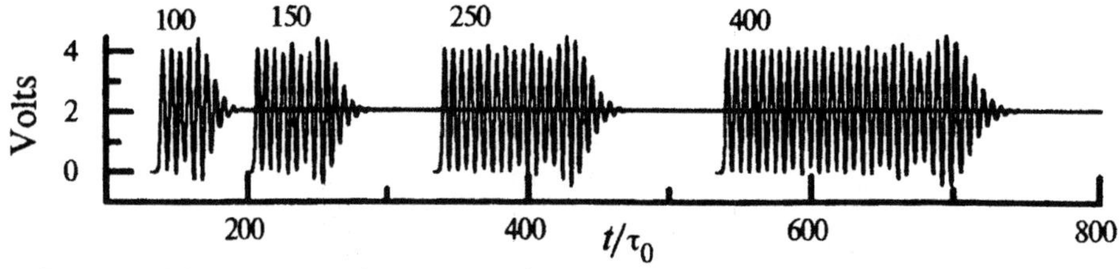

Figure 4. Voltage waveforms at 100, 150, 250 and 400 sections of NLTL.

The research described in this publication was supported by Russian Scientific Program "Physics of solid state nanostructures" (project N 01-030).

REFERENCES

1. D.W.van der Weide (1994) *Appl. Phys. Lett.* **65**, 881.
2. Ikezy H., Wojtowich S.S., Waltz J.S., deGrassie J.S., Baker D.R. (1991) *J.Appl.Phys.* **64**, 3277.
3. Belyantsev A.M., Climin S.L. (1993) *Izv. VUZov. Radiofizika*, **36**, 1011.
4. Rydberg A., Gronqvist H., Kollberg E. (1990) *IEEE Electron. Lett.*, **11**, 373.

PICOSECOND SPECTROSCOPY STUDIES OF CuS AND $CuInS_2$ QUANTUM DOTS WITH CHEMICALLY MODIFIED SURFACE

A.M. Malyarevich*, K.V. Yumashev*, P.V. Prokoshin*,
M.V. Artemyev**, V.S. Gurin**, V.P. Mikhailov*
** International Laser Center, Belarus State Polytechnical Academy,*
F. Skaryna Av. 65/17, Minsk 220027, BELARUS, tel/fax: 375(17)2326286
*** Physico-Chemical Research Institute of Belarusian State University,*
Minsk 220080, Belarus

In this report we present the results of recent studies devoted to nonlinear optical properties of some semiconductors in the form of nano-sized particles with chemically modified (oxidized) surface. They are $CuInS_2$, CuS and Cu_2S particles embedded in polyvinyl alcohol thin film. The materials were studied using picosecond pump-probe technique at room temperature. The passive mode-locked $YAlO_3$:Nd laser generated pulses with duration of 15. The picosecond pulses with λ=1.08 μm as well as frequency doubled with λ=540 nm (KDP crystal) were used as the pump beams. The energy of pump pulse was ~ 80 mJ/cm^2. The white light continuum from a D_2O cell was used as the probe pulse.

Chemical modification of particles' surface leads to appearance of broad additional absorption band with maximum at 1100-1200 nm. We attribute this band to formation of CuO shell around the particles and believe the correspondent energy level inside the band gap has the donor of electrons character. Figure 1 demonstrates differential

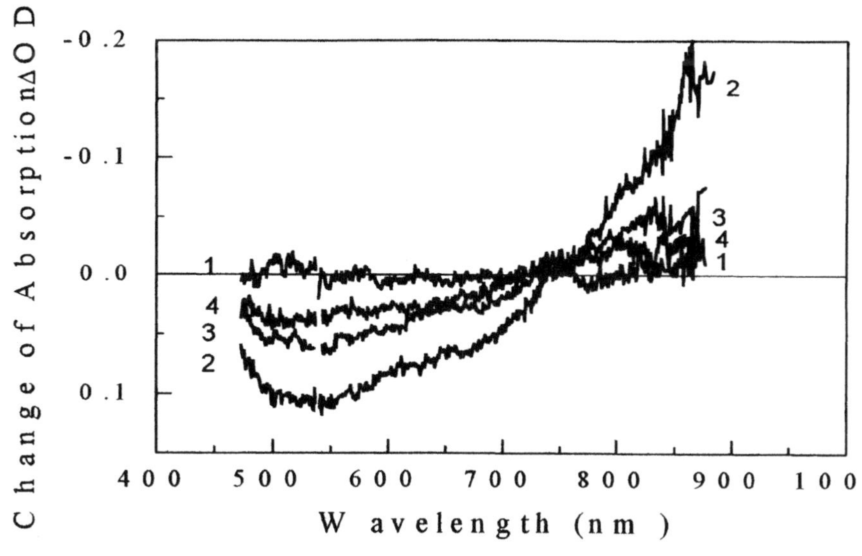

Figure 1. A series of differential absorption spectra for $CuS_{(ox)}$ samples after 1080 nm laser pulse pumping. Delay times between pump and probe pulses are: (1) -35 ps, (2) 10 ps, (3) 25 ps, (4) 55 ps.

M. Balkanski and N. Andreev (eds.),
Advanced Electronic Technologies and Systems Based on Low-Dimensional Quantum Devices, 69–70.

absorption changes spectra of CuS nanocrystals. As one can see pumping leads to the bleaching of additional absorption band and induced absorption in blue-green spectral region. This picture is the typical one for nonlinear bleaching and induced absorption effects in all studied samples under powerful laser excitation. Relaxation of nonlinear effects has short (ps) and long (ns) lived parts. Character time of decreasing of bleaching / induced absorption varies from ≈20 ps in Cu_2S to ≈50 ps $CuInS_2$ and ≈80 ps in CuS and depends on the type of semiconductor. Proposed energy level scheme for modified semiconductor nanoparticlesis shown at Fig. 2.

Additional absorption band concerning with level A is attributed to ligand-to-metal charge transfer transitions of copper (II) complexes which appear after thermal oxidation. Thus we suppose these transitions have the band-to-band nature and level A is actually an effective valence band (or has the character of an electron donor). Level B is attributed to the bottom of the conduction band or, at least, to shallow trap levels very close to respective band edge. Trap level B has long life-time (much more than 500 ps).

Short times of bleaching effect in studied samples make them very promising as saturable absorbers for mode-locking or Q-switching in pico- and femtosecond lasers with working wavelengths λ>800 nm.

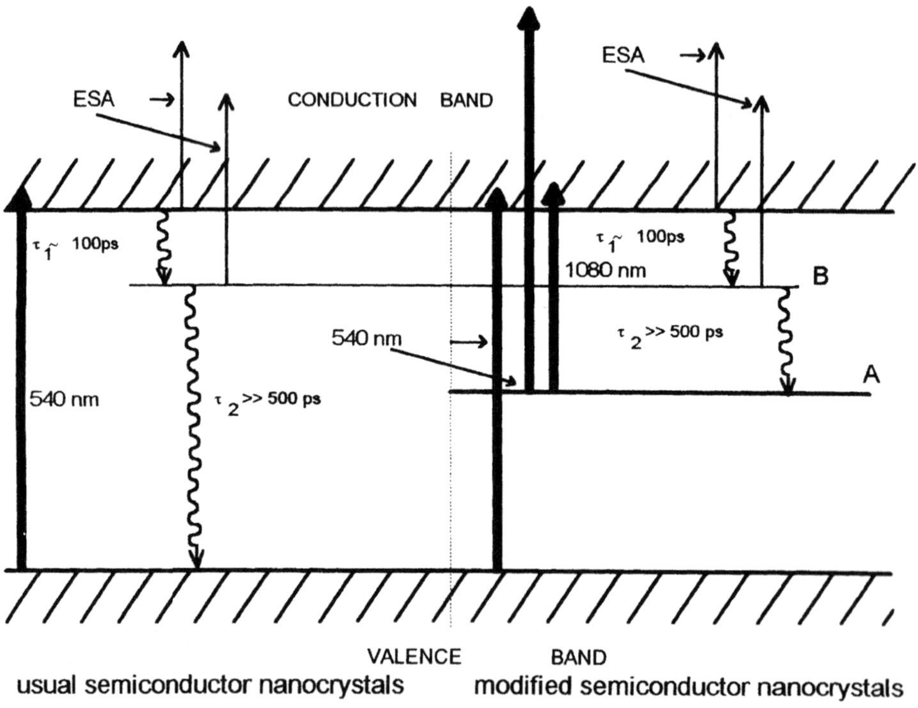

Figure 2. Proposed energy level diagram of modified CuS nanocrystals. Additional band appears due to chemical treatment (oxidation) of the microcrystal surface.

INTENSITY INDUCED POLARIZATION ROTATION DUE TO CASCADING IN IMBALANCED TYPE II NEARLY PHASE MATCHED FREQUENCY DOUBLING

S. SALTIEL, I. BUCHVAROV, K. KOYNOV, P. TZANKOV, CH. IGLEV,
University of Sofia, Faculty of Physics, Quantum Electronics Department, 5 J. Bourchier Blvd., 1164 Sofia, Bulgaria
Fax:359 2 9625 276 E-mail: ssv@phys.uni-sofia.bg

The effect of intensity dependent polarization rotation can be of great interest for nonlinear optical devices used for all-optical switching [1,2], self induced transparency and darkening [3], mode-locking and pulse shortening [4].

Intensity dependent polarization rotation due to nonlinear phase shift (NPS) in a crystal for type I SHG was predicted in [3]. It is shown there by numerical analysis that devices based on this process may posses self-induced darkening and self-induced transparency effect. Polarization rotation as a result of the nonlinear phase shift due to $\chi^{(2)}:\chi^{(2)}$ cascading in type II nearly phase matched SHG crystals has not been described in the literature by now.

This paper concerns the analytical study of the effect of intensity dependent polarization rotation in quadratic media suitable for imbalanced type II frequency doubling. It is shown that the nonlinear media can be considered as a wave plate with intensity dependent retardation effect. The analytical analysis shows that the phase retardation of this wave plate has stepwise dependence on the input intensity. The rotation angle depends on the angle between the input polarization and the main plain of the crystal and can reach 90°.

For description of the process of polarization rotation we used the derived in [4] analytical formulae for the intensities and the phases of fundamental waves involved in type II interaction for SHG. The formulae for the amplitudes have the following form:

$$a_j^2(z) = a_j^2(0)\left\{1 - N_j\left[\sin^2\left(\frac{\pi}{2}\frac{z}{l_{coh}}\right) + m^2 F \sin^2\left(\pi\frac{z}{l_{coh}}\right)\right]\right\}, \qquad (j=1,2) \qquad (1)$$

where $a_j(0)$ are the input values for the amplitudes of the two ortogonaly polarized fundamental waves. Their ratio is $a_2(0)/a_1(0) = \sqrt{r}$. The rest of the notations in (1) are:

M. Balkanski and N. Andreev (eds.),
Advanced Electronic Technologies and Systems Based on Low-Dimensional Quantum Devices, 71–75.

$N_1 = \sqrt{m.r}$, $N_2 = \sqrt{m/r}$, $l_{coh} = æK(m)$, $æ^2 = \dfrac{v - \sqrt{v^2 - 4u}}{2u}$,

$u = 4\sigma^4 a_1^4(0) r$, $v = 2\sigma^2 a_1^2(0)(1+r) + \dfrac{\Delta k^2}{4}$, $m = uæ^4$, $\Delta k = k_3 - k_2 - k_1$,

$F(r) = 0.66(r^2 + 1) - 1.064r$ $(r \in [0.1, 1])$. σ is the nonlinear coupling coefficient.

The function $K(m)$ in the expression for l_{coh} is the complete elliptic integral of the first kind and

$$K(m) = \left(p_0 + p_1(1-m) + p_2(1-m)^2\right) - \left(q_0 + q_1(1-m) + q_2(1-m)^2\right)\ln(1-m) + \varepsilon(m)$$

$p_0 = 1.3862944$, $p_1 = 0.1119723$, $p_2 = 0.0725296$; $q_0 = 0.5$,
$q_1 = 0.1213478$, $q_2 = 0.0288729$; $\varepsilon(m) \le 3 * 10^{-5}$.

The formulae for the phase shifts due to SHG $\Delta\varphi_j^{NL}$ (j=1,2) are as follows:

$$\Delta\varphi_j^{NL} = \varphi_j(L) - \varphi_j(0) = \frac{\Delta k l_{coh} c_j}{\pi(K_j^{(2)} - K_j^{(1)})}\left[\frac{K_j^{(2)} - 1}{\sqrt{K_j^{(2)}}}\operatorname{artg}\frac{p}{\sqrt{K_j^{(2)}}} - \frac{K_j^{(1)} - 1}{\sqrt{K_j^{(1)}}}\operatorname{artg}\frac{p}{\sqrt{K_j^{(1)}}}\right] - \frac{\Delta k L}{2} \quad (2a)$$

when $b_j^2 > 4c_j$ and

$$\Delta\varphi_j^{NL} = \varphi_j(L) - \varphi_j(0) = \frac{\Delta k l_{coh} c_j}{\pi}\left\{\frac{f_j}{2}\ln\left|\frac{p^2 + d_j p + \sqrt{c_j}}{p^2 - d_j p + \sqrt{c_j}}\right| + \right.$$

$$\left. + \frac{(1/\sqrt{c_j}) - d_j f_j}{2\sqrt{\sqrt{c_j} - d_j^2/4}}\left[\operatorname{artg}\frac{2p + d_j}{2\sqrt{\sqrt{c_j} - d_j^2/4}} + \operatorname{artg}\frac{2p - d_j}{2\sqrt{\sqrt{c_j} - d_j^2/4}}\right]\right\} - \frac{\Delta k L}{2} \quad (2b)$$

when $b_j^2 < 4c_j$.

In (3) the following notations are used:

$\varphi_j(0)$ - input values for the phases of the two ortogonaly polarized waves, $p = \operatorname{tg}(\pi L / 2 l_{coh})$, L is the length of the nonlinear media,

$$b_j = \frac{2 - N_j(1 + 4m^2 F)}{1 - N_j}, \quad c_j = (1 - N_j)^{-1}, \quad d_j = \sqrt{2\sqrt{c_j} - b_j}, \quad f_j = \frac{1 - \sqrt{c_j}}{2 d_j \sqrt{c_j}}, \quad (3)$$

$$K_j^{(1)} = \frac{1}{2}\left(b_j - \sqrt{b_j^2 - 4c_j}\right), \quad K_j^{(2)} = \frac{1}{2}\left(b_j + \sqrt{b_j^2 - 4c_{je}}\right).$$

The validity of formulae (2) is demonstrated on figure 1 where the analytical solution (2) (solid line) is compared with the exact numerical calculations (dashed line).

It can be seen that the agreement is excellent. The formulae (2) were used to study the effect of induced polarization rotation in SHG crystals with type II interaction.

Let us consider crystal for SHG oriented in such a way that the normal **N** to the plane formed by the wave vector **k** and the crystal axis is at angle α with respect to the input polarization (figure 2). After passing through the crystal the waves "1" and "2" collect both linear $\Delta\varphi^{LIN}$ and non-linear $\Delta\varphi^{NL}$ phase shifts. The polarization state at the crystal output will depend on the transmittance of the two waves and on the sum of the two phase differences

$$\Gamma = \Gamma_{LIN} + \Gamma_{NL} = \Delta\varphi_2^{LIN} - \Delta\varphi_1^{LIN} + \Delta\varphi_2^{NL} - \Delta\varphi_1^{NL}$$

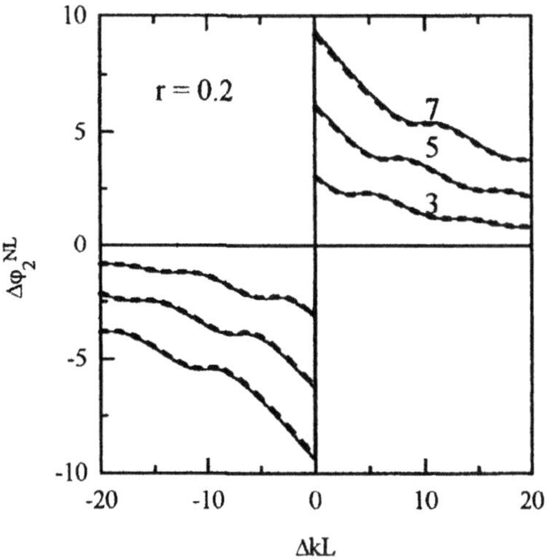

Figure 1. Non-linear phase shift $\Delta\varphi_2^{NL}$ of the weak wave "2" versus normalized phase mismatch for different values of the normalized input amplitude of the wave "1"-$\sigma\, a_1(0)L$. The solid line is the analytical solution. Dashed line represents the exact numerical solutions.

Γ_{LIN} is easy to compensate by inserting an additional phase corrector or by proper choice of the non-linear media length making it $\Gamma_{LIN} = 2n\pi$ (n-integer). Consequently the change of the state of the polarization will depend only on Γ_{NL}.

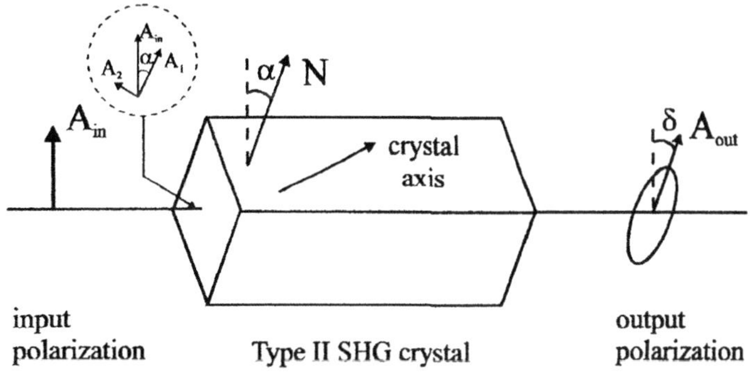

Figure 2. Schematic drawing of the arrangement considered for obtaining intensity-dependent change of the fundamental wave polarization state

Figure 3 shows the nonlinear part of the phase retardation $\Gamma_{NL} = \Delta\varphi_2^{NL} - \Delta\varphi_1^{NL}$ as a function of the input intensity. The middle of each plato of the curves shown on figure 3 are the points of full reconstruction of the pump intensities at the output of the nonlinear media. The first point of maximum intensity reconstruction corresponds to the fundamental intensities for which $L = 2l_{coh}$. The next points of maximum intensity reconstruction occur at higher intensities for which $L = 2nl_{coh}$ (n- integer). These points are the most interesting cases because the fundamental beam is transmitted through the crystal without losses.

The crystal for SHG for type II interaction oriented at $\alpha \le 43^o$ can be considered as an wave-plate with retardation effect that depends on the input intensity and the phase mismatch.

In general, output polarization will be elliptical with a big semiaxis rotated at angle δ with respect to the input polarization

$$\delta = \alpha - \frac{1}{2}\text{arctg}\frac{\rho_1\rho_2 \sin 2\alpha}{\rho_1^2\cos^2\alpha - \rho_2^2\sin^2\alpha}\cos\Gamma .$$

In this formula $\rho_j = a_j(L)/a_j(0)$ are the amplitude transmission coefficients for the two waves, $a_1(0) = a_{in}\cos\alpha$ $a_2(0) = a_{in}\sin\alpha$. The semiaxes "a" and "b" are defined by:

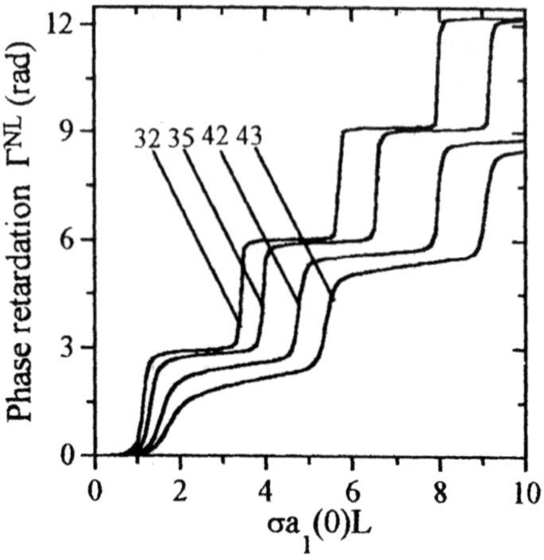

Figure 3. Non-linear phase retardation $\Gamma^{(NL)}$ versus normalized input amplitude $\sigma a_1(0)L$ of the strong wave for ΔkL =0.3 rad. The parameter is input angle α. Solid line is the analytical solution .Dashed line represents the numerical solutions.

$$\begin{aligned} a^2 &= \rho_1^2\cos^2\alpha\cos^2(\alpha-\delta) + \rho_2^2\sin^2\alpha\sin^2(\alpha-\delta) + \frac{1}{2}\rho_1\rho_2\sin 2\alpha\sin 2(\alpha-\delta)\cos\Gamma \\ b^2 &= \rho_1^2\cos^2\alpha\sin^2(\alpha-\delta) + \rho_2^2\sin^2\alpha\cos^2(\alpha-\delta) - \frac{1}{2}\rho_1\rho_2\sin 2\alpha\sin 2(\alpha-\delta)\cos\Gamma \end{aligned} \quad (4)$$

Since the transmission coefficients ρ_j and Γ_{NL} strongly depend on the input intensities, the rotation angle δ is intensity dependent as well. By using the above formulae we analyzed the rotation angle and the eccentricity, $e = \sqrt{a^2 - b^2}/a$, of the output

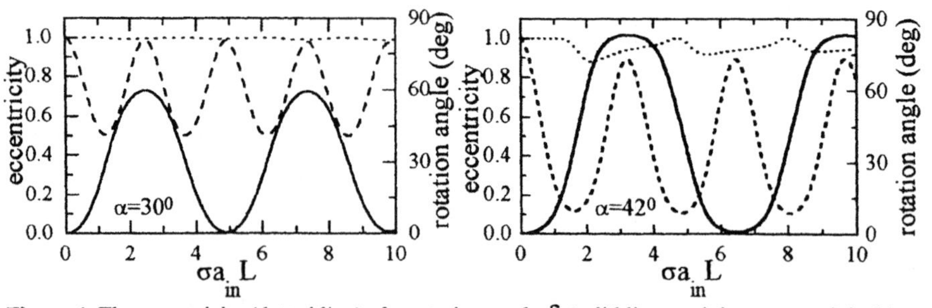

Figure 4. The eccentricity (dotted line), the rotation angle δ (solid line, and the square of the big axis, a^2, (dashed line) as a function of the normalized input intensity for two different input angles. The normalized phase mismatch $\Delta kL = 0.3$.

fundamental wave.

The angle of polarization rotation δ of the output polarization (with solid line) as a function of the input intensity is shown on figure 4 for different input angles α.

On the same figure are shown also the eccentricityand the square of the big axis (a^2). It is seen that the angle δ is changing periodically between 0 and δ_{max}. The maximum rotation angle δ_{max} is obtained on the point of minimum losses for the input intensity and its magnitude depends on the input angle α and the phase mismatch ΔkL. The dependence of the maximum rotation angle and the eccentricity on the input angle α is shown on figure 5. For small ΔkL the output fundamental wave has eccentricity close to one i.e. the output fundamental wave remains linearly polarized. The rotation angle in this case can reach the values close to 90°.

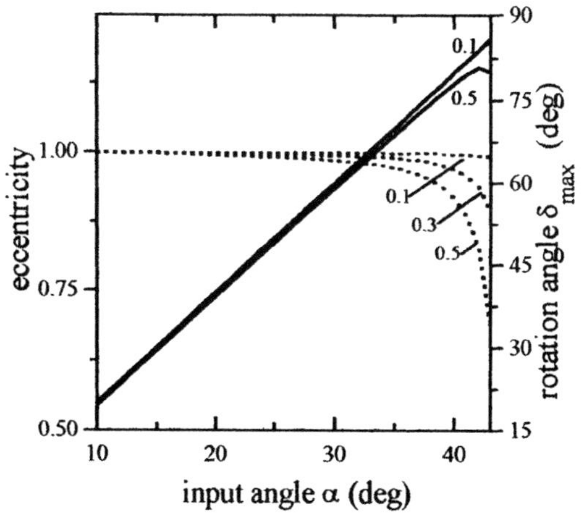

Figure 5. Maximum rotation angle δ_{max} (solid line) and eccentricity (dotted line) versus the input angle α. The parameter is the normalized mismatch ΔkL

In summary here we show that near phase matched type II SHG crystals can be used as "frequency doubling polarization rotator" with intensity dependent angle of polarization rotation. Possible applications of such kind of devices may be mode-locking, all-optical switching and sensor protections.

Acknowledgments

We would like to express our acknowledgements to the Bulgarian Science Foundation for the support under contracts F 601 and F405.

References

1. Lefort, L. and Barthelemy, A. (1995) Intensity-dependent polarization rotation associated with type II phase-matched second harmonic generation: application to self induced transparency, *Opt. Lett.*, **20**, 1749-1751.
2. Lefort, L. and Barthelemy, A. (1995) All-optical transistor action by polarization rotation during type II phase-matched second harmonic, *Elect. Lett.*, **31**, 910-911.
3. Saltiel, S., Koynov, K., Buchvarov, I. (1996) Self-induced transparency and self-induced darkening with nonlinear frequency doubling polarization interferometer, *Appl. Phys. B*, **63**.
4. Buchvarov, I., Saltiel, S., Iglev Ch., Koynov, K. (submitted) Intensity dependent change of the polarization state as a result of nonlinear phase shift in type II frequency doubling crystals, *IEEE J Quant. Elec.*

MODELLING QUANTUM WELL LASER DIODE STRUCTURES

PETER BLOOD, DAMIAN L FOULGER and
PETER M SMOWTON

Department of Physics and Astronomy
University of Wales Cardiff
PO Box 913
Cardiff CF2 3YB
UK

1. Introduction

Many of the attractions of quantum wells in lasers derive from the properties of the density of states function of the two-dimensional electron system. The abrupt edge of the density of states as a function of energy provides a very high differential gain above transparency leading to significant reductions in threshold current in appropriately designed devices compared with their bulk counterparts. The effective density of states of a two-dimensional system has a linear dependence upon temperature which leads directly to an intrinsic linear temperature dependence of threshold current compared with the stronger "three-halves" dependence of a bulk material. Because of the benefits of these characteristics of the two-dimensional system, much of the modelling of quantum well lasers has understandably concentrated on the intrinsic gain-current characteristic of the quantum well active region. These calculations usually use an ideal, square, potential well and extrinsic non-radiative currents are neglected. In devices operating at wavelengths below about 1μm, where intrinsic Auger recombination is negligible, this approach has been reasonably successful in predicting trends in the room temperature threshold current with respect to parameters such as well width or cavity length. This is particularly true for the GaAs/AlGaAs material system which can be grown free of significant concentrations of non-radiative recombination centres.

The intrinsic gain-current characteristics have not been successful in predicting the behaviour of the threshold current as a function of temperature, even in GaAs/AlGaAs devices [1]. In all quantum well lasers the threshold current follows the intrinsic linear dependence over a low temperature region but as the temperature is increased an approximately exponential increase in threshold with temperature is superimposed on the linear behaviour and this becomes dominant at sufficiently high

M. Balkanski and N. Andreev (eds.),
Advanced Electronic Technologies and Systems Based on Low-Dimensional Quantum Devices, 77–89.

temperature (figure 1). This additional current above the intrinsic linear component is often referred to as the "excess current". In good quality GaAs/AlGaAs devices the threshold current is linear in T to temperatures above 300K though by about 350K the excess current begins to appear. Experiments have shown that this current is chiefly due to non-radiative recombination via deep states in the AlGaAs barrier material forming the core of the waveguide [1] which contains a higher density of deep states than the GaAs material comprising the well. Because the carrier density in the barrier increases exponentially with temperature compared with that in the well, this excess current component has an exponential temperature dependence.

In visible-emitting GaInP/AlGaInP lasers, operating in the range 630nm to 690nm, the excess current is apparent at lower temperatures than in GaAs devices and in some cases, especially in 633nm lasers, it is significant even at room temperature [2]. At these wavelengths, calculations based only on the intrinsic behaviour of the quantum well are not particularly successful in predicting even the room temperature behaviour. The excess current is greater in these AlGaInP devices because the band offsets are smaller than in the AlGaAs system, giving less confinement of carriers in the well, so there is significant leakage of carriers through the p-cladding layer [2]. It has been suggested that drift of minority carrier electrons makes a significant contribution to the leakage current because the conductivity of the p-cladding is lower than in AlGaAs structures [2,3].

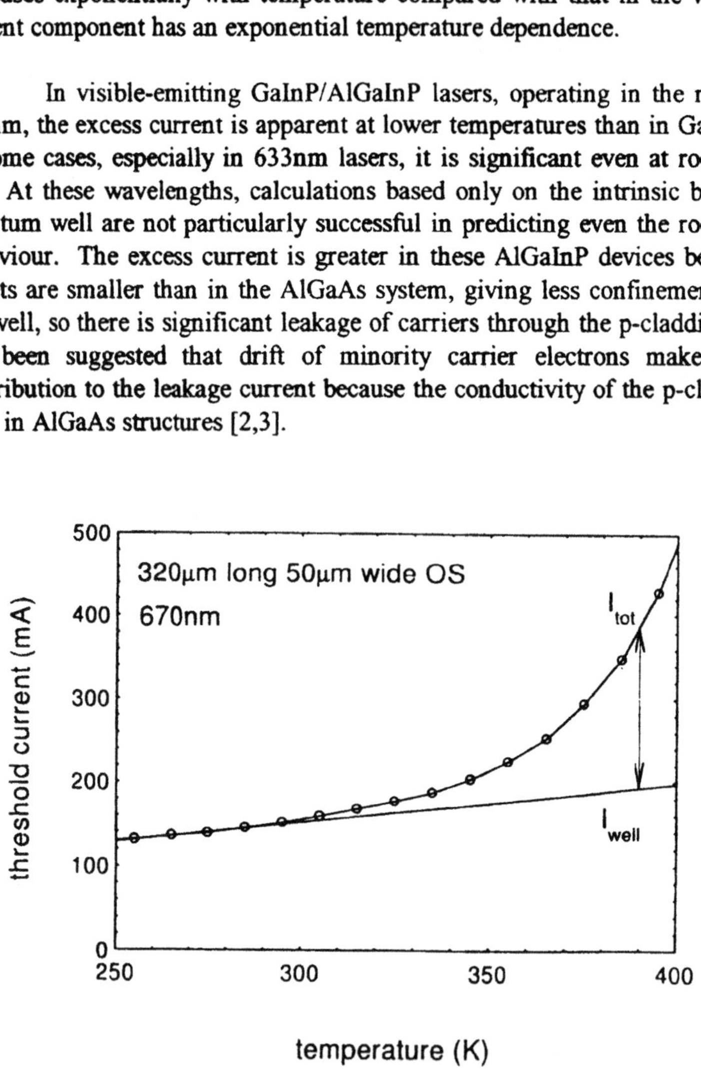

Figure 1. Temperature dependence of threshold current of a typical GaInP quantum well diode laser showing the intrinsic linear temperature dependence at low temperature (I_{well}) with the thermally activated "excess current" appearing at higher temperatures.

These examples show that it is often not sufficient to consider the active region of the quantum well in isolation from the rest of the device structure, especially for visible-emitting lasers where there can be significant leakage currents through the cladding layers. A realistic calculation of the threshold current, particularly its temperature dependence, should include current paths in other parts of the device structure. Four current paths have been identified in addition to the intrinsic spontaneous recombination in the well. These are illustrated in figure 2 and are as follows:

- radiative recombination in the barrier
- non-radiative recombination in the barrier
- leakage through the cladding layers by diffusion
- leakage through the cladding layers by drift.

We neglect radiative recombination in the cladding layers because these usually comprise indirect-gap material. Non-radiative recombination in the well is also negligible in good quality structures based on GaAs substrates. Of the above processes, radiative recombination in the barrier is usually only a few percent of that in the well and is negligible [3,4], though it is straightforward to include this in a calculation.

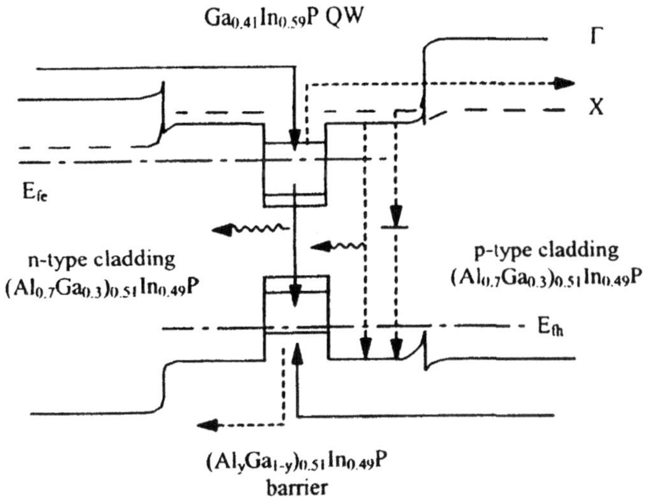

Figure 2 Schematic band diagram through a separate confinement quantum well laser illustrating the possible current paths outside the quantum well.

The currents due to all of the above processes can be calculated in a straightforward manner, provided there are no rate-limiting steps so that the carrier distributions throughout the whole structure are in dynamic thermal equilibrium at the lattice temperature. The current components are then simply additive [1]. If it is also valid to assume that the bands are flat and that the quasi-Fermi levels are constant across the core region of the waveguide/barrier region then the quasi-Fermi level positions are defined by the gain requirement for the well and all the above currents can be calculated analytically. This flat-band approach to the calculation of these additional currents is easy to implement, gives a reasonable description of the essential physics of the problem, and provides a much better representation of experimental data for the temperature dependence of threshold current than is given by the intrinsic current in the well alone [1]. However this flat-band approach does contain a number of inconsistencies and does not describe correctly the behaviour of real structures in detail. The chief difficulties are as follows.

(1) The assumption of a constant Fermi level across the well and adjacent barrier implies a charge imbalance in the barrier because the densities of states in the wells and barriers are different. This is inconsistent with the assumption that the bands are flat. Some progress can be made within a simple model by imposing the requirement of charge neutrality on the whole of the well and barrier region rather than the well alone and leaving all the bands flat [5], but this still leads to errors in the Fermi level positions and is not really satisfactory.

(2) The absence of charge neutrality in the well and barrier region also implies that the potential well is not square.

(3) The absence of charge neutrality in the active region shifts the quasi-Fermi level positions relative to the rest of the structure, affecting calculations of leakage current.

(4) The inclusion of a leakage contribution due to *drift* implies the presence of an electric field through the cladding layers and, while there is an analytic expression describing the current flow due to drift and diffusion combined [2,6], this is based on the assumption that the field is constant throughout the cladding region.

The conclusion which emerges is that the best approach to modelling the current through a laser is a self-consistent simulation of the current flow and potential by numerical solution of the current continuity equations and Poisson's equation throughout the complete structure, together with the solution of Schrodinger's equation for a non-square well. The purpose of this paper is to describe the results of such simulations for 670nm GaInP lasers, where carrier drift through the cladding is significant. These results give a good description of the temperature dependence of threshold current and provide tutorial illustrations of the inadequacies of the simple flat-band model.

2. Description of the calculation and the device structure

The simulation program is based upon computer code written by colleagues at the University of Wales in Swansea [7], originally for simulation of silicon devices, and which we have extended to include non-square quantum wells and have linked to a calculation of optical gain including carrier scattering effects, to enable us to simulate realistically the operation of quantum well laser diodes.

The model is a steady state, two-dimensional, finite element program in a plane perpendicular to the axis of the laser, which solves Poisson's equation for the electric field E

$$\nabla . E = \rho / \varepsilon$$

and the current continuity equations for the electron and hole current densities, J_n and J_p, where

$$\nabla . J_{n,p} = e(R - G)$$

The charge density is $\rho = e(p\text{-}n\text{+}N_d\ \text{-}N_a)$, and R and G are the recombination and generation rates per unit volume respectively. The electron current density is given by

$$J_n = e\{-nv + D_n \nabla . n\}$$

where n is the electron density, v the electron velocity, and D_n the diffusion coefficient. A similar equation applies for holes. The local electron densities in both the direct (Γ) and indirect (X) conduction bands are assumed to be in equilibrium, and the electron and hole densities are determined by the local quasi-Fermi levels and band edge positions using Fermi Dirac statistics and the lattice temperature. Dopant atoms are included with a specified ionisation energy with respect to the appropriate band edge.

In the bulk parts of the structure radiative band-to-band recombination is incorporated in the form

$$R(T) = B(T).n(T).p(T) \qquad \text{per sec per unit volume}$$

where B(T) is the radiative recombination coefficient. The coefficient B has a $T^{-3/2}$ temperature dependence. Non-radiative Shockley-Read-Hall recombination is specified by the non-radiative lifetimes for electrons and holes, τ_n and τ_p respectively (taken to be constant in a given layer) and the local values of n and p.

The quantum well is treated by solving Poisson's equation and Schrodinger's equation self-consistently assuming no lateral current flow along the well. The in-plane dispersion relation is taken to be parabolic and the net charge distribution across the well is determined by the probability distributions for electrons and holes defined by the respective envelope functions. With a self-consistent solution for the quasi-Fermi levels and sub-band energies in the well, the spontaneous recombination rate is calculated from first principles using Fermi's Golden rule. This calculation only requires the spectrally integrated recombination rate so it is not necessary to include spectral broadening at this stage.

With specified details of the compositions, doping densities and types, and layer thicknesses of the device structure, the program seeks a consistent solution for the potential distribution and current flow for a specified terminal voltage corresponding to the quasi-Fermi level difference at the two boundaries of the structure. Having arrived at the corresponding sub-band separation and quasi-Fermi level positions in the well, the optical gain spectrum is calculated including spectral broadening due to carrier scattering using a scattering time calculated as a function of energy within the sub-band as described in ref [8]. By this procedure we avoid a lengthy calculation of the broadened recombination spectrum at each iteration of the program and only do this calculation for the final self-consistent solution at each terminal voltage. The threshold current is defined as the current which flows when the optical gain is equal to the cavity losses. Since we are only interested in simulating the threshold behaviour, and we do not require a simulation of the light-current curve above threshold, it is not necessary to include a self-consistent treatment of the photon equations and stimulated emission. This again reduces the computational time significantly.

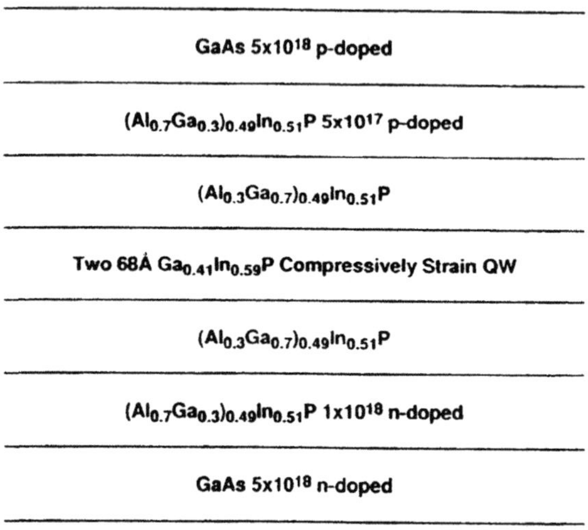

Figure 3 Schematic diagram of the layer structure of the 670nm GaInP/AlGaInP double quantum well laser used in the simulations.

In this paper we describe the results of simulations of a threshold behaviour of the 670nm GaInP/AlGaInP quantum well laser diode with layer structure given in figure 3. This comprises two compressively strained $Ga_{0.41}In_{0.59}P$ wells each 6.8nm wide embedded in $(Al_{0.3}Ga_{0.7})_{0.51}In_{0.49}P$ barriers. The waveguide core is nominally undoped and the n and p cladding regions are doped to 1.10^{18} cm^{-3} and 5.10^{17} cm^{-3} respectively, as measured for our structures and typical of many devices. The material parameters we have used for this structure are given in table 1.

TABLE 1. Parameter values used in the simulations.
[x defines the Al content in lattice matched $(Al_xGa_{1-x})_{0.51}In_{0.49}P$.]

Parameter		Value	ref
Direct Γ bandgap		$E_\Gamma = 1.91 + 0.61x$ eV	[9]
Indirect X bandgap		$E_X = 2.24 + 0.022x$ eV	[9]
Density of states effective mass:			
	Γ electrons	$m^* = (0.194 + 0.036x)m_0$	[9]
	X.electrons	$m^* = 0.6m_0$	
	holes	$m^* = (0.50 + 0.071x)m_0$	[9]
Relative permittivity		$11.72 + 1.01x$	
Non radiative electron lifetime in barrier		1 ns	[10]
Cladding layer properties:			
Non radiative electron lifetime		1 ns	[10]
Non-radiative hole lifetime		1 ns	[10]
Electron mobility		28 $cm^2V^{-1}s^{-1}$	
Hole mobility		7 $cm^2V^{-1}s^{-1}$	[11]
Donor ionisation energy		15 meV	
Acceptor ionisation energy		54 meV	

3. Results

3.1 ENERGY BAND PROFILES

Figure 4(a) shows the simulated profile of the conduction and valence band edges and the electron and hole quasi-Fermi levels as functions of distance through the structure with zero applied bias. The diagram indicates clearly the wide gap cladding layers, the GaAs contact layers at each end of the structure, and the built-in electric field across the undoped barrier/waveguide region containing the quantum wells. With the application of a forward bias across the terminals of 2.25 volts, corresponding to the threshold condition, the diagram takes the form shown in figure 4(b) which is similar to the flat-band diagram which is commonly drawn to represent a laser diode.

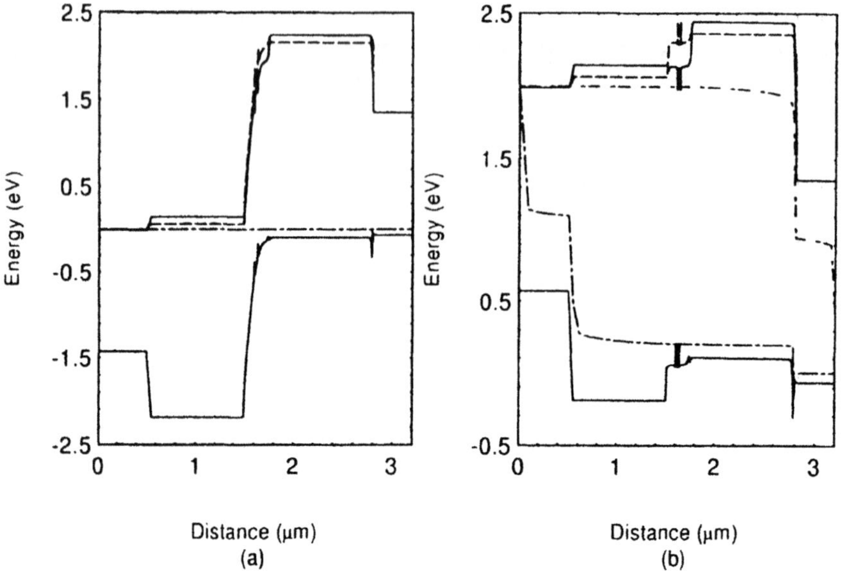

Figure 4. Simulated depth dependence of the energy band edges and the Fermi levels through the laser structure in figure 3, (a) with no applied bias and (b) with an applied bias of 2.25V.

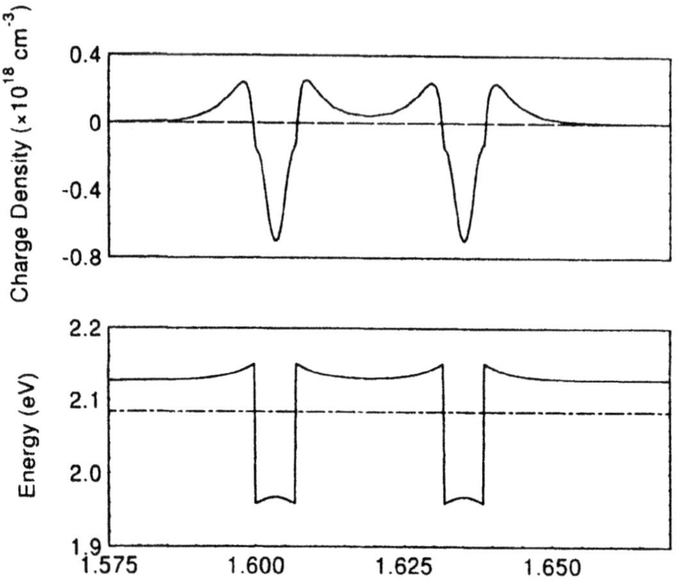

Figure 5. Simulated conduction band profile and net charge distribution in the vicinity of the quantum wells at V=2.25V, showing the accumulation of carriers in the well and the associated band-bending.

The detail of the band diagram in the vicinity of the quantum wells is shown in figure 5 where the band-bending in and alongside the well can be seen. This arises primarily from the differences between the densities of states in the well and barrier rather than from differences in the shape of the envelope functions for electrons and holes. There is a net transfer of electrons into the well region which induces an electric field such that in the steady state the diffusion of electrons into the well is balanced by the drift of electrons into the barrier alongside the well. The band-bending in the well is about 2 meV and the energy of the barrier at the edge of the well is raised by 12 meV relative to the flat-band region away from the well. The band bending in the well has a very small effect on the sub-band energies, and there is only a small change in the intrinsic recombination rate and optical gain due to the departure from the n = p condition.

3.2 BARRIER RECOMBINATION

To test the some of the details of the calculation we have measured the ratio of the spontaneous emission rates at threshold from the well and barrier as a function of temperature by means of a window in the top contact of a GaInP laser [3,12]. Figure 6 is a comparison of this data (shown as points) and the simulation (continuous line). The dash line is a calculation of this ratio using a simple flat-band model. Both calculations assumed that the optical losses were independent of temperature, and used a value of B in the barrier material of 1.0×10^{-10} cm^3s^{-1} at 300K.

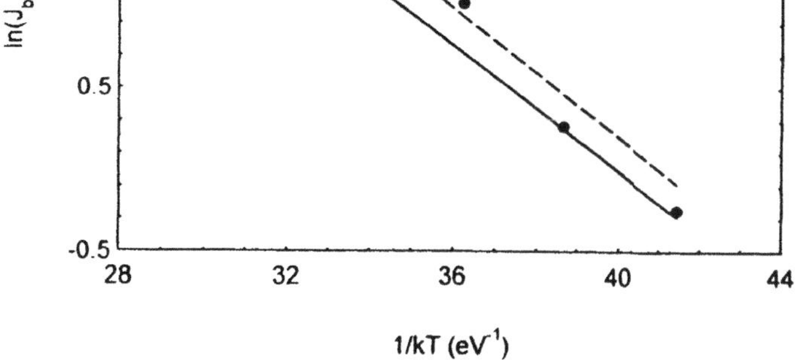

Figure 6. Plot of the logarithm of the ratio of spontaneous emission from the well and barrier regions of the 670nm laser at threshold as a function of reciprocal temperature. Experimental data is shown as points and the results of a flat-band calculation and the simulation are shown as dash and continuous lines respectively.

Both calculations give an acceptable description of the data. There appears to be a small systematic difference between the simulation and the flat-band calculation, associated chiefly with the absolute rates rather than the activation energy as defined by the energy band diagram. Neglect of band-bending does not introduce significant errors into the calculation of the intrinsic gain-current relation, but it does affect calculation of extrinsic currents which depend upon the position of one of the quasi Fermi levels. The radiative current depends on the (np) product and is therefore most sensitive to the Fermi level separation rather than the absolute Fermi level positions.

3.3 THRESHOLD CURRENT

A major challenge in modelling the operation of 670nm strained layer GaInP quantum well lasers is achieving an accurate description of the temperature dependence of threshold current. Experiments have shown [3] that the activation energy of the excess current corresponds to the loss of carriers via the X conduction band minima in the barrier and cladding layers. It is difficult to account for the magnitude of this current by non-radiative recombination in the barrier region alone (as has been shown to occur in GaAs/AlGaAs quantum well lasers [1]) and it is supposed that carrier loss through the cladding layers must be important in these devices.

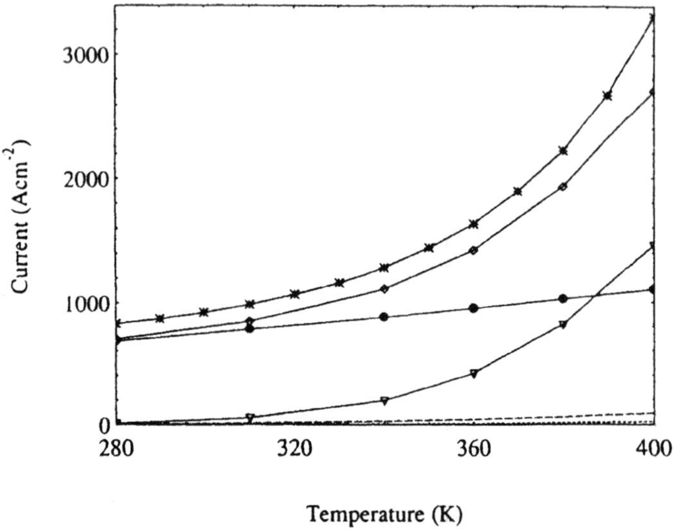

Figure 7. Experimental (stars) and simulated (diamonds) data for the temperature dependence of threshold current of the double well GaInP laser structure illustrated in figure 3. The device is 250μm long and the threshold gain is taken as 1371 cm^{-1} per well. Simulated results using the parameters given in table 1 are shown for the intrinsic spontaneous recombination current in the wells (solid circles), the non-radiative and radiative current in the barrier/waveguide (dash lines), and leakage of electrons through the p-cladding layer (triangles).

Figure 7 shows experimental and simulated data for the threshold current as a function of temperature for the double well GaInP structure illustrated in figure 3. At each temperature the simulations were performed to obtain a local gain of $1371 cm^{-1}$ per well corresponding to the estimated threshold gain for these lasers. The device length is 250μm, the optical confinement factor was calculated to be 0.02 per well, and the modal waveguide loss was taken to be $10 cm^{-1}$.

The parameter values used in the simulation are given in table 1: the carrier mobilities are taken from available published data and the non-radiative lifetimes for electrons and holes in both the waveguide core and cladding layers were taken to be 1ns. This value represents a reasonable "worst case" and is supported by results of luminescence decay experiments on isotype double heterostructures in this material system [10]. The radiative recombination coefficient used for the barriers is the same as that used to simulate the data in figure 6, $B=1.0\ 10^{-10}\ cm^3\ s^{-1}$. The energy level of the zinc acceptors in the p-cladding layer influences the conduction band alignment between the core and cladding layers on the p-side of the structure and hence the electron leakage current. A value of 54meV, together with an X-electron mobility of $28\ cm^2V^{-1}s^{-1}$, gives good agreement with experiment as shown in figure 7. This ionisation energy is in agreement with a published result of 90meV for zinc acceptors in $(Al_{0.7}Ga_{0.3})_{0.51}In_{0.49}P$ [13] reduced by impurity banding at high concentrations as described in ref [14]. The electron mobility is consistent with data scaled by the effective mass from published results in ref [11]. These are the only two parameter adjusted in our simulation to achieve a fit to the data. The experimental results are for 50μm oxide isolated lasers and no account has been taken of current spreading which may account for the difference in absolute current between simulation and experiment.

Figure 7 also shows the temperature dependence of the various contributions to the threshold current. At 280K the intrinsic recombination current in the wells is responsible for 98% of the current whereas at 400K this falls to only 41%. At 400K the threshold current is $2713\ Acm^{-2}$ and leakage of electrons through the p-cladding layer accounts for 54% of this. Recombination in the barrier/waveguide core is responsible for the remaining 4% of the current at this temperature. The electric field across the p-cladding layer is about $4\ kVcm^{-1}$ accounting for a voltage drop of 0.4 V.

4 Implications

The calculations above show that the flat-band model of the active region comprising well and barrier is satisfactory for calculation of the intrinsic properties of this region alone. The band-bending in the well produced in the full simulation is very small. Thus for GaAs/AlGaAs lasers flat-band calculations of the gain-current relations are reasonably accurate if the barrier material is of good quality. However, in devices where there is significant current flow outside this active region, the flat-band model is poor because it distorts the line-up of the Fermi levels with the bands in the barrier and it neglects any electric fields in the structure.

We have shown that a full simulation gives very good agreement with the measured temperature dependence of threshold current for 670nm laser diodes and that in these devices there is a significant temperature-dependent current due to electron leakage through the p-cladding layer. The magnitude of this current depends critically upon the p-doping density because this controls the band line-up at the barrier-cladding interface and the strength of the electric field which extracts the electrons.

It is commonly assumed that the stimulated emission process above threshold has an internal differential quantum efficiency of unity due to the pinning of the quasi Fermi levels in the active region. Our calculations show that although the Fermi levels may pin in the vicinity of the well, this is not necessarily so in the cladding regions. As the current drawn by the well increases above threshold, so the electric field in the cladding layer increases, increasing the leakage current. Consequently, an increment in the external current above threshold results in an increment in leakage current in addition to an increment in the stimulated radiative current in the wells. Even though this latter process remains internally 100% efficient, this leads to a reduction in the external differential quantum efficiency. This reduction in efficiency is clearly not due to any deficiency in the well and barrier material and is not associated with any other non-radiative recombination process at defect centres. It is a characteristic of the *device structure* and is critically dependent upon the contribution made by drift to the leakage current. The nature of the current paths outside the active region therefore have a significant influence on the external differential efficiency of the laser above threshold and its temperature dependence.

5 Conclusions

We have obtained an excellent description of the measured temperature dependence of threshold current of 670nm GaInP laser diodes using a full, self-consistent, simulation of current flow and potential distribution through the complete device structure linked to a state-broadened optical gain calculation. These simulations show the importance of electron drift through the p-cladding layer for the temperature dependence of threshold current and suggests that the conventional interpretation of measurements of external differential quantum efficiency is not appropriate for these devices.

6 Acknowledgements

Damian Foulger and Peter Smowton are supported by EPSRC. The epitaxial structures used for the devices described here were grown by EPI Ltd with funding provided by the EPSRC/DTI LINK scheme, and the lasers were fabricated in Cardiff by Paul

Hulyer. We are grateful to Phil Mawby in Swansea for making the basic simulation program available to us and for his help and advice throughout this work.

References

1. Blood, P., Fletcher, E.D., Woodbridge, K., Heaseman, K.C., and Adams, A.R., (1989) Influence of the barriers on the temperature dependence of threshold current in GaAs/AlGaAs quantum well lasers, *IEEE Journ Quantum Electron* **QE 25** 1459-1467.
2. Bour, D P, Treat, D W, Thornton, R L, Geels, R S, and Welch, D F. (1993) Drift leakage current in AlGaInP quantum well lasers", *IEEE Journ Quantum Electron* **QE29** 1337-1343.
3. Smowton, P.M., and Blood, P. (1995) GaInP-(AlGa)InP 670nm quantum well lasers for high temperature operation, *IEEE Journ Quantum Electron* **QE-31** 2159-2164.
4. Blood, P., Tsui, E. S-M., and Fletcher, E.D., (1989) Observations of barrier recombination in GaAs-AlGaAs quantum well structures, *Appl Phys Letts* **54** 2218-2220
5. Tsui, E.S-M., Blood, P., and Fletcher, E.D., (1992) Electroluminescent processes in quantum well structures, *Semicond Sci Technol.* **7** 837-844.
6. Wolfe, C.M., Holonyak, N., and Stillman, G.E., (1989) *Physical Properties of Semiconductors,* Prentice-Hall, Englewodd Cliffs, New Jersey, USA. p253.
7. Gault, M, Mawby, P, Adams, A R, and Towers, M. (1994) Two dimensional simulation of constricted mesa InGaAs/InP burried heterostructure lasers. *IEEE Journ Quantum Electron* **QE-30** 1691-1700.
8. Hamilton R A H, and Rees, P, (1993) Line broadening due to carrier-carrier scatering in quantum wll heterostructures, *Semicond. Sci and Technol.* **8** 728-734
9. The band gap data used in this work originates from several sources and is summarised in: Smowton P M, and Blood P, Visible emitting (AlGa)InP laser diodes, to be published in *Strained quantum wells and their applications* ed M O Manasreh, Gordon and Breach Science Publishers SA.
10. C H Molloy, University of Wales Cardiff, private communication
11. Ohba H, Ishikawa M, Sugawara H, Yamamoyo M, and Nakansisi T, (1986) Growth of high quality InGaAlP epilayers using methyl metalorganics and their applications to visible semiconductor lasers, *Journ Crystal Growth* **77** 374-379.
12. Smowton P M, and Blood P, (1995) Threshold current temperature dependence of GaInP 670nm quantum well lasers, *Appl Phys Letts* **67** 1265-1267.
13. Honda H, Ikeda M, Mori, Y, Kaneko K and Watanabe N, (1985) The energy levels of Zn and Se in (AlGa)InP *Japan Journ Appl Phys* **24** L187-189
14. Schubert E (1993) *Doping in III-V Semiconductors* Cambridge University Press.

SILICIUM CRYSTAL PHOTOLUMINESCENCE AS TRANSDUCER FOR BIOSENSORS

N.F.Starodub*, L.L.Fedorenko[1], V.M.Starodub, S.P.Dikij[1]
Institute of Biochemistry of Nat.Acad.of Sci., 9 Leontovicha St., Kiev 252030, Ukraine;
Fax:380-44-2296365, E.mail:prof@progress.FreeNet.Kiev.UA
[1)] Institute of Semiconductor Physik of Nat.Acad.of Sci., Ukraine

1. Introduction

The solving of the problem of direct, sensitive and very rapid registration of biomolecule interactions is very important both for fundamental investigations and for solving numerous practical tasks in the field of medicine, biotechnology and environmental monitoring. A number of approaches have recently been proposed for this purposes. The approach based on the utilization of biosensors is the most promising. The success in the field of biosensors depends largely on overcoming difficultiess with combining biomolecules and physical surfaces as well as with finding of effective transducers. The most perspective among them are optical and optoelectronic tranducers. In comparison with the electrochemical transducers, they allow to prevent influence of the medium of samples to be analyzed. At the same time modern wellknown optical transducers are complicate (for example, surface plasmon resonance), or can not provide direct registration of the formation of specific biomolecule complexes. That is why search for new effective transducers is a pending task.

Today the porous silicon (PS) is freaquntly implemented into designing of optoelectronic elements [1]. This material has recently been shown to posses photoluminescence (PL) and electroluminescence in visible spectrum field. The nature of PL is the subject of discussion, but a great role of quantum-dimensional effect and surface properties of material are emphasized in the articles [2]. PS is characterized by high chemical activity and great specific square surface that give possibility to immobilize biological material by such a simple way as adsorbtion.

To attract attention to transducer properties of PS is the main aim of this article. We concentrate our attention on investigation of the influence of biochemical reaction of antigen- antibody (Ag-Ab) complex formation on the PL of PS. For the first time, we demonstrate that PL is reduced after specific antigen-antibody interraction. Obtained results provide basis for recomendation of using PS as tranducers for construction of immunosensors in which direct registration of the Ag-Ab or Ab-Ag complexes may be accomplished by estimation of their PL level.

2. Experimental

2.1. PREPARATION OF PS AND MEASURING PL

The PS layers were formed by stain etching of laser modified boron doped monocrystalline silicon. PL spectra of PS samples placed into a quartz cell was excited by He-Cd laser (l = 0,44 mkm, P = 0,001 W) and was measured by standart monochromator.

2.2. FABRICATION OF THE SELECTIVE BIOLOGICAL LAYER ON THE PS SURFACE

At first the surface of the PS crystals was clenead by consequtive alcohol and distilled water washing. For immobilization of Ab or Ag on the surfaces of the crystals (1x1 mm), they were immersed

---

* To whom correspondence should be addressed.

M. Balkanski and N. Andreev (eds.),
Advanced Electronic Technologies and Systems Based on Low-Dimensional Quantum Devices, 91–92.

in solutions (100 mkg/ml) of these substances for 1 h at the room temperature. After washing by tris-HCl buffer (pH 7.3) containing 140 mmol/l sodium chloride. Rabbit IgG was used as Ag and anti-IgG sheep to rabbit IgG as Ab.

2.3. RESULTS AND DISCUSSION

The best wave length for observation of the visible PL of PS was experimentyally established to be about 650 nm. The time decay of PL of the PS is described by a "stretch" exponent. Typical spectrums of the PL of PS at the contact of the PS with distilled water, or 0.01 mmol/l tris- HCl buffer (ph 7.3), or 0.14 mmol/l buffered sodium chloride solution, or solution of Ag, or Ab prepared in the above mentioned buffer containing sodium chloride were analyzed. It was found that in the absence of specific Ag-Ab interractions, the intensity and shape of PL spectrums remain almost immutable after the exposition of PS silicon to analysed solutions during no less than 2-2,5 hrs. It is also worth noting that above indicated effect was observed not only in case of immersion of PS silicon crystals in distilled water, buffer, buffered sodium chloride solution but also in solutions of Ag or Ab (in wide range of concentration, from 10 to 1000 μg/ml) prepared on the above indicated buffered salt. On the other hand, great extinguishing of PL intensity was observed when Ab or Ag were first immobilized on the PS surface and then allowed to contact with corresponding Ag or Ab solution to be analyzed. Degree of the reduction of PL intensity depended on the time of exposition and on the concentration of components in analysed solution. It was found that after completion of the Ab-Ag reaction, intensity of PL was reduced more than 3 times. There were not observed any other deformations of the PL spectra shape. The sensitivity of immunoanalysis using the biosensor based on the PL effects of PS was studied in preliminary experiments. In particular, it was shown that about 10 ng/ml of anti-IgG rabbit in solution may be detected by this method.

According to existing ideas pertinent to the nature of the visible PL of PS, extinguishing of the PL can be explained by a dehydrogenization of PS surface after specific immunocomplex formation. Hydrogen is released from silicon bonds and subsequently captured by the immunocomplex. In turn, torn Si bonds are known to intensify nonradioactive channel of recombination [3]. Therefore, effective activation of the above indicated channel is the most probable cause of PL extinguishing after Ag-Ab interractions on the PS surface.

So, the PL of PS can be effective indicative factor for direct interactions of biological molecules, especially those occuring between Ag and Ab. According to the obtained results we believe that discovered effect can be effectivelly used for solving actual problems of creating new biosensors.

3. Conclusion

The effect of the PL of PS extinguishing after the immunocomplex formation on the surface of PS has been observed for the first time. The use of visible PL of PS for immunosensor creation is suggested. This will intensify the detection of specific interections between Ab and Ag. The sensitivity of immunoanalysis with the usage of PL effects of PS is comparable with that of ELISA-method, but the overall time needed for fulfilment of the presented method is sufficiently less.

4. References

1. Vial J., Billet S., Bsiesy A, Fishman G., Gaspard F., Herino R, Legion M., Madeare R., Michalcesku T., Miller F, and Romestaine F. (1993) Bright visible light emission from electro-oxidised porous silicon, *Physica B, 185*, 593-602.
2. Canham L. (1990) Silicon quantum wire array fabrication by electrochemical dissosiation of wafers, *Appl. Phys. Lett., 57*, 1046-1048.
3. Koch F. (1993) Models and mechanisms for the luminescence of porous Si, *Mat.Rec.Soc.Symp.Proc., 298*, 319-329.

INJECTION LASERS BASED ON VERTICALLY COUPLED QUANTUM DOTS

A.E.ZHUKOV, V.M.USTINOV, A.Yu.EGOROV, A.R.KOVSH,
N.N.LEDENTSOV, M.V.MAKSIMOV, A.F.TSATSUL'NIKOV,
N.Yu.GORDEEV, S.V.ZAITSEV, P.S.KOP'EV
A.F.Ioffe Physico-Technical Institute,
Politekhnicheskaya 26, 194021, St.Petersburg, Russia

Using quantum dots (QDs) as an active region of a semiconductor laser should lead to ultra low threshold current density (J_{th}) due to delta function density of states [1]. The method based on a self-organization process at the initial stages of strained layer heteroepitaxy is currently widely used for QD formation. The quantum islands formed under appropriate growth conditions allowed us to realize lasing via the ground state of QDs at low temperatures [2]. We will show that using the concept of vertically coupled quantum dots (VECODs) leads to remarkable reduction in J_{th}, ground state lasing up to 300 K, and continuous wave (CW) operation of quantum dot injection laser.

The samples studied were grown by solid-source MBE. (In,Ga)As/GaAs VECODs are formed as the result of successive deposition of (In,Ga)As quantum dot sheets and thin GaAs spacers due to the effect of inhomogeneous strain fields [3]. It is clearly seen in TEM image of 3-period VECOD structure that QDs of subsequent sheet are formed just above the dot of the previous one, Fig. 1. It is important that no dislocations or large relaxed clusters are formed.

Since GaAs spacers are typically very thin (< 100 Å), the wavefunctions of neighboring islands are essentially overlapped and each VECOD can be considered as a one unified quantum-mechanical object. Increasing the number of QD sheets, N, leads to the red shift of PL peak position due to increase in carrier localization energy in VECODs. It is very important for laser applications since the population of upper states (wetting layer (WL) and GaAs matrix) at the expense of QD levels is one of the main reasons for the increase in threshold current density at elevated temperatures [2]. Moreover, the larger N, the less pronounced is the effect of PL intensity saturation under high excitation densities.

GRIN SCH AlGaAs/GaAs laser structures with a VECOD active region were studied. In the case of single-sheet QD the lasing wavelength shifts toward WL-related emission at higher observation temperatures, whereas in the case of the VECOD (N = 3) it remains within QD-related PL line in the entire 77÷300 K range, i.e., in VECOD structures lasing proceeds via the ground state of quantum dot up to room temperature.

M. Balkanski and N. Andreev (eds.),
Advanced Electronic Technologies and Systems Based on Low-Dimensional Quantum Devices, 93–95.

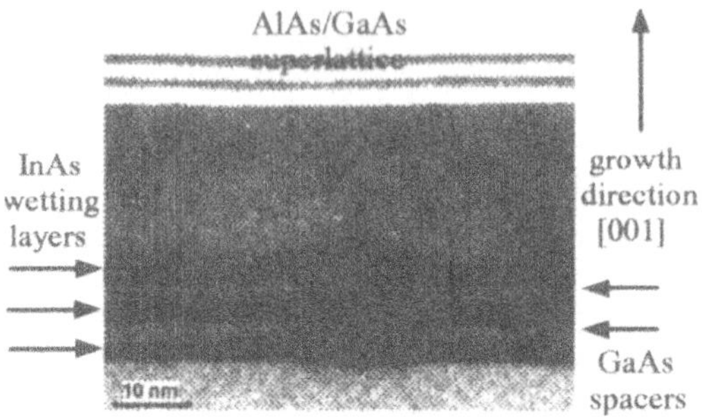

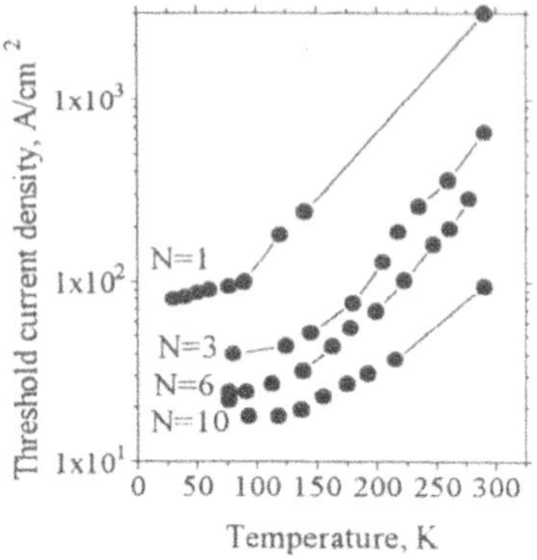

Fig. 1. Cross section TEM images of the 3-period VECOD composed of 5 Å InAs and 50 Å GaAs layers.

Fig. 2. Temperature dependences of threshold current density for the VECOD lasers with different number of QD sheets, N.

J_{th} values in a VECOD laser decreases with the number of QD sheets by one order of magnitude from 950 A/cm^2 (N = 1) to 98 A/cm^2 (N = 10), Fig. 2. The range of J_{th} thermal stability is extended up to 150 K. The reason for this is presumably the increase in optical confinement factor and carrier localization energy in VECODs [3]. It is worth mentioning that J_{th} as low as 18 A/cm^2 was observed at low temperatures (120 K).

Considerable reduction in threshold current density allowed us to realize RT CW operation of a QD injection laser. We used laser diodes based on InGaAs/GaAs VECOD (N = 10). Maximum power per facet obtained was 160 mW (W = 60 μm) and differential efficiency was 45 %. Lasing spectrum only slightly shifted toward short wavelengths with increasing pumping current. Consequently, lasing proceeds via the states of quantum dots up to $I = (6\div7)\times I_{th}$.

In conclusion, remarkable reduction in threshold current density (20 and 100 A/cm^2 at 77 and 300 K, respectively) and RT CW operation via QD states have been demonstrated for lasers based on InGaAs/GaAs vertically coupled quantum dots.

This work is supported by INTAS (Grant 94-1028), Russian Foundation for Fundamental Research (Grant 96-02-17824), and Program of Ministry of Science of RF "Nanostructures" (Project 2-001).

[1] Arakawa, Y. and Yariv, A. (1986) QW lasers: Gain, spectra, dynamics, *IEEE J. Quantum Electron.* QE-22, 1887-1899.

[2] Kirstaedter, N., Ledentsov, N.N., Grundmann, M., Bimberg, D., Ustinov, V.M., Ruvimov, S.S., Maximov, M.V., Kop'ev, P.S., Alferov, Zh.I., Richter, U., Werner, P., Gosele, U., and Heydenreich, J. (1994) Low threshold, large T_0 injection laser emission from (InGa)As quantum dots, *Electron. Lett.* **30**, 1416-1417.

[3] Ustinov, V.M., Egorov, A.Yu., Zhukov, A.E., Ledentsov, N.N., Maksimov, M.V., Tsatsul'nikov, A.F., Bert, N.A., Kosogov, A.O., Kop'ev, P.S., Bimberg, D., and Alferov, Zh.I. (1995) Formation of stacked self-assembled InAs quantum dots in GaAs matrix for laser applications, *Proc. MRS*, Nov.27-Dec.01, 1995, Boston, USA, EE3.6.

KEYWORDS/ABSTRACT: injection laser / semiconductor heterostructures / quantum dots / continuous wave operation

We have fabricated and studied injection lasers based on vertically coupled quantum dots (VECODs). VECODs are self-organized during successive deposition of several sheets of (In,Ga)As quantum dots separated by thin GaAs spacers. Increasing the number of periods in the VECOD leads to remarkable decrease in threshold current density. Lasing proceeds via the ground state of quantum dots up to room temperature. Room temperature continuous wave operation have been demonstrated.

INFRARED EMISSION OF HOT HOLES IN STRAINED MULTI-QUANTUM-WELL HETEROSTRUCTURES InGaAs/GaAs UNDER REAL SPACE TRANSFER

V.Ya.ALESHKIN, A.A.ANDRONOV, A.V.ANTONOV, N.A.BEKIN, V.I.GAVRILENKO, D.G.REVIN, *E.R.LIN'KOVA, *I.G.MALKINA, *E.A.USKOVA and *B.N.ZVONKOV
Institute for Physics of Microstructures of Russian Academy of Sciences GSP-105, Nizhny Novgorod, 603600, Russia
**Physical-Technical Institute of Nizhny Novgorod State University*

The paper presents results of the first experimental study of lateral transport and far and middle infrared (FIR&MIR) emission of hot holes in the strained multi-quantum-well (MQW) InGaAs/GaAs heterosructures (HSs). $In_xGa_{1-x}As$/GaAs HSs ($0.03 \leq x \leq 0.2$, $d_{InGaAs} = 50 \div 100$Å, $d_{GaAs} \approx 600$Å, $n_{QW}=20$) were grown by MOCVD technique on semiinsulating GaAs(001) substrates. Two δ-layers of carbon were introduced at 50 Å from both sides of each InGaAs QW in GaAs barrier layers. Typical values of 2D holes concentration and mobility were $p_s=(1\div3)\cdot10^{11}cm^{-2}$, $\mu_{4.2K}\approx3000$ $cm^2/V\cdot s$. The lateral pulsed electric field was applied to the structure via electric contacts deposited on the sample surface. Spontaneous emission of hot holes was studied in the sensitivity bands of FIR detector Ge:Ga with teflon filter ($\lambda=50\div120$ μm) and of MIR detector Si:B ($\lambda=20\div28$ μm).

Fig.1a,b represent emission-voltage and current-voltage (I-V) characteristics of two samples. I-V characteristics demonstrate pronounced saturation in strong electric fields. In some samples in the electric fields corresponding to the saturation current oscillations were observed, the amplitude and the decay time of the oscillations being increased with electric field (*Fig.2*). The nonlinear parts of I-V characteristics correspond to the remarkable features in emission-voltage characteristics both in FIR and MIR ranges (*Fig 1a,b*).

Both the current saturation and the nonmonotonous dependences of the emission intensities on the electric field may be explained by real space transfer (RST) effects [1]. Application of strong lateral electric field results in the heating of holes in QWs and in substantial population of the barriers. Doping of barriers provides the barrier wells (BW) where holes are trapped after RST. Since the hole mobility in BWs is much less then in QWs the trapping results in the saturation of I-V characteristic and in the drop of emission intensity. For the sample #1721 (x=0.18) FIR emission results from intrawell hh3→hh2 and hh2→hh1 transitions (*Fig.3*) while MIR one results from the upper to lower subband hh3→hh1 transitions and transitions between

M. Balkanski and N. Andreev (eds.),
Advanced Electronic Technologies and Systems Based on Low-Dimensional Quantum Devices, 97–98.

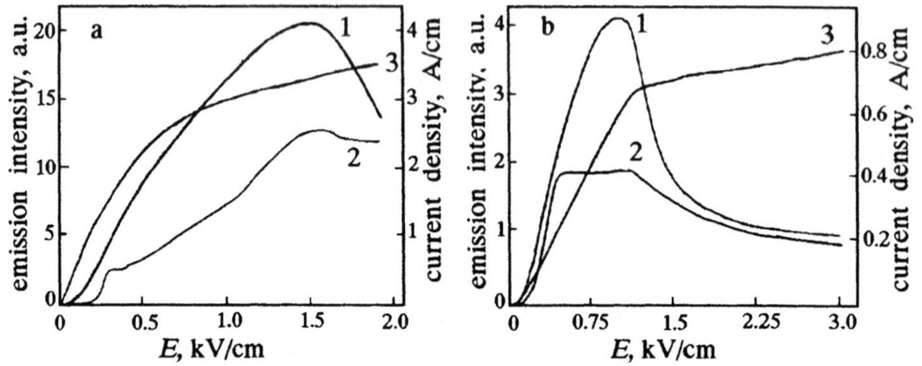

Figure 1. Emission-voltage (1,2) and current-voltage (3) characteristics of the HSs $In_xGa_{1-x}As/GaAs$; T=4.2K. Detectors: 1 - Ge:Ga, 2 - Si:B.
a) sample #1721 (x=0.18, d_{QW}= 100 Å) b) sample #1845 (x=0.1, d_{QW}= 77 Å)

barrier and QW states. In the sample #1845 (x=0.1) the band offset is smaller: $\Delta\mathcal{E}_v$=39 meV. Therefore there are only two hole subbund hh1 and hh2 in QW and FIR emission may be observed at hh2→hh1 transitions while MIR emission may result only from

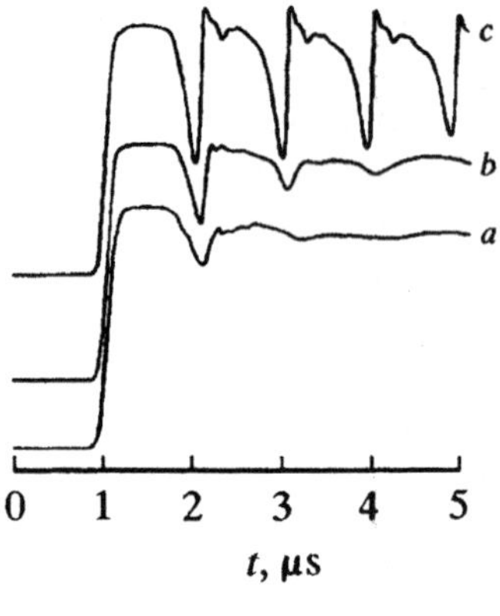

Figure 2. Oscillograms of the current pulses, sample #1721. T=4.2 K. E= a) 1.2 kV/cm, b) 1.3 kV/cm, c) 1.5 kV/cm

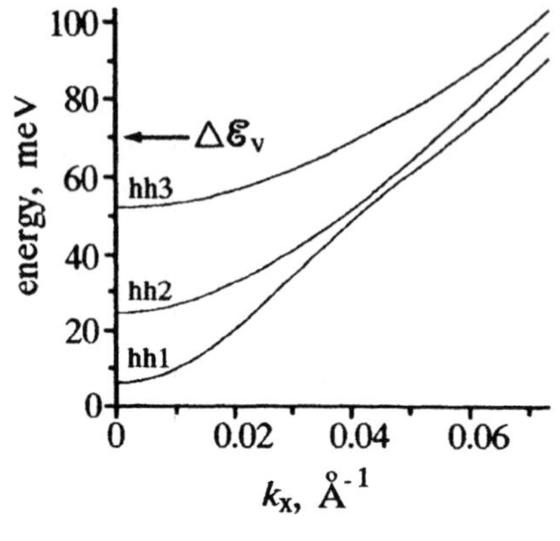

Figure 3. E(k) dependences for 2D hole subbunds in sample #1721 (x=0.18, d_{QW}=100Å); k||[100]

transitions between barrier and QW states. Correspondingly maximum of emission intensity in the sample #1845 (*Fig.1b*) is reached at less electric fields than in the sample #1721 (*Fig.1a*) that confirms the existence of RST under lateral electric field.

The research described in this publication was made possible in part by NATO Linkage Grant HTECH.LG.960931, CDRF Grant #1713 and Grant #94-842 from INTAS.

References

1. Gribnikov, Z.S. Hess, K. and Kosinovsky, G.A. (1995) Nonlocal and Nonlinear Transport in Semiconductors: Real Space Transfer Effects, *J.Appl.Phys.*77, N 4, p.1337.

MICROCAVITY SEMICONDUCTOR LASERS

John G. McInerney*, Damien P. Courtney and Peter M. W. Skovgaard
Optronics Ireland/Physics Department,
National University of Ireland, University College, Cork, Ireland

Brian Corbett
Optronics Ireland/National Microelectronics Research Centre
National University of Ireland, University College, Cork, Ireland

* also with the Optical Sciences Center, University of Arizona, Tucson, AZ 85721, USA

Abstract

We review the theory and recent experimental developments in microcavity semiconductor lasers and their implications. Particular attention is paid to microdisk lasers which support whispering gallery modes. InGaAsP/InP microdisks have recently been fabricated in our laboratories and have been pumped optically, resulting in the first achievement of CW room temperature lasing in these devices.

1. Introduction

The future viability of optoelectronics as a mainstream technology in communications, computing, data storage and consumer products is contingent on developing efficient, flexible and controllable photonic emitters, detectors, filters, amplifiers, memory elements and logic devices. It is intriguing and exciting to consider utilising the quantum nature of light itself in designing and studying photonic devices of all types. A critical part of this effort is the study of wavelength scale structures for control and selection of photon modes; structures such as microcavities and the closely related photonic bandgap materials.

A microcavity laser may be defined operationally as one whose cavity length is comparable to the emission wavelength, in at least one dimension [1,2]. A simple example is the planar microcavity laser whose thin ($\sim\lambda$) active region is bounded by two parallel, highly reflecting mirrors (Figure 1a). A related structure is the vertical cavity surface-emitting laser (VCSEL) [3] in which a planar microcavity is modified (by implantation, oxidation, etching or optical pumping) so that the gain is of limited spatial extent: the cavity is only a few wavelengths long but much larger in lateral extent (Figure 1b). Both the planar microcavity and VCSEL are one-dimensional microcavities since they are wavelength scale only in one dimension. Such structures have been made possible by advances in high-reflectivity semiconductor Bragg reflectors, placing the quantum well gain elements at the antinodes of the cavity standing wave [4] and by highly effective materials growth, doping and fabrication techniques.

From a physical point of view, a microcavity may be considered as an atom or atom-like emitter inside a cavity. In a true microscopic system the dimensions are comparable to the size

M. Balkanski and N. Andreev (eds.),
Advanced Electronic Technologies and Systems Based on Low-Dimensional Quantum Devices, 99–115.

of the atom and are thus much smaller than the (optical) wavelength. In a macroscopic system the dimensions are much greater than the wavelength. A microcavity as defined here is therefore a *mesoscopic* system, in that its scale is intermediate between those of microscopic and macroscopic objects. Because of this scaling, it is often possible to adopt a semi-classical point of view, in which the active atom or atom-like species (in this case the coupled system consisting of an electron in the conduction band of the semiconductor and the corresponding hole in the valence band) is treated quantum mechanically and the electromagnetic field is considered classically. Only when the quantum optical properties of the light field are to be studied is it necessary to adopt a fully quantum mechanical approach. A further complication is that useful microcavities, including microcavity lasers, are open (non-conservative, *ie* non-Hermitian) systems whose radiation should more properly be described in terms of quasi-normal mode bases [5]. This question, and associated issues related to theoretical treatments of microcavity lasers, will be discussed in Section 2.

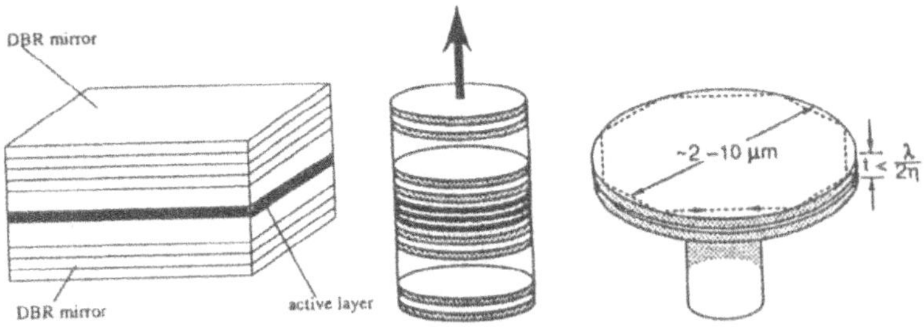

Figure 1: Schematic of (a) planar microcavity, (b) vertical cavity surface-emitting laser (VCSEL), (c) microdisk laser. (Adapted from [2])

From an engineering or applied-physics perspective, a microcavity may be considered as a filter or distributor of the radiation from the atom. Since the longitudinal mode spacing of a Fabry-Perot cavity of length L is the free spectral range $\Delta n = c/2nL$, where n is the refractive index of the material filling the cavity, 1-D microcavities have very sparse mode spectra which enable them to select single longitudinal modes from the broad gain spectra of bulk or quantum well semiconductor active media. The essential function of the 1-D microcavity is therefore spatial and spectral control of photon modes by filtering and distributing the spontaneous emission. Such spectral narrowing may be accompanied by increased power efficiency but often this is not the case due to non-ideal processes such as carrier and photon scattering, optical absorption, mode curvature due to thermal and carrier effects *etc*. The standard microcavity laser theory presented here predicts the maximum enhancement of the radiative transition rate due to a 1-D microcavity to be a factor of 3, but observed enhancements have been less than this - with the notable exception of the work of Jin *et al.*[**6**], who observed an order of magnitude enhancement in bulk GaAs at room temperature from carriers generated using picosecond pulses. Explanation of these data requires detailed considerations of sub-picosecond carrier kinetics and is not attempted in this review.

To counteract such difficulties, higher dimensional microcavities and photonic bandgap materials have been developed, but their fabrication is fraught with difficulty. A good compromise is the microdisk laser, in which light undergoes total internal reflection along the perimeter of an isolated circular disk, where two-dimensional mode confinement is provided via so-called "whispering gallery modes" (Figure 1c). Microdisk lasers have been pioneered by McCall, Slusher *et al.* at AT&T Bell Laboratories [7-9] and by Corbett *et al.* at the National

University of Ireland - University College Cork **[10]**, both in III-V semiconductors, and by collaborating groups at Brown and Purdue Universities **[11]** in II-VI semiconductors.

In this chapter we concentrate on these semiconductor microdisk lasers. Section 2 outlines some theoretical and fundamental considerations, particularly regarding the predicted effects of 1-D and 2-D microcavities on the density of photon modes. Section 3 describes fabrication of typical microdisk lasers. In Section 4 we describe recent optical pumping experiments in our laboratories culminating in the first demonstration of CW room temperature lasing in a semiconductor microdisk laser. Section 5 contains some conclusions and discussion.

2 Theoretical description of microcavity lasers

As mentioned in Section 1, microcavity lasers are open mesoscopic systems. The degree of "openness" of the system is given by the characteristic time τ associated with the (assumed exponential) decay of the optical field amplitude, also known as the photon lifetime of the system. An equivalent parameter is the cavity loss rate $\gamma = 1/\tau$ which appears operationally as the spectral linewidth of the optical mode at frequency ω. To determine an appropriate theoretical description of the microcavity, we must compare τ with the characteristic time T of the atomic transitions, or equivalently compare γ with the coupling or transition rate $\Gamma = 1/T$ as given for example by Fermi's golden rule (if applicable). Γ appears operationally as the spectral extent of the transitions - in our case the gain spectrum of the semiconductor.

The strong coupling regime is defined by $\gamma << \Gamma$ and has been implemented in atomic physics by placing long-lived atomic states (Rydberg atoms) inside high finesse cavities **[12]**. Recently the concept has been extended to semiconductors by studying excitons in quantum wells placed in high finesse resonators similar to VCSELs **[13]**. The strong coupling regime belongs in the domain of cavity quantum electrodynamics, attended by such phenomena as normal mode coupling, anti-crossing between the Fabry-Perot mode and the exciton photoluminescence peak. Fermi's golden rule, based on a perturbation approach, is not valid here. In this regime there are many parallels between atomic and semiconductor physics, for example observations of vacuum field Rabi splitting. However, there are also significant and interesting differences in that exciton transport properties differ markedly from those of atomic beams, and in the strong influence of many-body effects in semiconductor samples **[14]**. The semiconductor microcavity lasers we have demonstrated do not operate in the strong coupling regime.

In the weak coupling regime, $\gamma >> \Gamma$ and Fermi's golden rule may be used. Because of this weak coupling, and if the losses are not too high, we may consider as a first approximation that the system is Hermitian, with real eigenvalues and orthogonal eigenfunctions: this simplifies the counting of modes and hence the determination of the most important density of photon states **[15]**. We also assume that the "dressing" of the atomic transition frequency by the cavity is a small perturbation, and direct our attention to the most important influence of the microcavity, that is the geometric modification and quantisation of the photon modes **[16]**.

2.1 Mode counting techniques and densities of states

The most important advantage of using a microcavity is the modification of the spontaneous emission spectrum and decay time, resulting in better efficiency and noise performance and possibly realisation of "thresholdless lasing" **[17,18]**. We will consider this phenomenon theoretically by describing an excited atom in a microcavity using Fermi's Golden Rule. For simplicity consider a cubic box having 100% reflecting walls of side length L as in Figure 2. The electromagnetic field inside will obviously be quantized, and only photons with appropriate wavelengths can exist there. The mode density will be a series of narrow peaks at the resonance frequencies. If the atom has a single transition frequency which is not close to one of the cavity resonance frequencies, the atom will in principle stay excited inside the box forever. If, however, the transition frequency fits one of these frequencies, the atom will decay more quickly than in free space. This phenomenon has the potential to enhance dramatically the performance of diode lasers and LEDs and has been demonstrated using several types of microcavities [**19-23**].

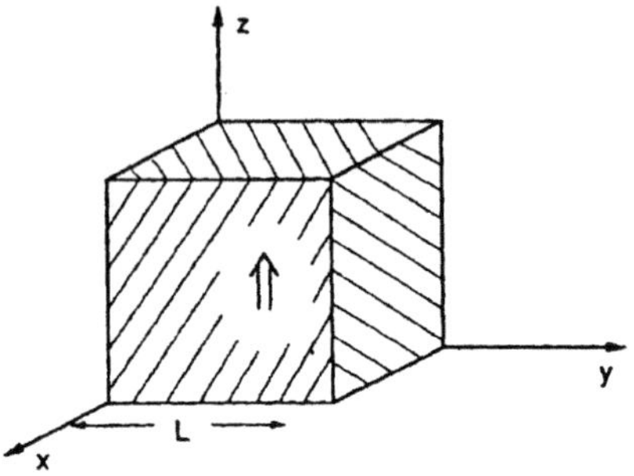

Figure 2: A cubic box having perfectly reflecting walls containing an oriented atomic dipole.

To calculate the modified decay time, we use Fermi's Golden Rule with a new density of photon states. This altered density of states is calculated using a method called *mode counting*. We then compare the results to the free space mode density to obtain the effect of the microcavity.

2.2 *Finding the transition rate from the mode density.*

An excited, two-level atom in free space has the Hamiltonian

$$H = H_a + \hbar\Omega(a^{+}a + \frac{1}{2}) - e\boldsymbol{d}\cdot\boldsymbol{E}$$

where H_a is the atomic Hamiltonian, Ω is the atomic resonance frequency, $-e\boldsymbol{d}\cdot\boldsymbol{E}$ is the dipole-field coupling. Assuming that Ω does not change due to the presence of the cavity, the transition rate Γ is then given by Fermi's Golden Rule

$$\Gamma = \frac{1}{\tau} = \frac{2\pi}{\hbar^2 c}\sum_{\boldsymbol{k}}|<\psi_g|ed\cdot E|\psi_e>|^2\,\rho(\boldsymbol{k})$$

where the sum is taken over all available $\boldsymbol{k}$ vectors satisfying $|\boldsymbol{k}| = k = \Omega/c$. $\rho(\boldsymbol{k})$ is the mode density of the field in the direction of $\boldsymbol{k}$. It is convenient to express Fermi's Golden Rule as

$$\Gamma = \frac{2\pi}{\hbar^2 c}|M|^2\,g(k)$$

where M is the matrix element describing the dipole-field coupling, and $g(k)$ the effective mode density including the angular dependence of M

$$g(k) = \frac{1}{V}\sum_{\boldsymbol{k}} P(\boldsymbol{k})\sigma(\boldsymbol{k})$$

for a cavity volume V, with the sum taken over all available $\boldsymbol{k}$ vectors satisfying $|\boldsymbol{k}| = k = \Omega/c$. $P(\boldsymbol{k})$ is the polarization factor containing the angular dependence of M. $\sigma(\boldsymbol{k})$ describes the modes available in the $\boldsymbol{k}$ direction on the surface of a sphere of radius k. In the case of a microcavity, $\sigma(\boldsymbol{k})$ will normally depend not only on the length of $\boldsymbol{k}$, but also on its direction.

2.3 *Finding the mode density of free space*

The allowed k-states for the emitted photon lie on the surface of a sphere with radius k and thickness dk. $\sigma(\boldsymbol{k})$ will be $(L/2\pi)^3$, the inverse of the unit volume in k-space, since there are no preferred directions in free space. Assuming that the atom is oriented vertically, the polarization factor is $P(\boldsymbol{k}) = \sin^2\theta$, where θ is the angle between $\boldsymbol{k}$ and the dipole moment $e\boldsymbol{d}$. Integrating over all the modes using the k-space volume element $d^3k = \sin\theta\, k^2\, d\theta d\phi dk$ gives

$$g(k)dk = \frac{1}{V}\int_0^{2\pi} d\phi \int_0^{\pi} d\theta \sin\theta\, k^2 dk \left(\frac{L}{2\pi}\right)^3 \sin^2\theta$$

resulting in the density of photon states in free space

$$g(k) = \frac{1}{3\pi^2}k^2$$

for the assumed vertical polarisation of the atomic dipole.

2.4 *Density of states in a 1-D planar microcavity*

Now we place the excited two-level atom inside a planar cavity consisting of two perfectly reflecting mirrors placed a distance L apart, so that the field is only modified in one dimension, assumed to be the vertical (z) direction (see Figure 3).

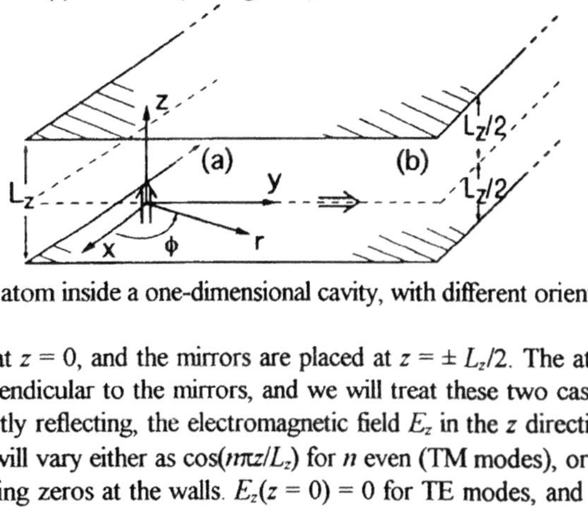

Figure 3: An excited atom inside a one-dimensional cavity, with different orientations.

The atom is situated at $z = 0$, and the mirrors are placed at $z = \pm L_z/2$. The atom can be oriented either parallel or perpendicular to the mirrors, and we will treat these two cases separately. Since the mirrors are perfectly reflecting, the electromagnetic field E_z in the z direction inside the cavity will be quantized. E_z will vary either as $\cos(n\pi z/L_z)$ for n even (TM modes), or as $\sin(n\pi z/L_z)$ for n odd (TE modes), having zeros at the walls. $E_z(z = 0) = 0$ for TE modes, and has a maximum for TM modes. In a *dielectric* waveguide the walls are partially reflecting and E_z is much greater at the boundaries for TM modes, which consequently suffer far higher losses in thin waveguides or in those which are clad by lossy media such as metal contacts. For this reason the microdisk lasers studied should be predominantly TE-polarised.

Only photons with $k_z = n\pi/L_z$, n an integer, can be emitted from the atom. This can be seen in Figure 4, where the allowed k-values are planes perpendicular to the z-axis. Since $k = \Omega/c$, the allowed modes will be the intersections of a sphere with a radius Ω/c and the planes $k_z = n\pi/L_z$.

To find the effective mode density, we count the number of allowed modes in a spherical shell of radius k and width dk; these are rings having radii

$$k_{ring}^{(n)} = k \sin\theta_n,$$

where θ_n is the angle between the ring in the n'th plane and the z-axis. The width of each ring is

$$dk_{ring}^{(n)} = \frac{dk}{\sin\theta_n}$$

We integrate over each ring separately and then sum over all the rings. Since we only integrate in the xy plane, the volume element in k-space is $d^2k = k_r d\phi\, dk_r$. Also, $\sigma(k)$ contains a factor $(L/2\pi)^2$ to account for the two unconfined dimensions x and y.

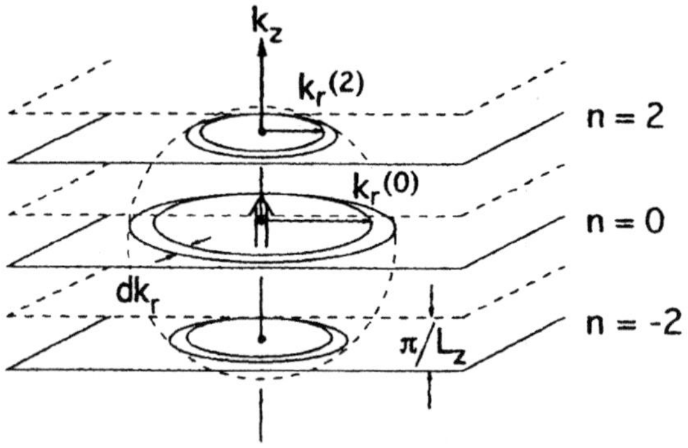

Figure 4: Allowed modes for an excited atom inside a planar mirror cavity

TM modes

The polarization factor for a dipole oriented *parallel* to the z-axis is $P_\parallel = \sin^2\theta_n$. For such a dipole situated at z = 0, only the *even* order modes (TM) are summed, since only the dipole couples to E_z:

$$g_\parallel(k)dk = \frac{1}{V}\sum_{\text{even } n}\int_0^{2\pi} d\phi\, k_{ring}^{(n)}\, dk_{ring}^{(n)} \left(\frac{L}{2\pi}\right)^2 \sin^2\theta_n$$

summing over all rings with even n ($|n| < k\,L_z/\pi$). The volume of the cavity in this case is $V = L^2 L_z$.

Now, defining the "cut-off" wave number $k_c = \pi/L_z$, the radii of the rings are

$$k_{ring}^{(n)} = \sqrt{k^2 - \left(\frac{n\pi}{L_z}\right)^2} = \sqrt{k^2 - (n k_c)^2}$$

from which the mode density is found to be

$$g_\parallel(k) = \frac{1}{2\pi^2}\, k_c\, k \sum_{\text{even } n}\left(1 - \left(n\,\tfrac{k_c}{k}\right)^2\right)$$

which can be further reduced to

$$g_\parallel(k) = \frac{1}{2\pi^2}\, k_c k \left[1 + \left(\tfrac{2k_c}{k}\right)^2 2\sum_s \left[\left(\tfrac{k}{2k_c}\right)^2 - s^2\right]\right],\quad 0 < |s| \le \left\lfloor \frac{k}{2k_c} \right\rfloor$$

This is a finite sum that can be expressed as:

$$g_\parallel(k) = \frac{k_c k}{2\pi^2}\left[1 + 2\left[p - \frac{1}{6}\left(\tfrac{2k_c}{k}\right)^2 p(p+1)(2p+1)\right]\right],\quad p = \left\lfloor \frac{k}{2k_c} \right\rfloor$$

In the two expressions above, $\lfloor k/2k_c \rfloor$ denotes the largest integer less than or equal to $k/2k_c$. $g_\parallel$ is plotted as a function of *k* in Figure 5 alongside the free space mode density for comparison.

TE modes

Now consider the dipole oriented perpendicular to the z-axis. In this case, the dipole couples only to odd-order modes (TE modes), and the polarization factor $P_\perp$ found by rotating $P_\parallel$ by 90°:

$$P_\perp = \sin^2\theta_n \cos^2\phi_n + \cos^2\theta$$

Counting modes as before, we get

$$g_\perp(k)dk = \frac{1}{V}\sum_{odd\,n}\int_0^{2\pi} d\phi\, k_{ring}^{(n)}\, dk_{ring}^{(n)} \left(\frac{L}{2\pi}\right)^2 (\sin^2\theta_n)$$

which can again be reduced to a much simpler form:

$$g_\perp(k) = \frac{1}{4\pi^2} k_c\, k \sum_{odd\,n}\left(1 + \left(n\,\frac{k_c}{k}\right)^2\right)$$

This is also a finite series, and can be reduced to (see Figure 5):

$$g_\perp(k) = \frac{k_c k}{2\pi^2}\left[q + \left(\frac{k_c}{k}\right)^2\left[\frac{4}{3}q^3 - \frac{1}{3}q\right]\right],\quad q = \left\lfloor \frac{k}{2k_c} + \frac{1}{2} \right\rfloor$$

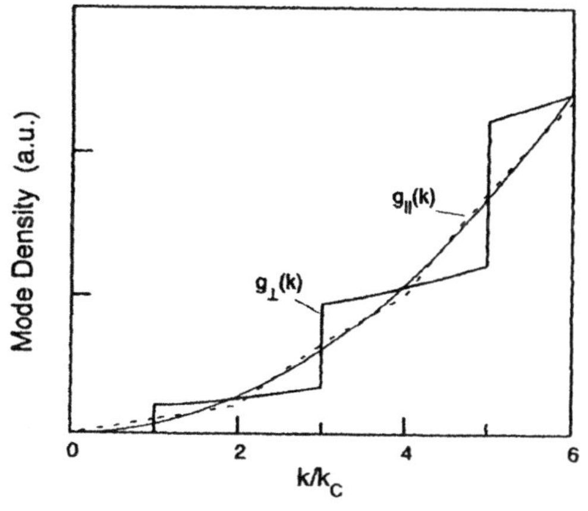

Figure 5: The mode density inside a planar mirror cavity versus the transition frequency for the dipole oriented both parallel ($g_\parallel$) and perpendicular ($g_\perp$) to the z-axis.

To see how much the presence of the cavity will alter the spontaneous transition rate, we divide the mode density in the cavity by that in free space (see Figure 6).

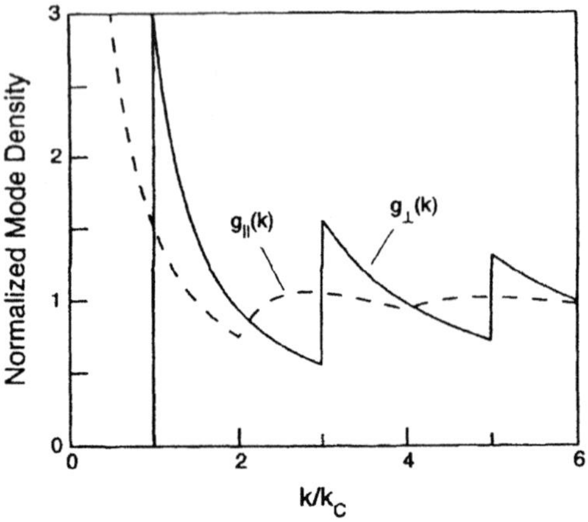

Figure 6: The ratio of the mode density inside the cavity to the free space mode density for the two possible orientations of the dipole.

2.5 2-D microcavity with rectangular cross section

We now place the excited atom in a two dimensional microcavity consisting of a waveguide with rectangular cross section having perfectly reflecting walls (see Figure 7).

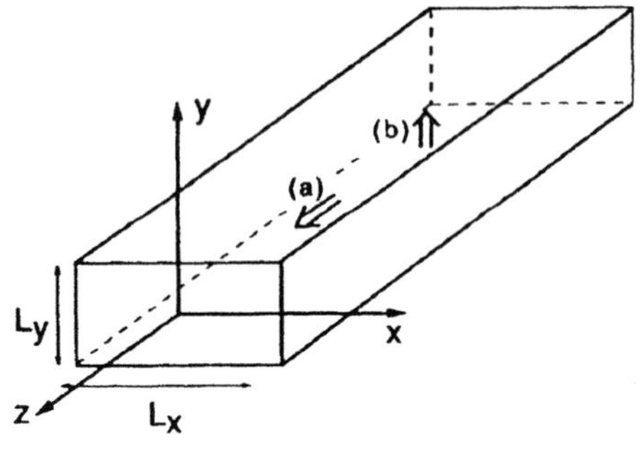

Figure 7: 2-D cavity with rectangular cross-section.

The field will now be quantized in both the *x* and *y* directions, so that the atom can only emit photons with $k_x = n\,\pi/L_x$ and $k_y = m\,\pi/L_y$ where *n* and *m* are integers. The allowed cavity modes are now lines parallel to the k_z axis, and the allowed modes for photons are the intersections between those lines and a sphere of radius $k = \Omega/c = 2\pi/\lambda_a$.

We assume that the excited atoms will be positioned exactly in the centre of the waveguide cavity. The dipole can have three orientations: one parallel and two perpendicular to the *z*-axis. As for the planar mirror cavity, the dipole couples to odd or even *n*'s and *m*'s, depending on orientation.

When the dipole is oriented parallel to the *z*-axis, only odd order modes for both *n* and *m* are excited, since the dipole only couples to waveguide modes for which E_z has a maximum at $z = 0$. In this way the 2-D cavity can be considered as a superposition of two orthogonal 1-D planar cavities.

Instead of integrating over each ring and summing all rings as in the 1-D case, here we sum over both n and m, multiplying by 2 since the modal lines intersect the sphere twice. We label the angle between the z-axis and the intersection with the line n,m as $\theta_{n,m}$. The polarization factor $P_{\|}$ is given by $P_{\|} = \sin^2\theta_{n,m}$. The mode counting sum now reads:

$$g_{\|}(k)dk = \frac{1}{V}\sum_{n\,odd}\sum_{m\,odd} 2\left(\frac{L}{2\pi}\right)\sin^2(\theta_{n,m})\, dk_z^{(n,m)}$$

Defining the cut-off k-values $k_{cx} = \pi/L_x$ and $k_{cy} = \pi/L_y$ and $k_\perp$ as the distance from line n,m to the k_z axis, we see that

$$k_z = k\cdot\cos(\theta_{n,m}),\quad k_z = \sqrt{k^2 - m^2k_{cx}^2 - n^2k_{cy}^2}$$

$$k_\perp = k\cdot\sin(\theta_{n,m}),\quad k_\perp = \sqrt{m^2k_{cx}^2 + n^2k_{cy}^2}$$

This can be written as:

$$\cos(\theta_{n,m}) = \sqrt{k^2 - m^2k_{cx}^2 - n^2k_{cy}^2}$$

$$\sin(\theta_{n,m}) = \frac{\sqrt{m^2k_{cx}^2 + n^2k_{cy}^2}}{k}$$

Inserting $dk_z^{(n,m)} = dk/\cos(\theta_{n,m})$ and the above expression in the mode counting sum, we get:

$$g_{\|}(k)dk = \frac{k_{cx}\cdot k_{cy}}{\pi^3}\sum_{n\,odd}\sum_{m\,odd}\frac{m^2\frac{k_{cx}^2}{k^2} + n^2\frac{k_{cy}^2}{k^2}}{\sqrt{1 - m^2\frac{k_{cx}^2}{k^2} - n^2\frac{k_{cy}^2}{k^2}}}\,dk$$

which must be taken over all the lines intersecting the sphere *ie* over all n and m satisfying

$$k^2 \geq (mk_{cx})^2 + (nk_{cy})^2$$

which must be evaluated numerically (see Figure 8 for the case $L_x = L_y$)

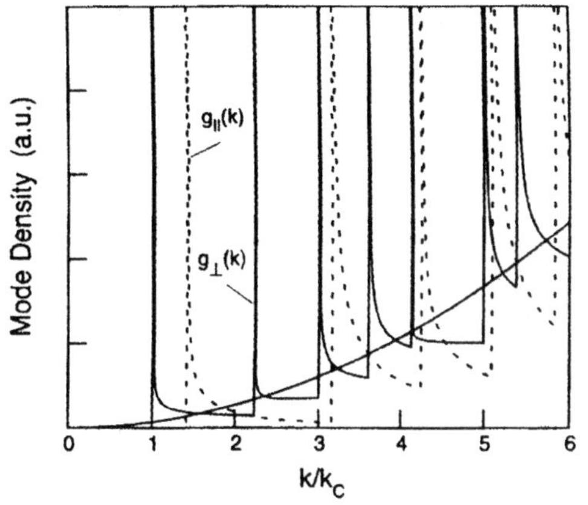

Figure 8: The mode density inside a rectangular waveguide cavity for both orientations considered.

Now let the dipole be oriented parallel to the y-axis. (For the dipole oriented parallel to the x-axis, we exchange x for y, n for m). The polarization factor is again rotated by 90°:

$$P_\perp = \sin^2\theta_{n,m}\cos^2\phi_{n,m} + \cos^2\theta_{n,m}$$

where $\phi_{n,m}$ is the angle between $k_\perp$ and the k_x-axis. It is easy to see that:

$$\cos\phi = \frac{k_x}{k_\perp} = \frac{m k_{cx}}{\sqrt{m^2 k_{cx}^2 + m^2 k_{cy}^2}}$$

from which the mode density becomes

$$g_\perp(k)dk = \frac{k_{cx} \cdot k_{cy}}{\pi^3} \sum_{n\,even} \sum_{m\,odd} \frac{k}{\sqrt{k^2 - m^2 k_{cx}^2 - n^2 k_{cy}^2}} \left(1 - \left(\frac{n k_{cy}}{k}\right)^2\right) dk$$

which must again be evaluated numerically (see Figure 8).

2.6 *Spatial modes in microdisk lasers*

There are many experimental difficulties in measuring the spatial modes and non-equilibrium field patterns in microdisk lasers, and to our knowledge no such reliable measurements have yet been performed. (They have recently been performed for dielectric microspheres using near field coupling to an optical fibre [24].) However, theoretical considerations point strongly to whispering gallery modes which propagate by total internal reflection along the disk periphery [25]. For such a mode to be stable requires that the total optical path length be an integer number of wavelengths. The mode wavelength can be estimated from geometrical optics; a cavity of length l supports those wavelengths λ that satisfy the condition $l = m(\lambda/2n)$. For a mode that undergoes many reflections as it propagates along the disk boundary, the cavity length can be approximated by the circumference of the disk. The angular frequency of the mode can then be expressed as

$$\omega_m = mc / nR$$

where R is the disk radius and m is the azimuthal mode number (*ie* half the number of field maxima encountered along the circumference of the disk). The mode number of a whispering gallery mode can be determined from the mode spacing of a microdisk laser spectrum. The mode numbers m and m+1 of two adjacent modes that have wavelengths $\lambda+\delta\lambda$ and λ respectively, can then be determined by

$$m \approx \lambda / \delta\lambda$$

The full vector Maxwell theory of such modes is generally quite complicated, but can be simplified here to provide physical intuition and to appreciate the experimental requirements of future investigations of the spatial modes of these lasers. Neglecting the effects of the supporting post, one starts from the 3D Helmholtz equation for the scalar field components ψ

$$\nabla^2 \psi + k^2 \psi = 0$$

which in cylindrical coordinates becomes

$$\frac{\partial^2 \psi}{\partial r^2} + \frac{1}{r}\frac{\partial \psi}{\partial r} + \frac{1}{r^2}\frac{\partial^2 \psi}{\partial \phi^2} + \frac{\partial^2 \psi}{\partial z^2} + k^2 \psi = 0$$

Using the cylindrical symmetry to express these components as

$$\psi(r,\phi,z) = F(r)\exp(\pm im\phi)\exp(ik_z z)$$

where m is an integer, we obtain the following equation for the radial field dependence:

$$\frac{d^2 F}{dr^2} + \frac{1}{r}\frac{dF}{dr} + (q^2 - \frac{m^2}{r^2})F = 0$$

The lateral wavenumber q is given by $q^2 = k^2 - k_z^2$. For each m this equation has a spectrum of eigenvalues q_{mn} determined by the boundary conditions on the fields at the disk periphery [26]. For disks such as ours, with thicknesses ~ $\lambda/4n$, TM modes and unguided waves can be neglected [7]. We also require that the mode be a traveling wave along the periphery and a stationary wave in the z direction. The eigenvalues $k_{z,p}$ are given essentially by the solutions to the planar step-index waveguide problem defined solely by the disk thickness and index steps [27]. A suitable approximation for the eigenmodes F_{mn} is

$$F_{mn}(r) = a_{mn} J_{mn}(n_{eff}\omega r / c)$$

where $J_m(x)$ is the mth order Bessel function of the first kind and a_{mn} is a normalisation factor. The mode frequencies can be deduced using the approximate boundary condition $F_{mn}(R) = 0$:

$$\omega_{mn} = x_{mn}\, c/Rn$$

where x_{mn} is the nth positive zero of $J_m(x)$. Ideally the eigenmodes radiate outwards in the disk plane and do not generate radiation at large out-of-plane angles. Such radiation does in reality occur due to scattering, and it is this scattered light which is measured experimentally due to the extreme (and to date insurmountable) difficulty in obtaining clear access to the lasing output of a working microdisk. Since there is no preferred location for whispering gallery modes in a perfectly symmetric disk, the entire disk periphery may be filled with these modes, some of which may rotate due to thermal and carrier induced lensing. The reader should realise that the posts separating the microdisks from the substrate are only ~1 μm high, far smaller than an optical fiber or other suitable optical probe: only at the wafer edges can one hope for ready access to the disk output.

2.7 *Advantages of microcavities summarised*

The primary advantages of a microcavity are (i) to enhance the radiative transition rate of an excited atom or atom-like species, and hence improve the radiative efficiency of the system, (ii) to increase the fraction β of the total spontaneous emission rate which is emitted in a lasing mode, (iii) to select particular modes (directions, frequencies and field patterns) which are useful. From the theory presented, we can summarise the potential of 1-D and 2-D microcavities as follows.

Even with the simplifying assumptions used above, we see that 1-D microcavities (which are relatively simple to fabricate) can provide only a modest alteration of the spontaneous emission rate (up to a factor of 3) relative to the free space value. 2-D microcavities, however, can provide very large modifications of the transition rate, but are generally extremely difficult to fabricate. Micro-disk lasers allow us to use standard photolithography and etching techniques to define a 2-D cavity.

3. Fabrication of semiconductor microdisk lasers

The initial approach **[7,9,11,28]** to fabricating semiconductor microdisks was to use dry etching to form the disk perimeter and selective wet etching to form the support pedestal. This approach leads to a number of difficulties: (a) the damage created by the dry etch process adjacent to the active area, (b) the reduced thermal dissipation due to the lasing disk being surrounded by air and heat sinking occurring via the pedestal and substrate, (c) restriction to disklike geometries. At UCC we have extended **[10,29,30]** the range of microdisk configurations to optimise thermal dissipation and integrability. These are based on separate confinement heterostructure InGaAsP layers (λ=1.3-1.6 μm) grown on an InP substrate but the fabrication concepts are easily extendable to other material systems. The base epilayers were grown by MOVPE at Optronics Ireland, UCC and transferred to a host substrate to generate alternative processing possibilities **[31,32]**. A key fabrication step is the exclusive use of wet chemical etching which yields smooth surfaces, thereby reducing scattering and optimising the Q of the lasing modes which propagate close to the surface. The disadvantage is with the submicron control of disk dimensions. A variety of techniques was used to fabricate 1 mm^2 arrays of microdisks.

We have demonstrated three different structures: (1) disks suspended on InP pedestals bonded with Au to a silicon substrate, (2) disks on glass substrates and (3) microrings on glass substrates. Each of these is intended for optical pumping. For (1) a conventional InP-based laser structure is used, consisting of (from the top down) an InGaAs contact layer, an InP p-doped cladding, an undoped active region with InGaAs quantum wells and InGaAsP spacers, an n-doped InP cladding and substrate. Both this material and a Si substrate are coated with Au. The chips are brought into contact and annealed at 410°C for 15 minutes to form a strong metallic bond helped by the Au diffusion into both substrates. Following removal of the InP substrate and n-cladding a smooth surface containing the active region is exposed. Conventional lithography is used to define disks of various diameters which are wet etched using a non-selective HBr solution. The etching is allowed to proceed to give an undercut of ~1 μm which causes the active layer to straighten up. The solution also attacks the Au base but importantly not the Si. The suspended disk is completed by the formation of a pedestal by selective etching of the InP p-cladding (Figure 9). In the second approach, the active layer is isolated from the top of the structure using selective etchants and lifted off onto a glass substrate. Microdisks and rings are then patterned as above. Pedestals are not necessary in this case as the glass provides both strong optical confinement and structural and thermal support.

Newly fabricated microdisks are assessed by photoluminescence (PL) using CW optical pumping with the 514.5 nm line from an Ar^{3+} laser at temperatures from 77 to 300 K. An array of disks is butt coupled to one end of a multimode optical fibre splitter and immersed in liquid nitrogen. The step-index fibre illuminates a 50 μm diameter spot with nearly uniform intensity, so that several disks are pumped simultaneously. The incident power on a single disk is estimated from the fractional area. The absorbed power is much less. The scattered emission is collected in the same fibre and measured by a spectrometer with a cooled Ge detector.

Figure 9: SEM of 6 μm diameter microdisks fabricated using wafer bonding technique.

7.8 μm disks formed by the metal bonding technique oscillate primarily single mode. Figure 10 shows the emission spectrum of an active region consisting of four quantum wells just below and at threshold which occurs with 1.2 mW in the fibre corresponding to an absorbed threshold power of 20 μW per disk. As the pump energy is three times the band gap, lower thresholds were expected with a more suitable diode laser pump source. The measured Q of the mode is greater than 2000 with the measurement limited by the spectrometer resolution.

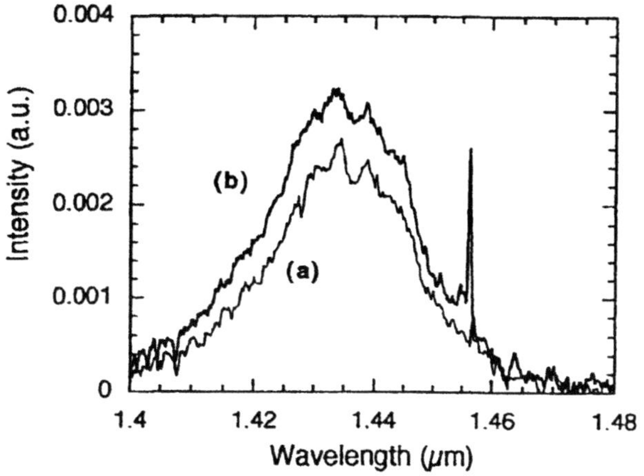

Figure 10: CW PL at 77 K from a 7.8 μm diameter microdisk with incident power on the disk of (a) 22 μW and (b) 29 μW with the appearance of the M=39 whispering gallery lasing mode at higher power. The measured linewidth is limited by the spectrometer resolution.

4 Optical pumping experiments on microdisk lasers

The spectra of two-dimensional arrays of microdisks were investigated when the disks were optically pumped under pulsed and CW conditions. Initial pumping experiments were conducted using a CW Ar^{3+} laser at 77 K using the same arrangement as for PL measurements, and low lasing thresholds were observed **[30]**. The mode spacing of the primary lasing modes was seen to vary inversely with the disk diameter, confirming that these modes were probably whispering gallery modes (WGM). Figure 11 shows the lasing spectra (just above threshold) for rings with outer diameter of 14 μm (annular region 4 μm). The material comprises a single quantum well. Several WGMs are measured with spacing 14.5 nm. This is in reasonable agreement with a calulated n_{eff} = 2.76 for the layers along with a dispersion contribution due to the wavelength size of the layers.

The spectra of individual disks on glass substrates were investigated at room temperature. It should be noted that CW operation of microdisk lasers had never been observed at room temperature. This was probably due to the insufficient thermal dissipation provided by the small supporting pedestal. A conventional 980 nm semiconductor laser delivering ~100 mW was used to pump the disks. Since the excitation energy of this laser is more closely matched to the bandgap of the InGaAs material than the Ar^{3+} laser, it provided a more efficient pumping source and there was less heat to dissipate. A coupled fibre arrangement was used both to deliver the pump light and to collect a portion of the emitted light. The 2x1 coupled multimode fibre had a 50:50 splitting ratio and a core diameter of 62.5 μm. The pump light was coupled into one of the arms on the double end of the fibre. The microdisk array was butt coupled to the single end of the fibre - no focusing arrangement was used. However the separation of disks on the array was such that a single disk could be illuminated. The disk array, which was contained on a glass substrate of area ~ 1 mm^2, was mounted on a three-way translation stage. Microdisk lasers emit into a narrow range of angles about the disk plane and so it may not be expected that a fibre placed above the disk array would collect the emission. However some of the light that is emitted by a disk is scattered from neighbouring disks or from disk imperfections into the vertical direction. A portion of the scattered emission from the microdisk was collected by the fibre and analysed using an optical spectrum analyser (Ando AQ-6312B) placed at the remaining end of the coupled fibre.

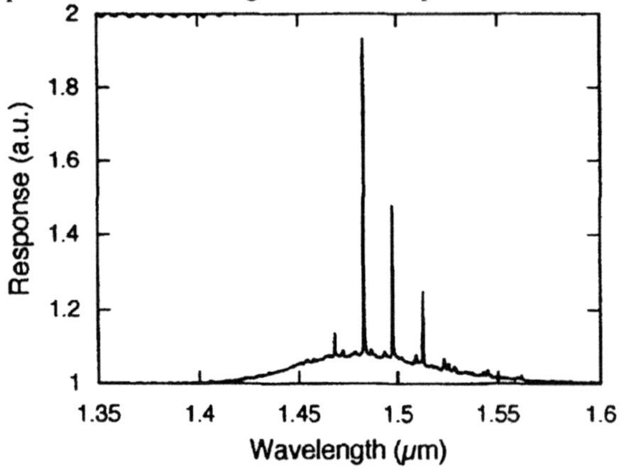

Figure 11. Emission spectra from a single quantum well, 14 μm diameter ring with 4 μm annular region with pump power 33 μW per ring, showing several whispering gallery modes.

Room temperature spectra of a 7 μm diameter microdisk were obtained when the disk was pumped under pulsed conditions. Measurements were performed using optical pulses of width 3 μs and period 100 μs. A series of spectra for a variety of pump powers is shown in Figure 12. The pump power was adjusted using a continuous neutral density filter placed between the pump laser and the coupled fibre. At low pump powers the spectrum is single mode. As the pump power is increased a second, higher wavelength mode appears and begins to dominate the spectrum. The mode wavelengths are 1579 nm and 1608 nm.

The light output power from the lasing mode was investigated as a function of pump power. The power in the lasing mode is defined to be the integrated area under the spectral line. The power in the lasing mode as a function of average incident power on the disk array is shown in Figure 13. The threshold in the curve is estimated by extrapolating from the linear lasing region to be 0.56 mW. The absorbed pump power is taken to be 0.6 times the incident power, assuming that 30% is lost by reflection and 85% of the remaining light is absorbed [7]. When the incident power on the individual disk is calculated from the fractional area and assuming a uniform pump intensity, the threshold corresponds to < 5 μW average absorbed power. It is not clear why the linearity of the curve is not preserved at pump powers greater than twice the threshold. However in this region the spectral lines were seen to broaden, possibly due to index variations across the wafer or to the onset of new loss mechanisms.

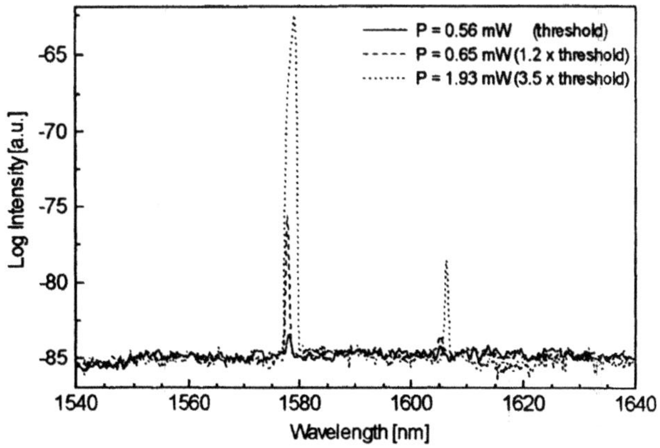

Figure 12 Series of spectra of a 7 μm diameter microdisk for a variety of pump powers. The indicated power is the average incident pump power on the microdisk array. The resolution of the spectrum analyser is 0.5 nm.

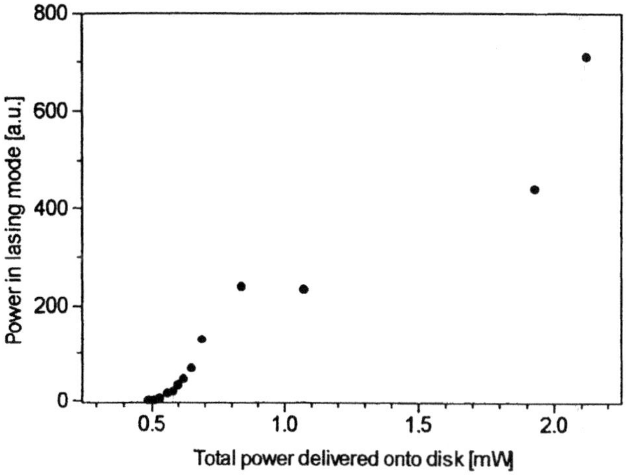

Figure 13 Integrated lasing mode power versus total power incident on the disk array.

The spectrum of another disk from the same disk array obtained when it was operated CW at room temperature is shown in Figure 14. At low pump powers the spectrum is single mode. As the pump power is increased, a second, longer wavelength mode appears and begins to dominate the spectrum. The mode spacing is 41 nm. The threshold characteristic of the disk is shown in Figure 15. The threshold is estimated to be 15 mW total incident pump power on the disk array. This corresponds to a pump power of 110 μW absorbed by the disk.

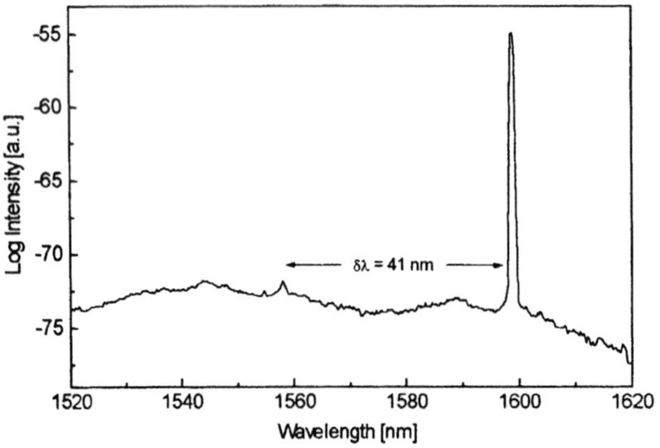

Figure 14 Spectrum of a microdisk from the 7 μm diameter disk array when pumped CW at room temperature. The spectrum displays two modes at 1558 nm and 1599 nm. The above spectrum was obtained at high pump power where the higher wavelength mode dominates the spectrum. The resolution of the spectrum analyser is 1 nm.

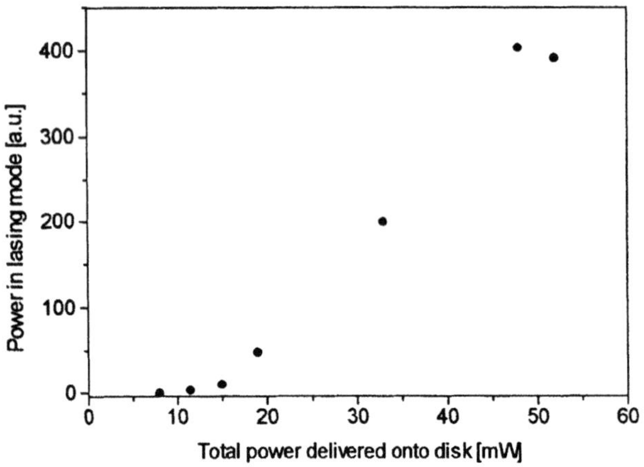

Figure 15 Power in the lasing mode as a function of the total incident power on the array.

5 Summary and conclusions

We have reviewed the theory of lasing modes in microcavities and presented some recent experimental developments in microcavity semiconductor lasers. Microdisk lasers supporting whispering gallery modes provide essentially two-dimensional microcavity effects and hence the possibility of large modification of the optical properties of the semiconductor. To date they have been operated only CW at low temperatures or pulsed at room temperature. By optimising the efficiency of the microdisk structure and its thermal environment, and by using a diode laser whose wavelength is chosen for most efficient optical pumping of the microdisk, we have succeeded in demonstrating room temperature CW operation for the first time in InGaAsP/InP microdisks. Future investigations will concentrate on mapping spatial mode spectra, dynamical behaviour and the study of spontaneous emission coupling factors.

6. Acknowledgements

We thank Forbairt (the Irish Science and Technology Agency) and the Commission of the European Union for financial support, Mr P O'Brien for experimental assistance, Ms S Walsh for MOVPE growth, Mr R Gillen and Mr J Sheehan for expert technical assistance.

7. References

1 R K Chang, A J Campillo (eds.), "Optical Processes in Microcavities" (World Scientific, 1996)
2 J Rarity, C Weisbuch (eds.), "Microcavities and Photonic Bandgaps: Physics and Applications" (Kluwer Academic Publishers, 1995)
3 See eg R A Morgan, Proc. SPIE **3003**, 14 (1997)
4 M Y A Raja, S R J Brueck, M Osinski, C F Schaus, J G McInerney, T M Brennan, B E Hammons, *IEEE J. Quantum Electron.* **25**, 1500 (1989)
5 E S C Ching, P T Leung, K Young, Chap. 1 in Ref. [1]
6 R Jin, M S Tobin, R P Leavitt, H M Gibbs, G Khitrova, D Boggavarapu, O Lyngnes, E Lindmark, F Jahnke, S W Koch, Ref. [2], p. 95
7 S L McCall, A F J Levi, R E Slusher, S J Pearton, R A Logan, *Appl. Phys. Lett.* **60**, 289 (1992)
8 A F J Levi, R E Slusher, S L McCall, T Tanbun-Ek, D L Coblenz, S J Pearton, *Electron. Lett.* **28**, 1010 (1992)
9 R E Slusher, A F J Levi, U Mohideen, S L McCall, S J Pearton, R A Logan, *Appl. Phys. Lett.* **63**, 1310 (1993)
10 B Corbett, L Considine, S Walsh, J Justice, W M Kelly, Proc. CLEO/EuropePaper, CThP2 (1994)
11 M Hovinen, J Ding, A V Nurmikko, D C Grillo, J Han, L He, R L Gunshor, *Appl. Phys. Lett.* **63**, 3128 (1993)
12 P Berman (ed.), "Cavity Quantum Electrodynamics" (Academic Press, 1994)
13 C Weisbuch, M Nishioka, A Ishikawa, Y Arakawa, *Phys. Rev. Lett.* **58**, 1320 (1992)
14 H Haug, S W Koch, "Quantum Theory of the Optical and Electronic Properties of Semiconductors", (World Scientific, 1990)
15 S D Brorson, P M W Skovgaard, Chap. 2 in Ref. [1]
16 E M Purcell, *Phys. Rev. Lett.* **69**, 681 (1946)
17 F DeMartini, J R Jacobovitz, *Phys. Rev. Lett.* **60**, 1711 (1988)
18 H Yokoyama, M Suzuki, Y Nambu, *Appl. Phys. Lett.* **58**, 2598 (1991)
19 H Yokoyama, K Nishi, T Anan, H Yamada, S D Brorson, E P Ippen, *Appl. Phys.Lett.* **57**, 2814 (1990)
20 H Yokoyama, S D Brorson, E P Ippen, M Suzuki, *Appl. Phys. Lett.* **58**, 998 (1991)
21 H-B Lin, J D Eversole, C D Merrit, A J Campillo, *Phys. Rev. A* **45**, 6756 (1992)
22 A M Vredenberg, N E J Hunt, E F Schubert, D C Jacobson, J M Poate, G J Zydzik, *Phys. Rev. Lett.* **71**, 517 (1993)
23 P M W Skovgaard, S D Brorson, I Balslev, C C Larsen, Ref. [2], p.309
24 J C Knight, N Dubreuil, V Sandoghdar, J Hare, V Lefevre-Seguin, J M Raimond and S Haroche, *Opt. Lett.* **20**, 1515 (1995)
25 R E Slusher, U Mohideen, Chap. 9 in Ref. [1]
26 M K Chin, D Y Chu, S T Ho, *J. Appl. Phys.* **75**, 3302 (1994)
27 M J Adams, "An Introduction to Optical Waveguides" (Wiley, 1981)
28 U Mohideen, W S Hobson, S J Pearton, F Ren, R E Slusher, *Appl. Phys. Lett.* **64**, 1911 (1994)
29 B Corbett, J Justice, L Considine, S Walsh, J Justice, W M Kelly, Proc. Int. Workshop on Compound Semicond. Dev. and ICs Europe (WOCSDICE) (1995)
30 B Corbett, J Justice, S Walsh, L Considine and W M Kelly, *IEEE Photonics Technol. Lett.* **8**, 855 (1996)
31 B Corbett, L Considine, S Walsh, W M Kelly, *IEEE Photonics Technol. Lett.* **5**, 1041 (1993)
32 B Corbett, L Considine, S Walsh, W M Kelly, *Electron. Lett.* **29**, 2148 (1993)

EXCITON ABSORPTION SATURATION AND CARRIER TRANSPORT IN QUANTUM WELL SEMICONDUCTORS

A. MILLER, T.M. HOLDEN, G.T. KENNEDY, A.R.CAMERON
and P. RIBLET
School of Physics and Astronomy
J.F. Allen Physics Research Laboratories
University of St Andrews,
North Haugh, St Andrews,
Fife, KY16 9SS, Scotland

1. Introduction

Optical nonlinearities [1-4] associated with excitonic absorption features in multiple quantum well (MQW) semiconductors [5] offer a number of useful functions for optoelectronics devices [6,7]. These include laser mode-locking elements, saturable elements for controlling the propagation of optical solitons in fibre transmission systems, all-optical bistable etalons, all-optical directional coupler switches and self-electro-optic devices for communications, signal processing and computing. The operation and optimisation of these devices rely on an understanding of the mechanisms which contribute to absorption saturation and the motion of optically generated electrons and holes in directions both parallel and perpendicular to the quantum wells. This chapter reviews measurements of exciton absorption saturation mechanisms and in-well transport processes relevant to new optoelectronic electro-optic and nonlinear optical devices.

2. Background

Resonant nonlinear and electro-optic interactions are particularly large in quantum wells at room temperature because of prominant excitonic features in their absorption spectra close to the band gap energy [5]. Absorption coefficients at the peak of the exciton absorption can be in excess of 10^4 cm^{-1}

M. Balkanski and N. Andreev (eds.),
Advanced Electronic Technologies and Systems Based on Low-Dimensional Quantum Devices, 117–135.

providing very efficient absorption in samples only a few microns thick. Optical generation of excitons and free carriers can bleach these absorption features by a number of mechanisms including phase space filling, Coulomb screening and lifetime broadening. Pump-probe measurements using ultrashort pulses of laser light with different linear and circular polarisations can be used to identify the relative magnitudes of the various contributions.

After creation of electron and hole pairs, a dynamical situation is produced whereby the bleaching will relax on timescales detemined by the time constants associated with the motion and recombination processes for the free carriers. Drift and diffusion properties normally determine the manner in which semiconductor devices operate, but we must consider many additional processes in MQW structures if we wish to assess the ultimate performance limits of low dimensional devices. These processes include thermionic emission from, tunnelling through, and trapping into the wells. The ability to control the motion of electrons and holes by designing the structures to make use of these processes gives opportunities for engineering devices with new and unique properties. Pump-probe and transient grating techniques can be used to monitor the motion and dynamics of the carriers on ultrashort timescales by employing excitonic nonlinearities as a probe.

3. Resonant optical nonlinearities in semiconductors

Here we are primarily interested in interactions between laser light and low dimensional semiconductors which cause changes of absorption coefficient and refractive index as a function of excess carrier density at room temperature. The optical generation of excess carrier populations produces changes in the real and imaginary parts of the susceptibility. Absorption and refraction are linked by causality via Kramers-Kronig relations. These familiar relations in linear optics may be modified [1] to account for optical nonlinearites when laser excitation of a quasi-equilibrium of electrons and holes results in renormalized material parameters. Thus, refractive index changes, Δn, and absorption changes, $\Delta\alpha$, may be connected by the integral expression,

$$\Delta n(\omega) = \frac{c}{\pi} P \int_0^\infty \frac{\Delta\alpha(\Omega)d\Omega}{\Omega^2 - \omega^2} \tag{1}$$

where P denotes the principal part of the integral. Note that *changes* in refractive index at a given frequency, ω, are induced by *changes* in absorption at *all* other frequencies, Ω.

The optical generation of free electrons and holes is essentially instantaneous, but the decay of the excess carrier population depends on the recombination rate, the diffusion coefficient (if there is a concentration gradient) or the sweep-out rate in the presence of an electric field. For instance, the excess carrier density as a function of position and time, *N(x,t)*, after excitation with an ultrashort pulse can be determined from the continuity equation (given here for one-dimensional motion in the absence of an electric field),

$$\frac{\delta N(x,t)}{\delta t} = -\frac{N(x,t)}{\tau_R} + D_a \nabla^2 N(x,t) \qquad (2)$$

where D_a is the ambipolar diffusion coefficient for electrons and holes and τ_R is the recombination time. The time for recovery to the equilibrium condition can range from microseconds to femtoseconds depending on the precise conditions.

A useful parameter is the diffusion length, i.e. the average distance a free carrier moves before recombining, defined by,

$$L_D = \sqrt{D_a \tau_R} \qquad (3)$$

The diffusion length in III-V semiconductors is typically on the order of a few microns.

4. Exciton absorption saturation in quantum wells

Excitons in quantum well semiconductors produce unique nonlinear optical properties [8]. Excitonic absorption features are clearly resolved at room temperature in quantum wells compared to bulk semiconductors because confinement leads to larger binding energies and enhanced oscillator strengths. The loss of degeneracy of light and heavy hole valence bands leads to an exciton doublet in MQWs.

Saturation of exciton absorption provides absorptive and refractive optical nonlinearities at moderate optical powers. Figure 1 illustrates room temperature, exciton saturation spectra for three GaAs/AlGaAs quantum well structures with different well widths, 4.4, 6.5 and 10nm. These spectra were recorded in a pump-probe configuration using linearly polarised, picosecond pulses for both the pump and probe. There was a fixed delay of a few picoseconds between the pump and probe pulses.

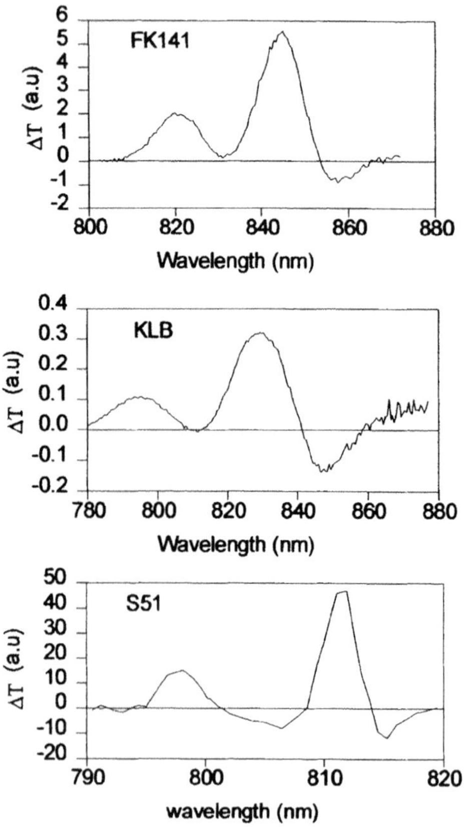

Figure 1. Comparison of differential transmission changes for three GaAs/AlGaAs MQW structures having different quantum well widths, (a) 10nm, (b) 6.5nm, and (c) 4.4nm.

Absorption saturation is clearly observed at below 1mW average power, for laser spot sizes on the order of 10-100 microns. This saturation occurs due to the optical generation of excitons or free carriers at 2-D densities in excess of 10^{11} cm^{-2}. Resonant excitation initially creates bound electron-hole pairs, i.e. excitons. When the density of excitons approaches that required to fill space (~10^{17} cm^{-3} based on an exciton diameter of 30nm for bulk GaAs), then the number of additional excitons which can be created is reduced by Pauli exclusion, thus decreasing the level of absorption. Knox et al [9] measured a room temperature ionisation time for excitons into free electron-hole pairs of 300fs. Therefore, on picosecond and longer timescales, it can be assumed that the exciton saturation results from free carriers. The increase in transmission results from a combination of phase space filling, Coulomb screening and broadening. The negative signals are caused by lifetime broadening of the exciton absorption features. Any reduction of the binding energy caused by Coulomb screening to produce a blue shift of the exciton is balanced by a red shift resulting from band-gap renormalisation, so that the wavelength of the exciton peak remains unaltered. Above 3mW, the absorption becomes progressively more difficult to saturate once the exciton feature has disappeared because band-gap renormalisation causes a red shift of the band gap energy and thus a progressively larger density of states needs to be saturated. At high optical powers, and thus very high carrier densities, band filling will cause the remaining absorption to saturate, but this is normally accompanied by a red shift due to lattice heating when using cw excitation or laser pulses at high repetition rates.

5. Selection rules and saturation mechanisms

Selection rules for transitions from the light and heavy hole valence bands to the conduction band for MQWs are shown in figure 2. It may be noted that, by employing circularly polarised light, 100% spin polarised electrons can be created by selecting the wavelength to be resonant with the heavy hole to conduction band transition [10].

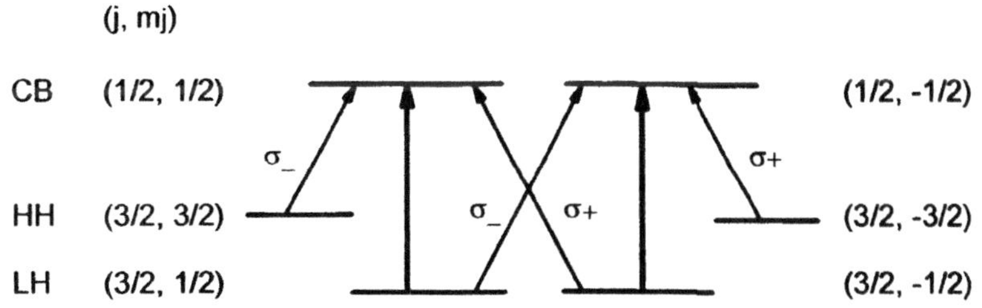

Figure 2. The selection rules for transitions from the heavy-hole and light-hole valence bands to the conduction band in MQWs. The (j, m_j) refer to the quantum numbers for angular momentum and its component along one direction. The σ_+ and σ_- refer to the transitions excited by each sense of circularly polarised light and correspond to $\Delta m_j = \pm 1$, respectively, where the propagation direction is used to define m_j.

The generalised Wannier equation describes the different nonlinearities responsible for exciton saturation associated with free carriers in room temperature MQWs,

$$[\hbar\omega - E_{e,k} - E_{h,k} + i(\Gamma_0 + \Gamma) + \Sigma_{k,\sigma}]\chi_{k,\sigma}$$
$$= (1 - f_{e,\sigma,k} - f_{h,\sigma,k})[d_k + \sum_{k'} V_s(k-k')\chi_{k',\sigma}] \qquad (4)$$

to calculate the polarisability,

$$\chi(\hbar\omega) = \sum_{k} d_k \chi_{k,\sigma}(\hbar\omega) \qquad (5)$$

whose imaginary part is proportional to the absorption coefficient. $E_{e,k}, E_{h,k}$ are the single-particle energies of electron and hole respectively. $V_s(k)$ is the Fourier transformed quasi-two dimensional screened Coulomb potential, d_k the dipole moment (proportional to the modulus of the interband momentum matrix element) and $f_{i,\sigma,k}$ (i={e,h}) are Fermi functions describing the distribution of electrons (holes) in their energy bands with spin σ. The real part of the self-energy is given by,

$$\Sigma_{k,\sigma} = \sum_{k'} (f_{e,\sigma,k} + f_{h,\sigma,k}) V_s(k-k') + \sum_{k'} V_s(k') - V(k) \qquad (6)$$

where V(k) is the Fourier transformed unscreened potential.

The first term of these summations is the “exchange hole” energy arising from Pauli exclusion which means that each fermion is surrounded by a region where the probability of the existence of another identical fermion is very small. This repulsive energy occurs for particles with equal spin and charge. The second term is the “Coulomb hole” energy which results from equally charged fermions avoiding each other because of Coulomb repulsion. This term is independent of the spin of the particles. A finite homogeneous linewidth Γ_0 is introduced to account for dephasing collisions (electron-phonon interactions) and is accompanied by the term Γ which represents the carrier-carrier scattering rate via the imaginary part of the self-energy.

The term $(1 - f_{e,\sigma,k} - f_{h,\sigma,k})$ is the filling factor describing blocking of transitions by the Pauli exclusion principle where a state occupied by a fermion is no longer available as a final state in an optical absorption process.

We can therefore distinguish three mechanisms responsible for the observed exciton absorption saturation [8]. (a) Phase-space-filling (PSF) results directly from Pauli exclusion and reduces the excitonic oscillator strength through the filling factor and the exchange hole energy. (b) Screening arises from the Coulomb interaction and will reduce the exciton oscillator strength through the screened potential V_s and the Coulomb hole energy. (c) Broadening is caused by carrier-carrier scattering and will increase the homogeneous linewidth of the exciton while retaining the same overall oscillator strength through the imaginary part of the self-energy.

The density-dependent modifications caused by PSF are similar to those caused by screening since both mechanisms lead to a reduced electron-hole attraction. The situation is different for broadening whose density dependence is complex and not well-established theoretically. Moreover, PSF will be spin-dependent since the filling factor and the exchange hole energy both depend on spin. In contrast, screening and broadening do not depend on spin in the low density regime. This difference has been exploited in pump-probe experiments at room temperature [10] where it is possible to separate experimentally the effect of PSF from the effect of screening and broadening in MQWs.

We noted above that circularly polarised light resonant with the heavy hole exciton generates 100% spin-polarised electron-hole pairs. The spin orientation is maintained by the electrons for tens of picoseconds at room temperature in GaAs/AlGaAs MQWs, even after rapid (300fsec) ionisation of the excitons by longitudinal phonons [9]. Hole spin is known to relax on subpicosecond timescales at room temperature because of the mixed spin character of the light hole valence states. Therefore, by using different polarisation combinations, it is possible to separate electronic PSF from Coulomb contributions using electron spin and to deduce the spin relaxation time of electrons by studying the change in transmission as a function of the time delay between pump and probe pulses.

6. Linear and circular polarisation pump-probe measurements

PSF is a spin-dependent optical nonlinearity since it depends on Pauli exclusion. Thus, circularly polarised light, which produces spin polarised electrons when in resonance with the heavy hole exciton, can be used to identify this component of exciton saturation. The principle is illustrated in figure 3.

The filling of the two conduction band states, electron spin-up and electron spin-down, are shown for linear and circular polarised pump pulses, based on the selection rules of figure 2. Three pump and probe polarisation combinations, OLP, SCP and OCP identify the spin-dependent phase space filling.

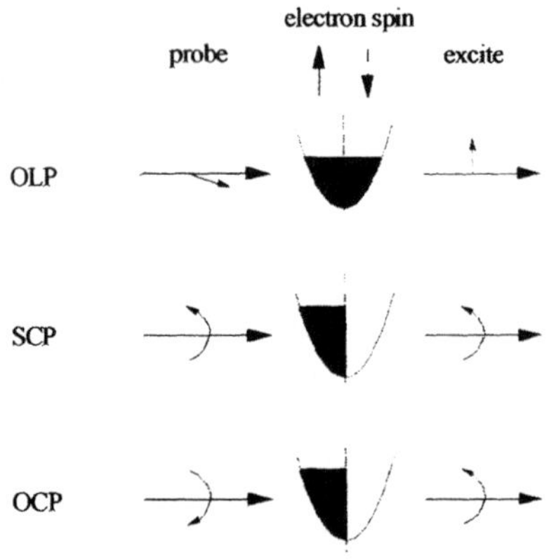

Figure 3. The effect of different polarisations of pump and probe beams (OLP: opposite linear polarisations, SCP: same circular polarisations, OCP: opposite circular polarisations).

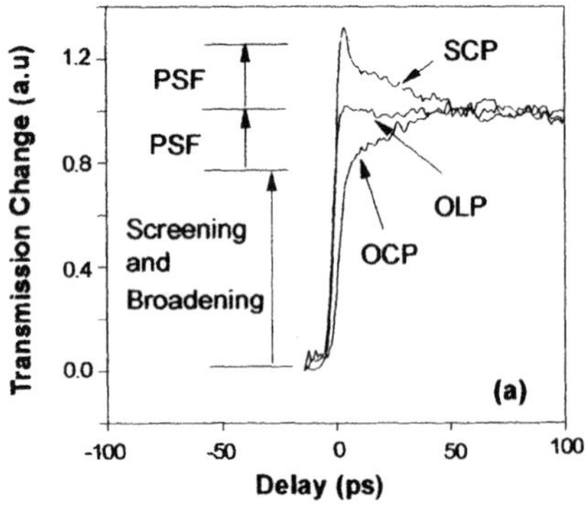

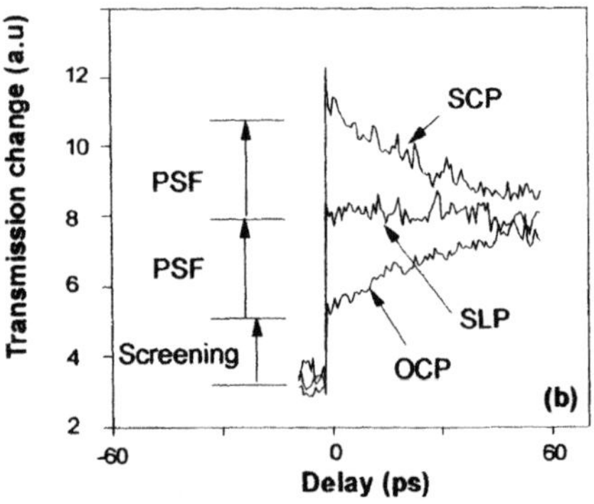

Figure 4. Probe pulse transmission changes as a function of time delay, (a) using 1 ps pulses and (b) using 100 fs pulses.

Figure 4(a) shows the results of time resolved saturation measurements carried out at the peak of the HH exciton absorption in GaAs/AlGaAs MQWs at room temperature using 1ps pulses with the three polarisation configurations, OLP, SCP and OCP of figure 3. One picosecond pulses have a spectral bandwidth much less than the hh exciton spectral linewidth. The transmission change is due to the combined effects of PSF, screening and broadening which all reduce the absorption at the centre of the line. The OLP configuration shows a rise in transmission due to exciton saturation, which recovers on nanosecond timescales by carrier recombination. SCP shows an initially enhanced saturation due to spin-dependent PSF. In this case, the probe interacts with electrons in spin-polarised states which are twice as full as those in the OLP case (for the same total number of excited carriers, figure 3), so initially there is a doubling of the spin dependent saturation. This transmission enhancement is observed to recover on the spin relaxation time of 32ps in this case. For the OCP condition, immediately after excitation, all of the electrons are in the opposite spin state to that being probed, so there is no PSF. This results in less saturation initially, but as the electron spins randomise, the OCP signal rises to the OLP level. The initial rise in the OCP signal is due to the combined spin-independent Coulomb effects of screening and broadening.

Figure 4(b) shows the results obtained using 100fs pulses for the same MQW sample [11]. (In this case the same linear polarisation, SLP, was employed instead of OLP, but this makes no difference to the interpretation). The measured 10nm bandwidth of these 100fs pulses was greater than the hh exciton absorption linewidth. Consequently the probe effectively measures the integrated absorption change rather than the change at the line centre. The striking difference between the two sets of results is the absence of lineshape broadening using the shorter pulses. Collisional broadening lowers the absorption change at the line centre and increases it in the wings but leaves the integrated absorption unchanged. A comparison of figures 4(a) and 4(b) demonstrate that broadening is the dominant contribution to absorption quenching using picosecond and longer pulses.

Higher time resolution measurements allow a comparison of exciton absorption saturation before and after exciton ionisation and an analysis of these results allows a comparison of exciton-exciton and free carrier contributions to spin-dependent and spin-independent optical nonlinearities [11]. An example of a set of higher time resolution pump-probe scans using 100fs pulses and recorded within 2ps of zero delay are shown in figure 5. A probe beam steering mirror was vibrated to eliminate any coherence artifacts. The enhanced transmission change at small delays in the SLP configuration results from the fact that the presence of excitons is more efficient in saturating the absorption than free carriers. Exciton ionisation produces the initial decay with a time constant on the order of 250fs, consistent with earlier studies. This transmission enhancement is observed to be larger for the SCP configuration than SLP confirming that PSF is larger for excitons than free carriers.

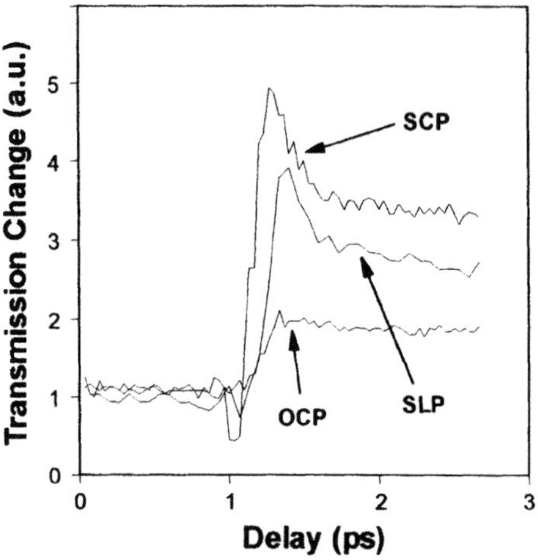

Figure 5. Probe transmission changes for SCP, LP and OCP as a function of time using 100fs pulses.

On inspection of the OCP trace in figure 5, the initial rise is seen to be determined by the optical pulse width after which the transmission will increase (beyond the scale of the plot) towards the SLP curve at the spin relaxation rate, figure 4(b). Surprisingly, there is no signature of exciton ionisation on short timescales in this OCP configuration. As PSF does not contribute in the OCP condition, this indicates that the Coulomb exchange effect is independent of whether the carrier pairs are bound or free. Theory predicts that the free carrier contribution should be a factor of 1.4 greater than the exciton contribution [8].

Given that the exciton ionisation time and pump pulse duration are comparable, some exciton ionisation will occur during excitation. Consequently the initial transmission change will have a significant contribution from both free carriers and excitons. This must be accounted for if we are to accurately resolve the exciton and free carrier contributions to the saturation. To this end, a simple rate equation was employed to describe the evolution of the particle sets. The initial density of excitons, and subsequently free carrier pairs, created by the excite pulse under the condition described above was estimated to be $10^{10}cm^{-2}$. At this density Boltzmann statistics apply and we can assume that the strength of the saturation is proportional to the instantaneous particle density. The transmission change for the SLP configuration is given by,

$$\Delta T_{LP}(t) = (\beta_{EX} + \gamma_{EX})n_{EX}(t) + (\beta_{FC} + \gamma_{FC})n_{FC}(t)$$

where β and γ are the spin dependent and spin independent saturation coefficients respectively, and n denotes the instantaneous particle densities with the subscripts EX and FC denoting excitons or free carrier pairs. The four coefficients, can be obtained by fitting the pump-probe traces corresponding to the different polarisation configurations for the same excitation intensities. The relative strengths of all contributing saturation mechanisms can then be deduced. The OCP fits give identical spin-independent coefficients, $\gamma_{EX} = \gamma_{FC} = \gamma$, reinforcing the observation that the Coulomb components of exciton-exciton saturation and free carrier-exciton saturation are identical. This makes γ a convenient parameter to make relative comparisons with the spin-dependent components. The relative strengths of the different mechanisms, as taken from the fits are summarised below.

$$\gamma_{EX}/\gamma_{FC}=1 \qquad \beta_{EX}/\gamma=3.5 \quad \beta_{FC}/\gamma=1.5 \qquad \beta_{EX}/\beta_{FC}=2.2$$

The PSF contribution from the excitons is 3.5 times greater than the spin independent contribution while for free carriers the PSF is only 1.5 times greater. Consequently the exciton PSF is more than twice as effective than the free carrier PSF contribution. These observations, particularly the equivalence of the exciton and free carrier spin independent saturation mechanisms differ from earlier theoretical predictions and suggests that the existing theory for exciton saturation is still incomplete.

Since the PSF and Coulomb contributions to exciton saturation can be distinguished by using circularly polarised light, the power dependence of these individual components can be measured. Figure 6 plots the initial change in transmission in the OCP configuration as well as the change in transmission due to PSF alone as a function of the average pump power for the three samples with different well widths (as used for the results shown in figure 1). Note that the Coulomb contribution is larger than PSF in all three cases. The PSF contribution is linear with pump power for all three well widths, however the sample with the narrowest wells, in which broadening dominates the Coulomb contribution, shows a nonlinear dependence of the transmission change.

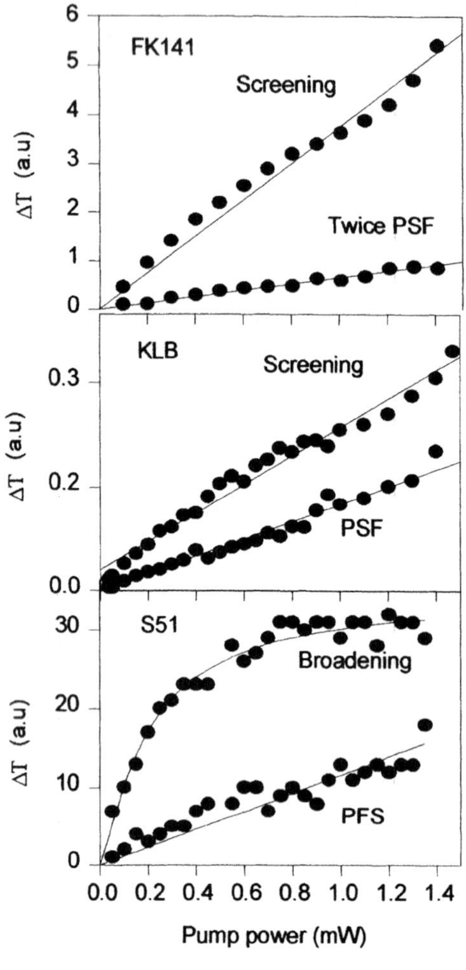

Figure 6. Change in transmission for Coulomb and PFS contributions at the peak of the heavy hole exciton as a function of pump power for three samples FK141 (10nm), KLB (6.5nm) and S51 (4.4nm).

7. Transient gratings

Four-wave mixing provides a useful means of determining the magnitudes of the nonlinear coefficients, response times and the characteristic parameters, which describe carrier motion. Time resolved experiments can be performed using three input pulses whereby two 'excite' pulses with the same linear polarisation overlapping at the sample in both space and time, create a temporary spatial modulation of the carriers, as illustrated in figure 7. A time delayed probe pulse

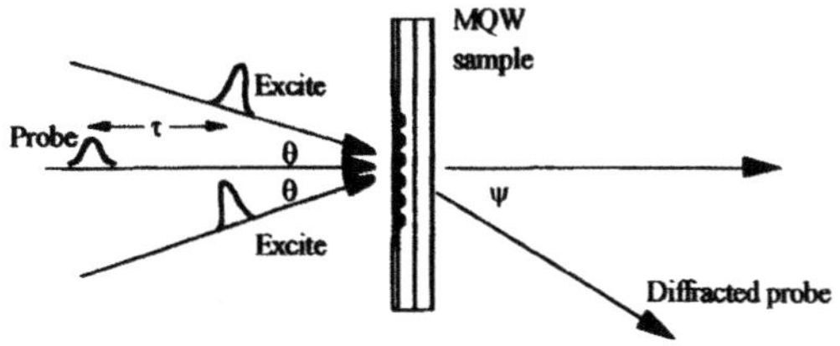

Figure 7. Transient grating configuration

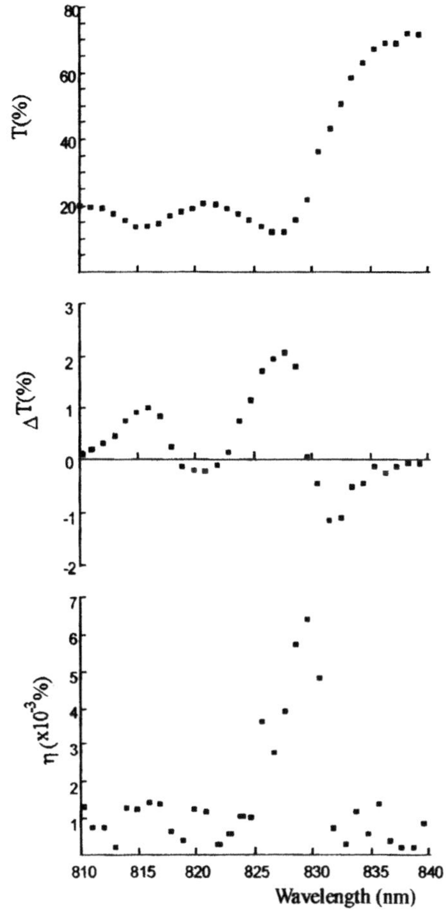

Figure 8. (a) Transmission of GaAs/AlGaAs MQW as a function of wavelength, (b) differential transmission in a pump-probe measurement with fixed delay of 15ps and (c) diffraction efficiency in three-beam transient grating measurement [12].

will diffract from this grating because of the carrier induced optical nonlinearity. The large excitonic optical nonlinearities in MQWs described above are very useful for this purpose [12].

Figure 8 shows the results of transient grating measurements as a function of wavelength at a fixed probe delay for 6.5nm MQWs at room temperture. The maximum diffraction efficiency is noted to occur on the long wavelength side (829nm) of the heavy hole exciton feature, where there is no transmission change, indicating that the grating is primarily refractive rather than absorptive. Information about the in-well motion of carriers can be obtained by measuring the decay of the diffracted probe signal as a function of time.

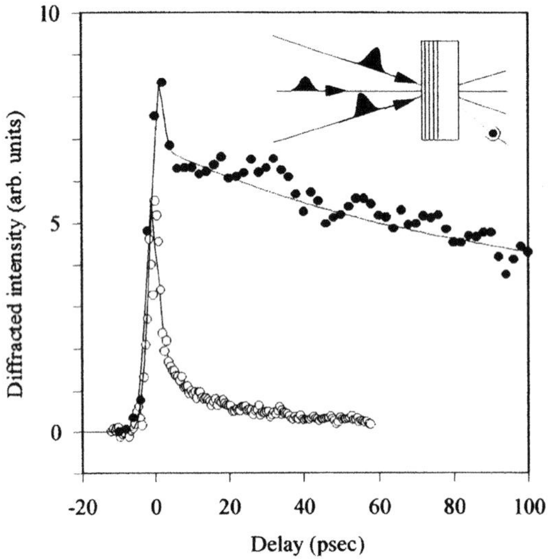

Figure 9. The measured diffracted signal decay rate for a 5μm amplitude grating (●) compared with the corresponding 5μm polarisation grating (○). Exponential decays with time constants of 220 and 13 ps are fitted to the two grating signals.

Figure 9 compares the time dependence of the diffraction probe when parallel and crossed linear polarisations are employed for the two pump beams. Interference of the parallel polarisations create an *amplitude* light modulation and thus a concentration grating of optically excited carriers. On the other hand, crossed linear polarisations produce no amplitude modulation, however, a polarisation grating is created [13]. The field modulation, E(x), along the polarisation grating direction, x, can be written in terms of two circular polarisations,

$$E(x) = E_0\left\{sin\left(\frac{\pi x}{\Lambda}\right)e^{-i\frac{\pi}{4}}\frac{\hat{x}+i\hat{y}}{\sqrt{2}} + cos\left(\frac{\pi x}{\Lambda}\right)e^{i\frac{\pi}{4}}\frac{\hat{x}-i\hat{y}}{\sqrt{2}}\right\} + c.c. \quad (7)$$

where Λ is the grating spacing, and $\hat{x}$ and $\hat{y}$ are unit vectors describing the polarisation directions of the incident beams.

The polarisation of the combined electric fields has the form shown in figure 10(a). Thus, for equal excite intensities, the polarisation changes from linear to circular to orthogonal linear to circular of the opposite sense and back to linear. The period of this modulation is identical to that produced for amplitude gratings and is defined by the angle between the two excite beams and the wavelength of light. Figure 10 also shows that crossed-linear polarisations provide a modulation in circular polarisation. An alternative description is obtained by separating it into two circularly polarised components with opposite directions of rotation, figure 10(b).

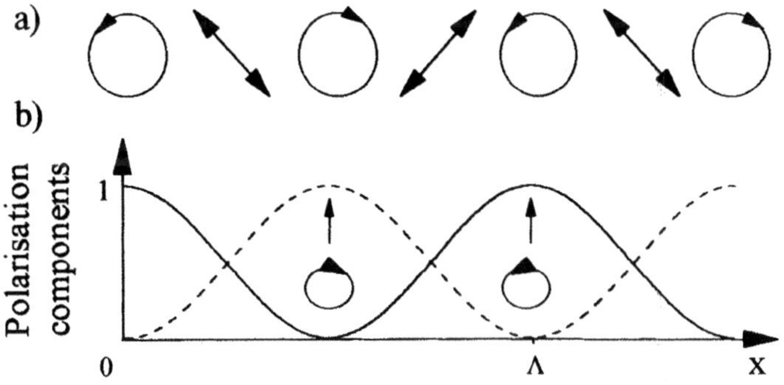

Figure 10. (a) The polarisation modulation produced when two orthogonally polarised light beams interfere. (b) Amplitudes for right and left circular components of the intensity as a function of distance in the plane of the grating.

A modulation of spin-polarised electrons can be created using this cross-polarisation configuration in resonance with the heavy hole exciton. Diffraction of the probe occurs via PSF, the spin-dependent free carrier contribution to exciton saturation as described in section 5 above. Since there is a π phase shift between the right and left circular polarisation components (spin-up and spin-down) of the grating, a 90^0 rotation of the linear polarisation of the diffracted probe beam was observed.

The decay rate of an amplitude grating of electrons and holes is given by,

$$\Gamma = \frac{1}{\tau_D} + \frac{1}{\tau_R} = \frac{4\pi^2 D_a}{\Lambda^2} + \frac{1}{\tau_R} \tag{8}$$

where D_a, is the ambipolar diffusion coefficient and τ_R is the recombination time for electrons and holes. The two terms arise from the recombination and diffusion of the carriers in the quantum wells. Since both electrons and holes are present, ambipolar diffusion is measured from an amplitude grating. The diffracted signal decays at twice the rate given in equation 8, since the diffraction efficiency depends on the square of the refractive index change (or excess carrier density). An exponential decay rate of 220ps is deduced for the amplitude grating shown in figure 9.

A much faster decay rate, at 13ps in figure 9, is observed for a polarisation grating with the same grating spacing, Λ. This is because the wash-out of a modulation of spin-polarised electrons will depend of the *electron* diffusion coefficient rather than the *ambipolar* diffusion coefficient. Electron diffusion is much faster than ambipolar diffusion because of the much higher mobility of electrons than holes. The spin-grating decay rate, Γ, is given by,

$$\Gamma = \frac{4\pi^2 D_e}{\Lambda^2} + \frac{1}{\tau_s} \tag{9}$$

where D_e is the electron diffusion coefficient and τ_s is the electron spin relaxation time.

Figure 11 plots measured decays for different grating spacings, Λ, produced by changing the angle between the pump beams for both concentration and spin-polarisation gratings. The slopes of the lines give the ambipolar and electron diffusion coefficients. For this 6.5nm well width sample, values of D_a = 13.3 and D_e = 125 cm^2/s were obtained. The ambipolar diffusion coefficient is approximately twice the hole diffusion coefficient.

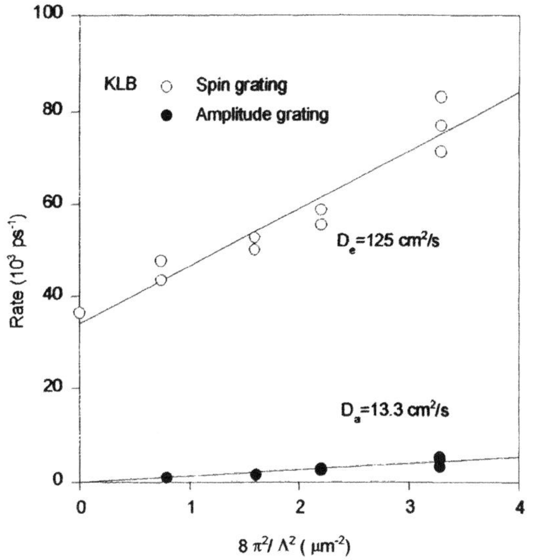

Figure 11. Measured decay rates of the diffracted grating signal against $8\pi^2/\Lambda^2$. Amplitude grating results (●) and spin grating results (O) are plotted together for comparison for the sample KLB.

The hole mobility of $\mu_h \sim 257$ cm^2/Vs, deduced from the Einstein relation, is in excellent agreement with values for bulk GaAs. The electron mobility deduced from this diffusion coefficient is $\mu_e = 4924$ cm^2/Vs. This value is lower than the typical room temperature value of 8500 cm^2/Vs for pure bulk GaAs. The difference may be attributed to interface scattering within the quantum well layers.

8. Conclusions

Excitons in low dimensional structures provide a powerful means of probing ultrafast carrier dynamics. Scattering phenomena, carrier thermalisation, in-well and cross-well transport can be monitored and time resolved via free carrier induced optical nonlinearities. In this chapter, we have emphasised the use of circular polarisations in pump-probe studies in distinguishing the various mechanisms contributing to exciton saturation and in transient gratings to make a unique measurement of both electron and hole mobilities within the plane of quantum wells.

Acknowledgements

This work was supported by the Engineering and Physical Sciences Research Council (EPSRC).

References

1. Miller, A., Miller, D.A.B., and Smith, S.D. (1981) Dynamical optical nonlinearities in semiconductors, *Adv. Phys.* **30**, 697-800.
2. Haug, H., ed. (1988) *Optical nonlinearities and instabilities in semiconductors*, Academic Press, San Diego.
3. Miller, A. (1993) Semiconductors, in R.W. Eason and A. Miller (eds.), *Nonlinear optics in signal processing*, Chapman & Hall, London, pp. 66-99.
4. Van Driel, H.M. (1995) Photoinduced refractive index changes in bulk semiconductors, in A. Miller, K.R. Welford and B. Daino (eds.), *Nonlinear optical materials and devices for applications in information technology*, NATO ASI Series, Volume 289, Kluwer Academic Publishers, Dordrecht pp. 141-181.
5. Chemla, D.S. and Miller, D.A.B. (1985) J. Opt. Soc. Am. B **2** 1155-1173.
6. Miller, A., (1995) Nonlinear optical devices, in M. Balkanski and I. Yanchev (eds.), *Fabrication, properties and applications of low-dimensional semiconductors*, Kluwer Academic Publishers, Dordrecht, pp. 383-413.
7. Miller, A., Cameron, A.R., Riblet, P. (1996) Physics and applications of exciton saturation in MQW structures, in M. Balkanski (eds.), *Devices based on low-dimesnional semiconductor structures*, Kluwer Academic Publishers, Dordrecht, pp. 199-225.
8. Schmitt-Rink, S., Chemla, D.S., Miller, D.A.B. (1985) Phys. Rev. B **32**, 6601-9.
9. Knox, W.H., Fork, R.L., Downer, M.C., Miller, D.A.B., Chemla, D.S., Shank, C. V., Gossard, A.C. and Weigmann, W. (1985) Phys. Rev. Lett. **54**, 1306-9.
10. Snelling, M.J., Perozzo, P., Hutchings, D.C., Galbraith, I. and Miller, A. (1994) Phys. Rev. B **49**, 17160-9.
11. Holden, T.M. Kennedy, G.T., Cameron, A.R. Riblet P. and Miller A. Appl. Phys.. Lett. in press 1997

12. Manning, R.J., Crust, D.W., Craig, D.W., Miller, A. and Woodbridge, K. (1988) J. Mod. Opt. **35** 541-551; Miller, A., Manning, R.J., Milsom, P.K., Hutchings, D.C., Crust, D.W. and Woodbridge, K. (1989) J. Opt. Soc. Am. B **6**, 567-578.

13. Cameron, A.R., Riblet, P. and Miller, A. (1996) Phys. Rev. Lett. **76**, 4793-4796.

INTEGRATED OPTOELECTRONICS-THE NEXT TECHNOLOGICAL REVOLUTION

A. S. POPOV
Physics Department, University of Sofia
Bulgaria

Materials science and advanced fabrication methods such as MBE, MOCVD and focused-ion-beam micromachining have enabled optical, electro-optic, and electronic components to be consolidated into monolithic circuits. One of the best examples of an optoelectronic integrated structures is the active-matrix LCD, in which thousands of electro-optic spatial light modulators are integrated with the same number of transistors into a large array. But other forms of optoelectronic integration are possible for example as diode lasers and their drive electronics; photodetectors and their amplification circuitry; switches, modulators, couplers, waveguides and other optical components.

One of the first experiences is the integration of diode lasers and their drive electronics. In Figure 1 [1]is illustrated a simple example of this kind of integration, a device that combines

M. Balkanski and N. Andreev (eds.),
Advanced Electronic Technologies and Systems Based on Low-Dimensional Quantum Devices, 137–153.

an AlGaAs buried-heterostructure laser and two

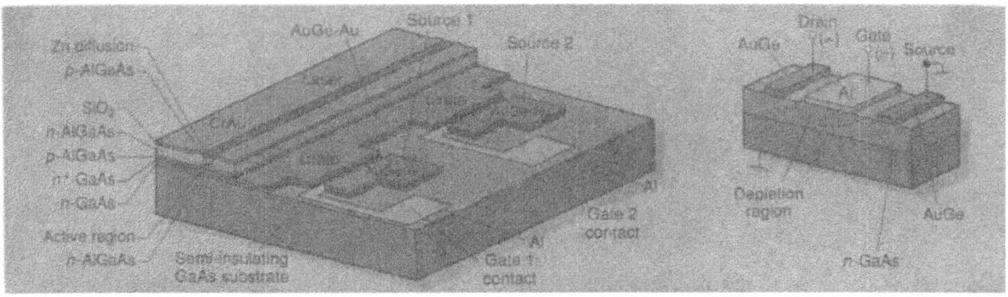

Figure 1. Integration of diode laser with drive electronics

MESFETs on a single substrate. One transistor controls a bias current to keep the laser above threshold, while the other provides a modulation current for direct modulation of the laser output. Both currents are separately controlled by voltages applied to the MSFET gates - 1 and 2.

The device architecture is constructed on the base of semi-insulating GaAs substrate. For MESFET structure a single layer of n- type GaAs is grown and capped with three separate metal contacts that serve as the source, gate and drain of the transistor. If the source is connected to ground and a positive voltage is applied to the gate, current will flow between the two contacts via the n- type GaAs layer, which acts as a channel. However, if a negative voltage is applied to the gate, the reverse bias will create a depletion region in the n- type channel below the gate, which will increase current flow between the source and drain.

In the process of integration a heavily-doped n^+- type GaAs layer is grown on top of the n- GaAs channel layer to help connect the drain electrode with the n- AlGaAs stripe directly below the active region of the laser. The rest of the laser structure is grown on the top of the n^+- GaAs. One of the disadvantages of this OEIC design is its nonplanar profile, which limits the smallest feature size attainable by photolitography. To achieve a planar architecture, a trench can be etched into the substrate at the location where the laser is to be grown. Than in the process of laser fabrication, the uppermost layers will lie at roughly the same height as the top layers of the transistors.

The integration of diode lasers and MESFETs has been a very successful approach to OEICs based on GaAs semiconductor compounds. But diode lasers based on AlGaAs/GaAs architecture emit light in the 800-nm region of the spectrum, which is fine for CD-ROMs and first generation fibreoptic transmitters only. But it is not applicable for second and third-generation fibreoptic communications at 1.3, and 1.55 μm.

Diode lasers for these regions are usually constructed from narrow-band-gap quaternary compounds of InGaAsP grown on InP substrates. But, MESFETs constructed from this material have high gate-leakage currents. As a result, InGaAsP/InP systems are used for OEIC with the other kind of transistors, for example heterojunction bipolar configurations. Such structure is shown in Figure 2, as combination of laser and bipolar transistor. Unlike MESFET and other field-effect transistors that function with a horizontal geometry, HBTs have a vertical geometry consisting of an emitter, base and collector arrayed in a stack. The n-doped emitter layer is made of a wider-band-gap material (n- InP); the base is formed by p-InGaAsP, and a collector - by n- GaInAsP. On the left side of Figure 2 [1] are shown the

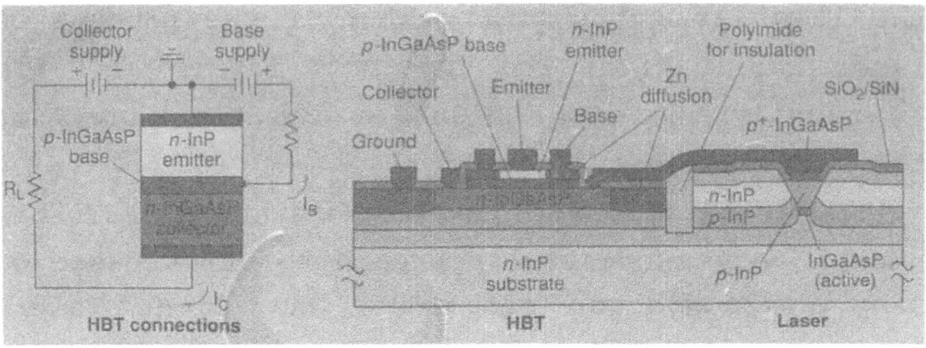

Figure 2. Integration of diode laser with bipolar transistor with different semiconductor material structures

connections of HBT. A forward bias is applied across the B-E heterojunction while between C-E a reverse bias is applied. The resulting current that flows through the B-E junction drows electrons into the thin base-layer, where they are attracted to the higher positive voltage on the collector. Thus, a relatively small current flowing through the E-B - circuit generates a much larger current through the E-C circuit via the base. Therefore, modulating the E-B current of the HBT modulates the current through the laser, which is connected to the collector in the OEIC.

Besides bipolar transistors certain FET designs have proved valuable for InGaAsP/InP OEICs, too. These include metal-insulator-semiconductor FET (MISFET), high-electron-mobility transistors (HEMT) and modulated-doped field-effect transistors (MODFET).

A number of different diode-laser designs have been successfully integrated in OEICs. Although buried-heterostructure and other-stripe geometry lasers work fine in an integrated environment. A better solution is to use DFB or DBR lasers with low-current quantum-well gain structures. Low current VCSEL also are ideal for high density OEICs.

Equally important for high speed fibreoptic communications are integrated optoelectronic receivers that combine photodetectors with electronics for amplification and signal processing. Two photodetector designs are particularly suitable for OEICs: p-i-n photodiodes and metal-semiconductor-metal (MSM) photodiodes. Both of these quantum detectors have very-high speed performance capacity. Figure 3 [1] depicts a p-i-n photodiode and HBT grown on single substrate of InP. This device is an example of a vertically integrated OEIC in which all of the semiconductor layers that define the various OEIC components are

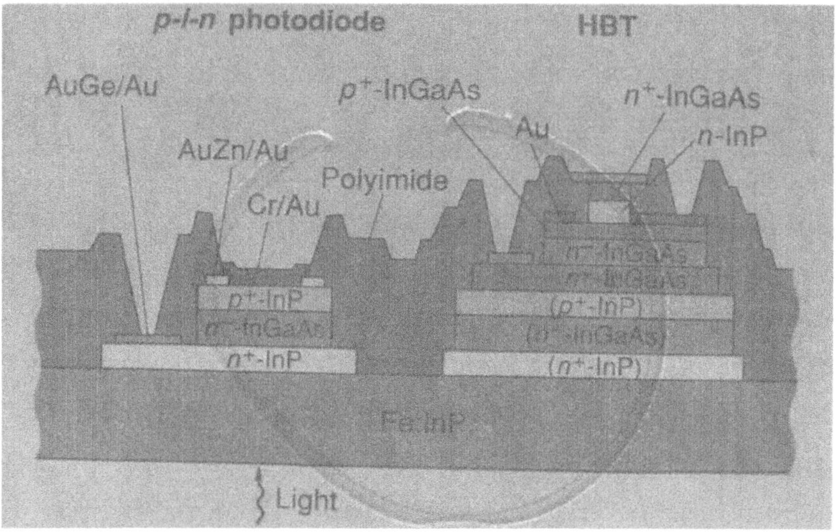

Figure 3. Vertical integration of p-i-n photodiode and heterostructure bipolar transistor grown on a InP substrate

first grown as a single vertical stack than selectively etched away to isolate distinct components. Especially in this case the photodiode layers are grown first, followed by the HBT layers. Selectively etching than sculpts the photodiode and transistors out of the stratified composite, after which the contacts are deposited. Finally, an insulating volume of polyimide is deposited to decrease a leakage current between photodiode and transistor. Electronic interconnections between them are made by separate metal depositions.

Such kind of OEIC can be integrated horizontally, if will use a photodetector with horizontal geometry, for example MSM-structures. This architecture is shown in Figure 4 [1], and it

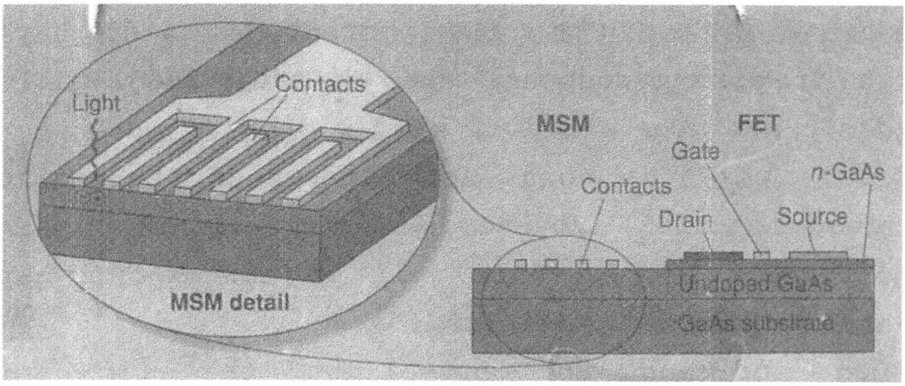

Figure 4. Horizontal integration of metal-semiconductor-metal photodetector and field-effect-transistor

consists MSM photodiode integrated with a MESFET in a GaAs system. The MESFET consists of a channel layer of n-type GaAs grown on layer of undoped GaAs and capped with source, gate and drain contacts. The enlarged perspective of the MSM region of the OEIC reveals a unique pattern of metal contacts deposited on a light-absorbing layer of undoped GaAs. In this view, the MSM structure is actually cut in half, and the front surface defines the cross-sectional plane of OEIC.

The contacts of the MSM photodiode form two integrated, fork-shaped electrodes that function as back-to-back Schottky diodes. When the electrodes are biased at potentials of 2 or 3 volts, the GaAs region between them becomes depleted and the structure behaves like a p-i-n photodiode. Separation between the "tines" of the forks is only about 2μm.

Besides their convenient horizontal geometry, which makes them excellent optoelectronic combination for FET, MSM photodiodes also have high frequency response due to their negligible capacitance. The receiver band-widths is more than 18 GHz, which is ideal for high-speed fibreoptic communication. However, conventional MSM photodiodes do not work well in narrow-band-gap systems such as InGaAs/InP.

In addition to controlling the flow of electrons between different components, OEICs also must control the flow of photons. It is known that if semiconductor materials are used to control photon flow - the technique is called photonic integration; if dielectric materials are used - it is called optical integration. In both types of integration, waveguides often play a crucial role in routing photons from one point to another on the OEIC, and a number of structures have evolved for different applications. Four common waveguide structures are the strip, embedded-strip, ridge, and strip-loaded. These waveguide structures are often completely buried beneath other layers, too, such as the buried-heterostructure laser. All of these structures can be fabricated into a variety of geometries for different OEIC functions such as straight-line links, offset links, beam splitting and combining, and wavefront interference and coupling.

If the waveguide is accompanied by an active element such as a gain source, absorption medium, or electrooptic device, it is called an active waveguide. Simple transparent waveguides are known as passive waveguides. The combination of active and passive waveguides on a single OEIC is what enable full optoelectronic integration.

In Figure 5 [1]are presented two OEICs designed

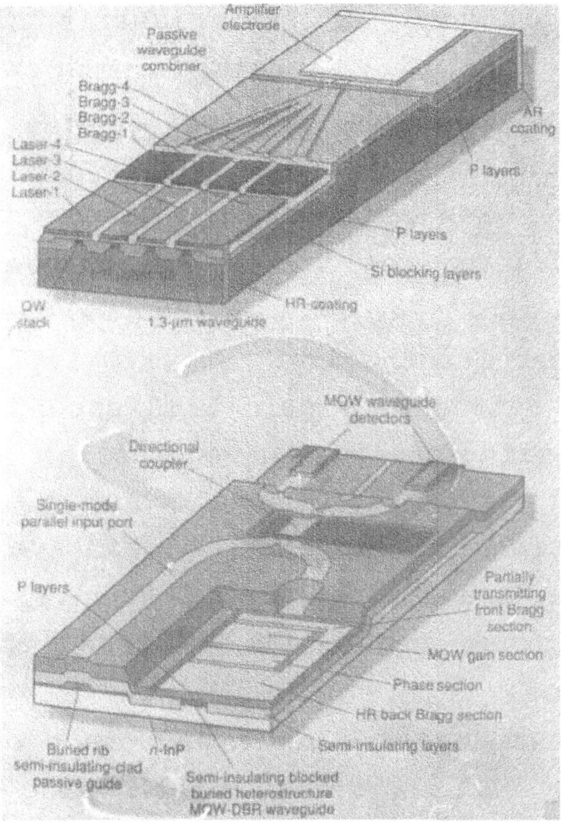

Figure 5. Full integration of active and passive waveguides

for fibreoptic communications, which incorporate both active and passive waveguides for full integration.

In the top of Figure 5 is shown an OEIC with active waveguide as a tunable transmitter for wavelength-division-multiplexing (VDM) at four different wavelengths in the 1.3 μm spectral region. It consists of four integrated multiple-quantum well DBR lasers linked by four passive waveguides to a MQW optical output amplifier. Separately controlled electrodes deposited directly over the laser structures are used for both wavelength tuning and direct intensity modulation. The four passive waveguides then combine the outputs of all four lasers into a single active waveguide that passes through the MAW amplifiers region. An electrode deposited above the amplifier section controls the gain of the four-wavelength output.

In the bottom of Figure 5 is shown second kind of OEIC with passive waveguide as a balanced heterodyne receiver for coherent fibreoptic communications. It consists of passive buried-rib waveguides, a directional couple, two MQW photodetectors, and a MQW DBR laser as the local oscillator.

To adjust the wavelength and phase of the local oscillator for proper heterodyning with the incoming light signal, the MQW laser is equipped with separately controlled Bragg-reflector and phase sections. Wavelength tuning is accomplished by varying the voltage to the electrode deposited over the front and back Bragg reflectors. The resulting electro-optic shift in the refractive index of the Bragg gratings changes their operating wavelength. Similarly, a voltage change to the electrode over the phase section of the laser changes the refractive index of the buried waveguide and hence the phase of the wavefront passing through it.

Light from the local oscillator and input signal are mixed at the directional coupler, which is electro-optically controlled by another electrode. Buried waveguides than direct the heterodyned light signal from the coupler to the two photodetectors, where it is converted into an electrical signal for further electronic processing.

The parallelism, noise immunity and speed of OEIC make them attractive solutions for high-speed fibreoptic transmitters, receivers, transreceivers, flat-panel displays, and optical storage. Perhaps most exciting of all is the explosive impact that OEIC will have on optical computing. The next generations of computers will need from two- and three- dimensional integration of bistable photonic switches, logic circuits and massively parallel optical interconnects. Because the duration of the lecture will be concentrated on the next generation of optical computers.

The best usual computers till now have multipurpose functionality at speeds up to 10^7-10^8

bit operations per second at clock rates 100 MHz. This parameter for second generation of optical digital computers is 10^{11}, and for third generation - 10^{14}, i.e. seven order higher.

Second generation of DOC is based on parallel architecture with N^2 interconnect technology, where N is the number of optical channels. The third generation of optical computers utilizes N^4 global interconnect technology. In a parallel configuration Figure 6 [2] two planes of spatial

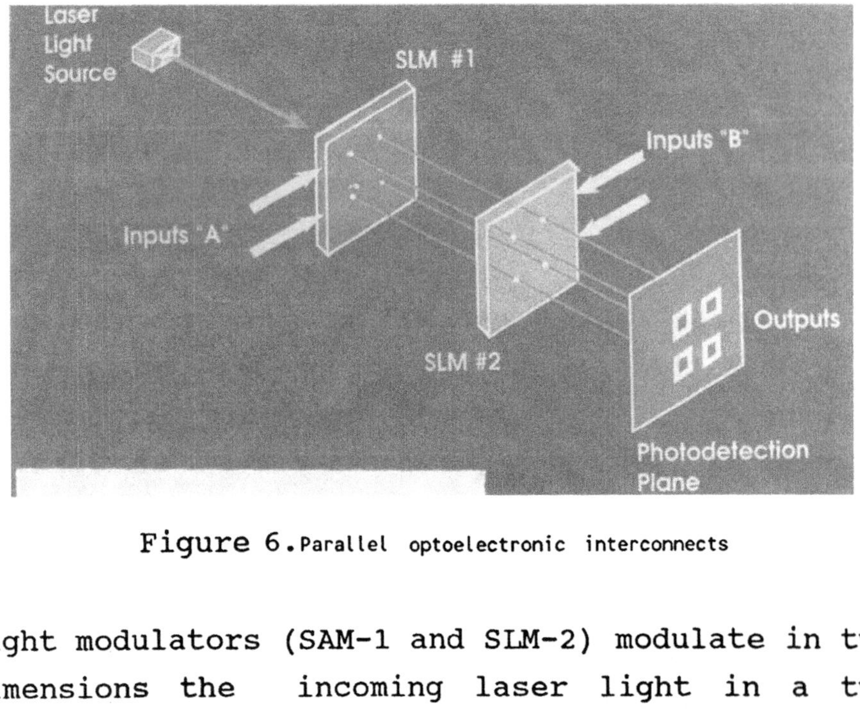

Figure 6. Parallel optoelectronic interconnects

light modulators (SAM-1 and SLM-2) modulate in two dimensions the incoming laser light in a two dimensional plane. Each point of plane **1** is connected to its respective point at plane **2** and the digital product **A AND B** is delivered optically to the photodetection plane or the output. This is referred as parallel interconnect because all interconnects are in fact parallel with respect to each other.

The Doc-II architecture is configured as a matrix/vector Boolean logic multiplier. As is shown in Figure 7.[2] each column of the

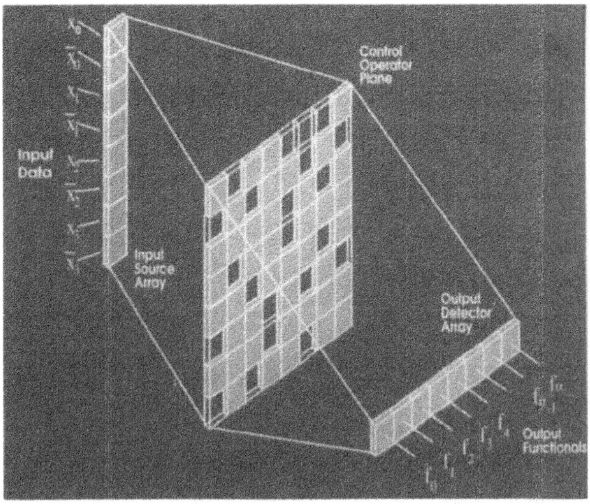

Figure 7. Parallel interconnects architecture used in a logic array for DOC-II

two-dimensional control operator is bit-wise multiplied against the input data vector. This products are Boolean summed on their respective output detectors. Parallel interconnects are used to connect an array of 64 laser diodes to a two-dimensional acousto-optical spatial light modulator which consists of 64 inputs, each with a band width of 400 MHz. The output cascade is realized onto a 128-element avalanche photodiode array.

With the global interconnect is constructed

third generation of digital optical computers. This scheme is shown in Figure 8 [2], where two

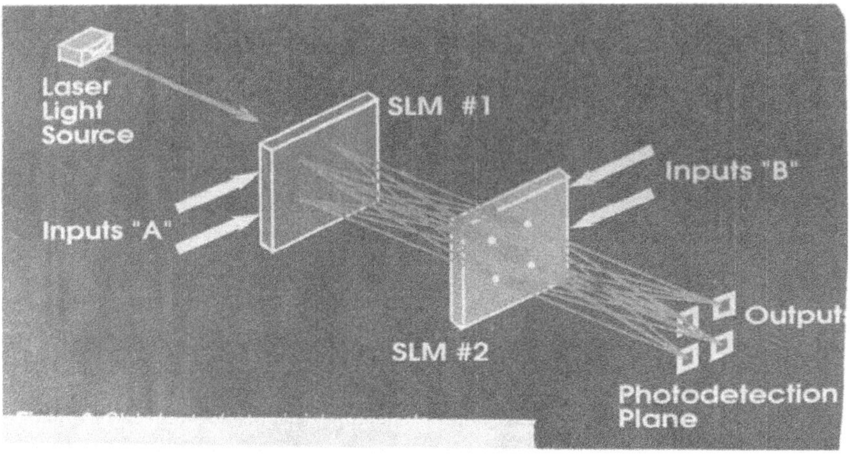

Figure 8. Global optoelectronic interconnects

planes of modulators modulate in two dimensions the incoming laser light in a two-dimensional plane, but the interconnects between the two planes are now global. Each point at plane 1 is selectively broadcast to all possible modulators at the second plane. The digital product **A**(m) **AND** **B**(n) is selectively broadcast to all possible detectors by optical delivery to the photodetector plane or the output. This is referred as global interconnect because all points are globally broadcast to all other points at each successive optical modulation plane or detection plane.

In Figure 9 [2] is shown a global architecture, where a double stage high performance computing module, interconnects between the input and the subsequent stage are globally broadcast by an array of thin planar holographic optical

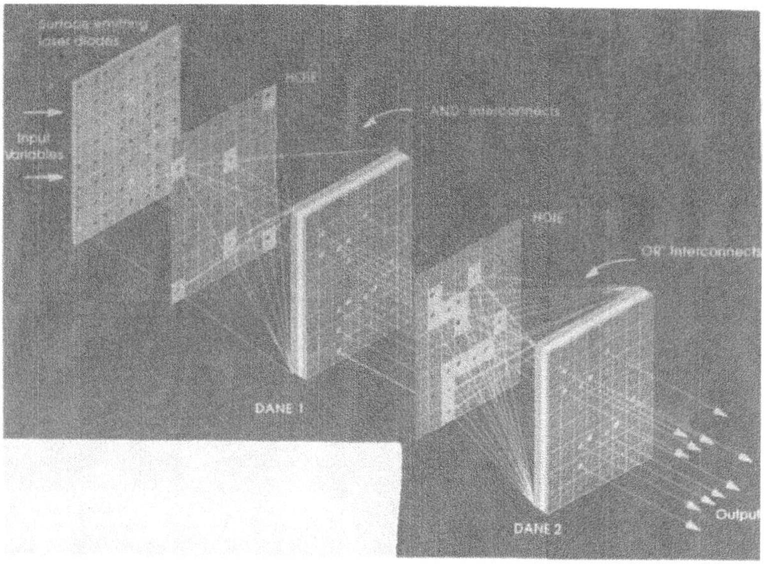

Figure 9. Scheme of global interconnects architecture of third generation of DOC

interconnect elements (HOIE).

"Smart pixel" technology is further developed in DOC-III within a GaAs smart pixel design. This improvement lowers the power consumption, reduces the package size and increases the throughput of the system. The concept of the smart pixel was originally utilized on the DOC-II platform and is extended to the DOC-III platform in global free space by folding the two-dimensional interconnect into four-dimensional interconnect. Figure 10 [2] shows the concept of the "smart pixel". Input light to the detectors is represented as negative logic. The data is ORed optically or threshold at the 0 to 1 level. The zero level is the no light

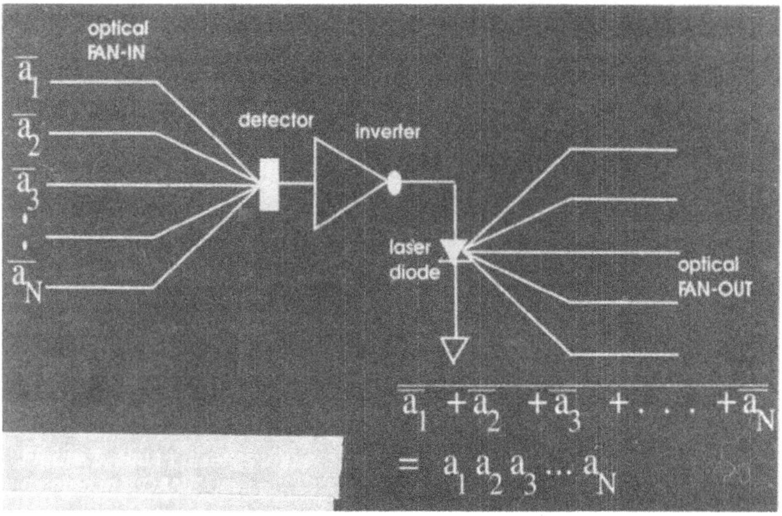

Figure 10. Smart pixel design in optoelectronic integration structure

condition or KT limit, and the one level is approximately 5000 photons or 1.2 fJ/b for switching energy.

After inversion, this represents the Boolean logical product computation $a_1, a_2, a_3 \ldots, a_n$. The cell is smart because it computed the n-bit **AND** product of effectively logical positive representation inputs $a_1, a_2, a_3 \ldots, a_n$. This represents a substantial benefit for optoelectronic computing systems.

The achieved successes in the optoelectronic logic schemes in a hybrid technology for constructing of different modules and nodes. In Figure 11 [2] is presented DANE-device (detection,

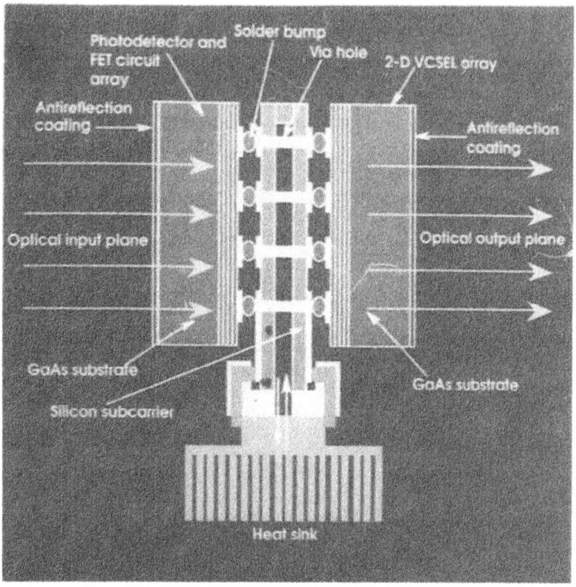

Figure 11. Hibrid structure of a DANE array on optoelectronic integrated structure

amplification, negation and re-emission). Each element of the DANE array consists of a simple detector, amplifier, inverter and output laser. The detector and amplifiers are fabricated on the back side of the GaAs-chip. On the other side is mounted VCSEL. For the all system the wavelength of 980 nm is selected.

References

1. Higgins,T.(1995) Optoelectronics:the next technological revolution,Lasr Focus World <u>31</u>,93-102

2. Guilfoule,P.(1993) A third generation DOC, Photonics spectra <u>27</u>,116-124

OPPORTUNITIES OF VERTICAL-CAVITY-SURFACE-EMITTING LASERS (VCSEL) IN DISPLAY AND OPTICAL COMMUNICATION SYSTEMS

A. S. POPOV
Physics Department, University of Sofia, Bulgaria

The diode laser is a semiconductor chip with an internal structure that makes it function like electrical diode.A simple diode however,makes a poor laser.Diode laser achieve high performance through their sophisticated internal structures, which control the flow of electrons and light through the layers of semiconductor materials. These structures can be fabricated in several families of materials :A3B5,A2B6,A4B6.

Early semiconductor lasers had many practical limitations for development of integrated circuits and systems.The first diode lasers could operate only in pulsed mode for microsecond pulses and required cooling to 77K. In fact,it wasn't until 1970 that continuous room-temperature operation becomes possible through the development of double heterostructures.

The most common type of diode laser with internal structure emits light from its edge.

M. Balkanski and N. Andreev (eds.),
Advanced Electronic Technologies and Systems Based on Low-Dimensional Quantum Devices, 155–173.

Smooth facets cut or polished on the edges of the chip reflect light back into the semiconductor; often the rear facet is coated for full reflection.The reflected light stimulates emission from recombining electron-hole pairs in the unboned active layer between the p- and n-regions. This process extracts energy that otherwise would be dissipated as heat,raising output power and efficiency far than LED's. In practice, drive current must reach a threshold before stimulated emission overcomes internal losses and produces laser output.At lower drive current the device emits as an LED.The critical physical parameter of laser diode is current density,measured in amperes per square centimeter of junction area.

The three major classes of diode lasers are defined by compositions of the layers next to active layer.The first one is homojunction diode laser which is made entirely of one compound, typically GaAs.The earliest diode lasers used this structure,are inefficient and they could operate only in pulsed mode at cryogenic temperatures.The second kind is single heterojunction laser which has an undoped active layer sandwiched between two layers of different compositions and different bandgaps,typically GaAs and AlGaAs.The difference helps confine light and excitation to the AlGaAs active layer,letting this lasers produce pulses with high power at room temperature.The third type are double heterostructure lasers with active layers sandwiched between two layers of different materials-for example,GaAs between AlGaAs layers. This confinement on both top end bottom of the active layer makes double-heterojunction lasers more efficient than single-heterojunction ones. It is the first type to emit continuously at room temperature.

The development of double-heterostructure lasers started many time ago.One step has been to limit the width of the laser region in the active layer.Early devices had active region width about 50 μm,but up to now it is 1-10 μm only.This design reduces drive current by concentrating it in a small area.Laser action can be limited to a narrow stripe either by restricting current flow through the laser or by creating an optical waveduide. Controlling current flow produces a population inversion only in the stripe and also creates wave guide effect called "gain guiding". Optical waveguides are formed by creating paterns in the junction plane.Differences in refractive index confine the light to a narrow stripe,called "index guiding". Index guided lasers are more efficient and have lower threshold currents.They are the usual choice for low power applications that require good quality beams or small focal spots. These lasers have many different constructions which are classified into five categories :

-buried-heterostructure lasers (Figure 1) they have the narrow stripe of the active region completely surrounded by material of lower refractive index.Such lasers can be formed from ordinary double heterostructure material by first etching away everything,but a mesa,often only 1-3 μm across which becomes the active stripe. This heterostructure stripe is buried by other material and deposited with either an insulating layer or reverse p-n junction to block current flow. This concentrates current in the stripe and forms waveguide [1];

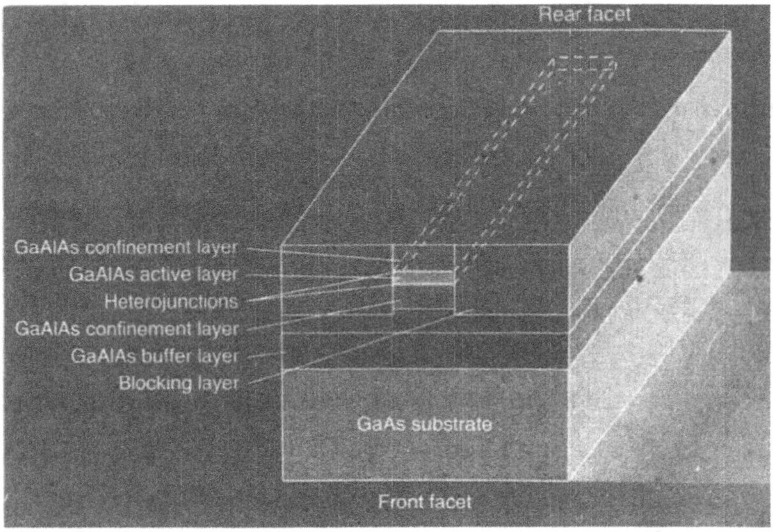

Figure 1. Horizontal-cavity diode laser with buried stripe geometry

-channeled-substrate planar lasers (Figure 2a) they have an active layer grown on a substrate with a channel etched into it.The substrate has a high refractive index,but is covered by low index cladding layer.The tick cladding fills in the etched channel,creating a planar surface for the active layer and isolates the active layer from the surface.So,the active layer has low loss in that region.But the loss is higher on the sides, where the cladding layer is thin, causing an effective decrease in the refractive index that is large enough to confine light laterally in the active layer;

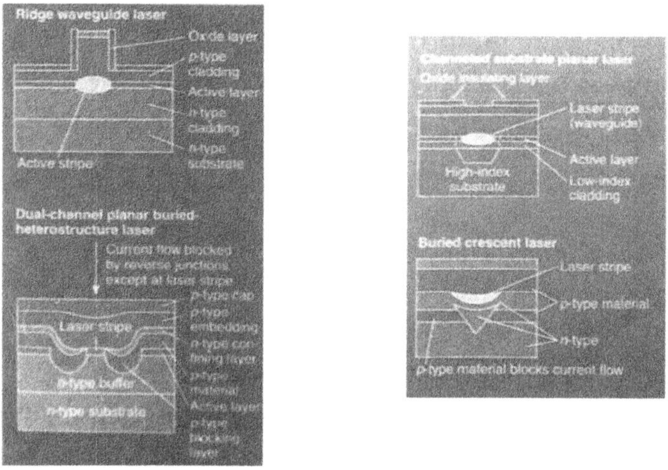

Figure 2. Index guided diode laser structures

-buried-crescent laser (Fig.2b) -it is like to the previous one and have a groove or channel etched into the substrate. The layer, however, is grown on the grooved surface using LPE and forms a stripe in the active layer with crescent-shaped cross section. This structure serves as a vaweguide to confine light in the active stripe. Additional layers cover the crescent and produce an overall planar structure;

-ridge waveguide laser (Figure 2c)-it have a ridge above the active stripe,while the surrounding areas are etched close to the active layer (about 0.2-0.3 μm).The edges of the ridge reflect light guided in the active layer, forming a waveguide.Insulating coatings on the surrounding areas help confine current flow to a path through the ridge and active stripe;

-dual-channel planar buried-heterostructure laser (Fig.2d)-it has an isolated active stripe in a mesa by etching two parallel channels around it. After that follows the growth of materials around to it to bury the heterostructure.This structure is attractive for high power lasers,made from InGaAsP.

The next generation of laser structures is based on the quantum wells.The concept of the quantum wells is logical continuation of the trend to shrink structures to smaller and smaller dimensions,made possible by sophisticated growth technics as MBE and MOCVD. When layers are thinner than about 20 nm,quantum mechanics becomes significant,changing the energy-level structure of the semiconductor in ways that improve laser operation.

A single quantum well is a thin layer sandwiched between two layers with larger band gaps.This confines electrons to the quantum well layer,which are so thin that energy states in the valence band and conduction band are quantized and do not form continuum in bulk semiconductors. The adjacent layers also can have grade refractive indexes,forming a grade index and separate optical and carrier confinement heterostructure.The high confinement of light and carriers improves efficiency giving higher output power and threshold current only about one-half to one-third of those of comparable double-heterostructure lasers.

Multiple quantum wells can also be produced by alternating layers with low and higher bandgaps.Their advantage is an increase in the volume of exited material,which allows higher power.The high-band gap barrier layers must be thick enough to prevent quantum tunneling. Adjustment of layer thickness and composition gives some control over wavelength.

Quantum wells are tightly confined in one dimension (Fig.3),but the scientists are studding structures tightly confined in two dimensions - quantum lines (Fig.4)[2], and three dimensions-quantum dots.This structures have quantized energy levels that promise useful properties.

All of the laser-diode structures discussed so far emit radiation from their edges and this feature makes many difficulties for construction of integrated and hybrid optoelectron circuits. More convenient structures for including into different integrated systems are surface emitting laser diodes. The optical cavity of this kind of laser comes in two basic variants: planar-cavity surface-emitting laser (PCSEL) and vertical-cavity-surface-emitting laser (VCSEL). A (PCSEL) (Figure 5.)[3] essentially consists of an

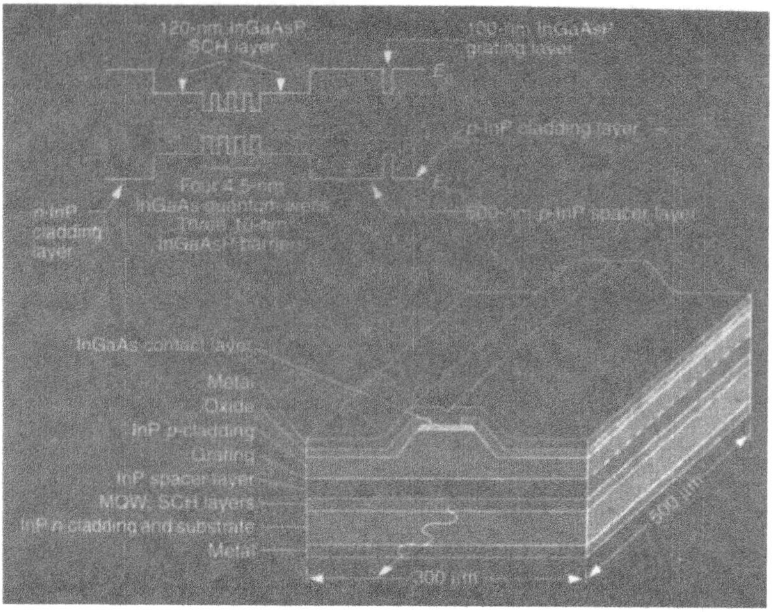

Figure 3. Diode laser structure with quantum wells

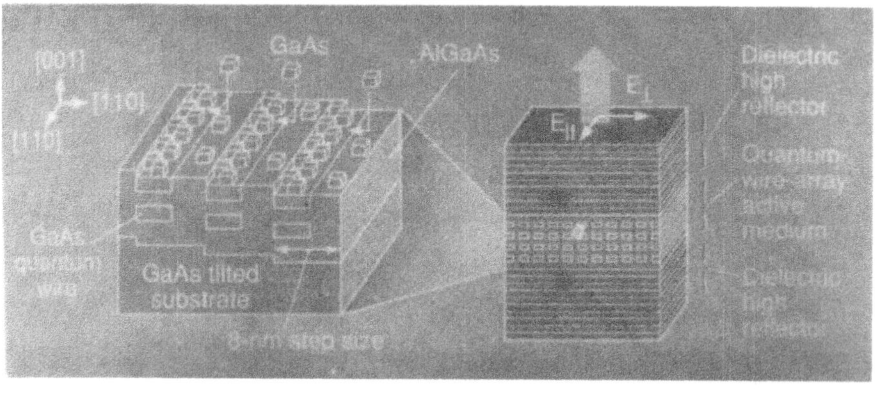

Figure 4. Quantum-wire microcavity laser with quantum wells

edge-emitting laser with an optical structure to

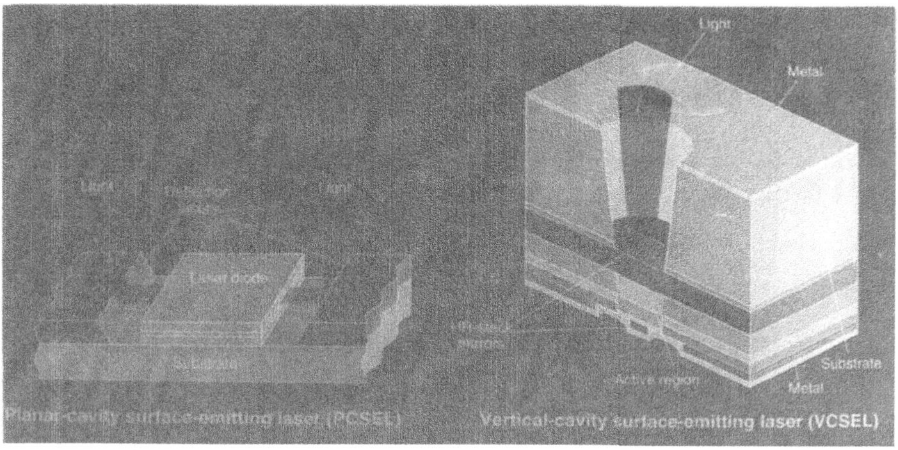

Figure 5. Principal structure of planar cavity and vertical cavity surface-emitting lasers

redirect the laser beam up through the chip surface. Therefore, these lasers are at least as long as the conventional edge-emitting variety and have the same elliptical shape of beam patterns.

The high efficiency of new diode-laser structures has been obtained by development of VCSEL. The resonant cavity of such kind of devices is in the active - layer plane in a traditional horizontal-cavity laser, but perpendicular to it in the vertical-cavity laser (Figure 6.)[1].Light resonates between mirrors on the top and bottom of the wafer, so photons pass through only a very short length (a micron or less) of active medium in which they can stimulate emission. Thus vertical-cavity lasers have low round-trip gain, rather than the high round-trip gain of horizontal

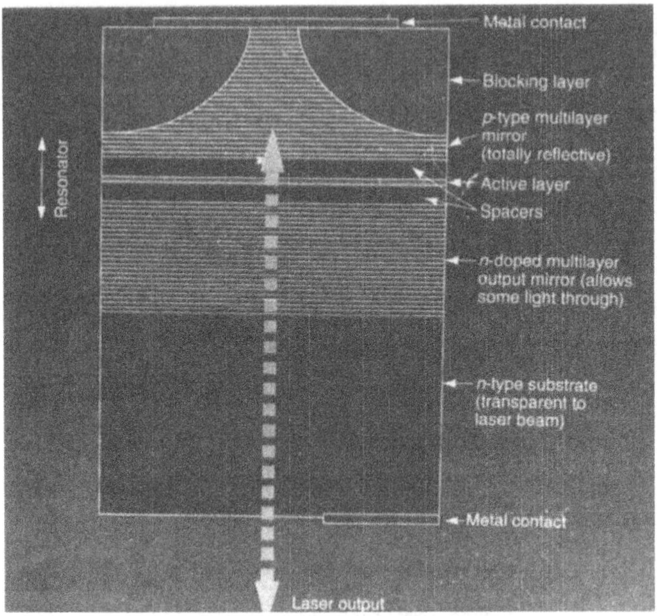

Figure 6. Vertical cavity diode laser with emission through the top and bottom surfaces

cavity lasers. All vertical-cavity lasers emit from their surface and this means that high reflectivity mirrors are needed to produce vertical-cavity-laser oscillation. VCSEL's have a fundamental advantage in beam geometry - they naturally emit clean, round beams. This property is fundamental. The edge-emitter puts out a highly elliptical beam with a minor-axis divergence that is about twice as large as the circularly symmetric divergence of VCSEL. Astigmatism is also present in the wavefront of edge-emitters and it becomes incorrectable during the age.

Circularly symmetric VCSEL beams have no inherent astigmatism. Emission perpendicular to the wafer surface means that microoptic lenslets can be integrated monolitically with the VCSEL, even on a wafer scale,as is shown in (Figure 7)[4].Among the

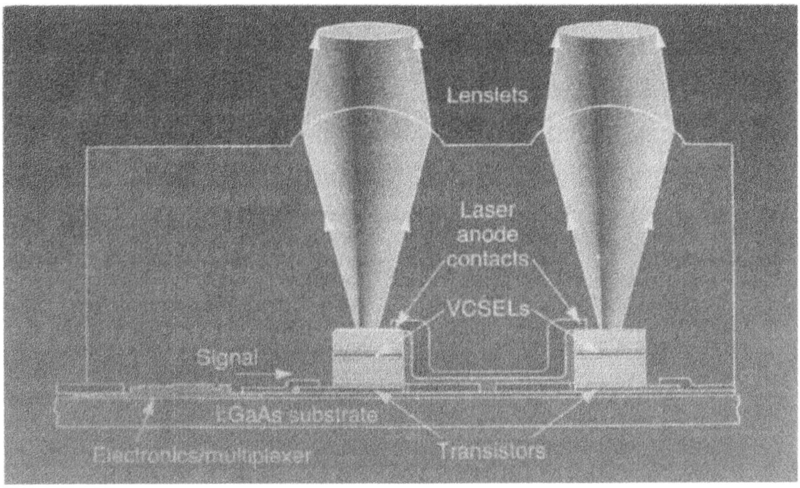

Figure 7. VCSEL integrated with microoptics system

technologies being pursued to advance interconnect techniques are VCSEL and the integration of high-density silicon-electronics with high-performance GaAs-optoelectronics. Direct optical communication between chips and stacked boards is designed to improve communication speed, eliminate electrical interference and crosstalk, and simplify packing requirements. To achieve submilliamper threshold current, VCSEL were made with an epitaxially grown AlAs oxide/GaAs distributed Bragg reflector (DBR) above the gain region and a conventional AlAs/GaAs DBR below is shown in (Figure 8.)[5]. The index-guided top emitters were fabricated with a technique that selectively oxidized an interfacial AlGaAs layer to provide current construction and served as both a vertical waveguide an current aperture. Lasing

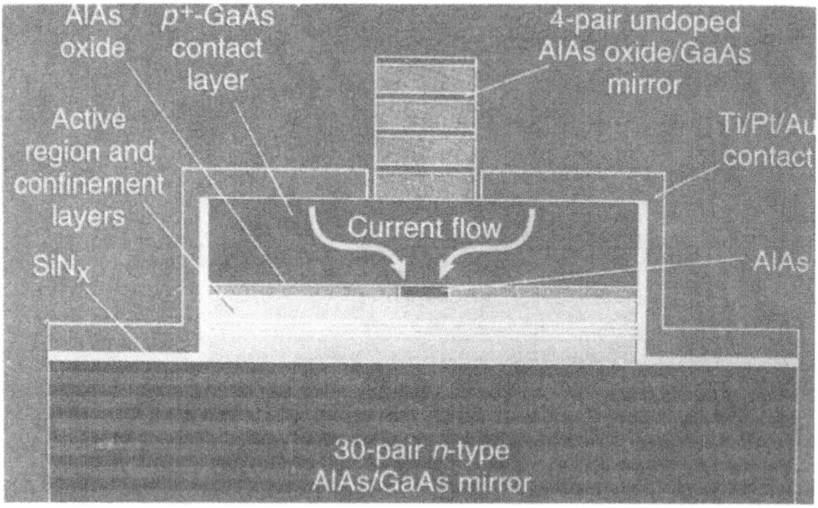

Figure 8. Structure of VCSEL with submiliampere threshold current

threshold current densities of 130 $\mu A/cm^2$ were obtained in VCSEL's with 2x2 μm mesa size.

Fabrication in one- and especially two-dimensional arrays is a VCSEL trade mark. By the end of 1989, a 32x32 laser array had already been demonstrated. High resolution miniature displays or two-dimensional scanners could use much larger arrays, on the order of 512x512. Such device requires 1024 individual connections to the outside world. To eliminate this situation is possible by integration with simple electronic circuitry - such as shift registers or demultiplexers. For example, an on-chip multiplexer could reduce the connections from 2N (for an NxN array) to 2 log 2N, or from 1024 to 36 for N = 512. VCSEL's can be integrated with on-chip electronics and microoptics to create optoelectronic integrated circuits (Figures 9, 7)[4]. If a 512x512 array is to be contained in a

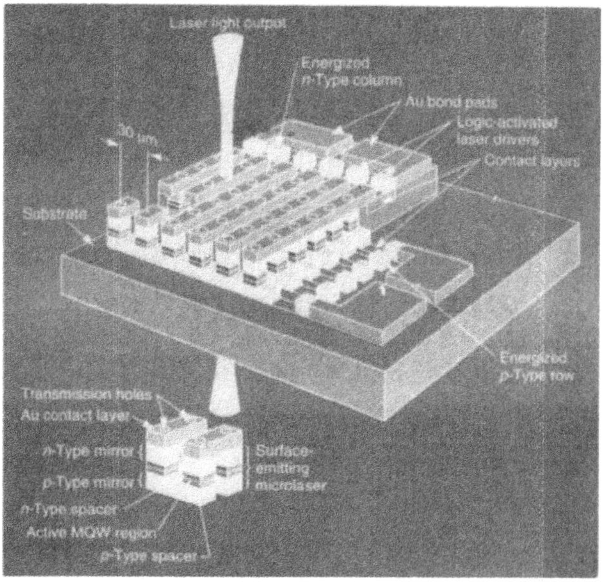

Figure 9. Two dimensional VCSEL array

1 cm^2, the device spacing must be 20 μm.

Two-dimensional VCSEL array could be used in helmet-mounted displays very compact and light. The video input port would receive data from TV or radio waves, or from helmet-mounted night-vision sensors. The sensor data are than translated to the appropriate format and communicated to the display. Generation of high resolution images can be accomplished in the near term with smaller two-dimensional arrays by using scanning technique. The other applications of VCSEL are optical communication systems. Circular beams are ideal for coupling into fibres - more than 90%. The communication bandwidth of long-distance fibres are accessed by multiplexing a number of data streams on different wavelength (Figure 12). A valuable component is monolithic source of evenly spaced wavelength-high speed lasers with beams combined by a simple monolithic device. System using multiwavelength VCSEL arrays with 140 evenly spaced wavelengths and one-piece holographic beam combiners that merged and coopled four beams into a single mode fiber is demonstrated on (Figure 10)[4]. For shorter

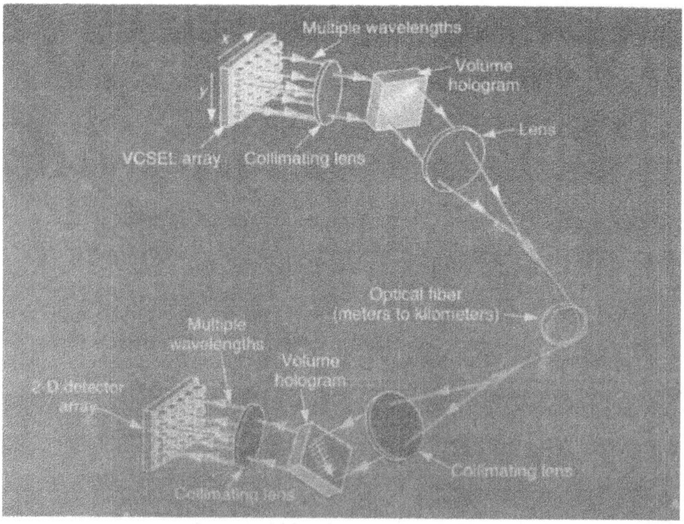

Figure 10. System for multiplex multiple wavelengths with volume hologram

distances extremely low loss and dispersion in fibre is less critical because applications of near IR and visible VCSEL is open. But the majority of such optical communications require high speed. The most-impressive VCSEL-speed demonstration is 80 Gbit/s produced by 2x8 flip-chip. This speed was based on the slowest device speed in the array, which was 5 Gbit/s with 10^{-9} bit-error rate. Each row of the VCSEL array had eight evenly spaced wavelengths. Until now such devices is possible to built on the base of VCSEL in which individual emission wavelengths in two dimensional laser arrays can be set predictably. For example, emission wavelengths are adjustable from 965 to 995 nm by a simple post fabrication method in which each cavity length accurately fixed by anodic oxidation. Such wavelength-tunable VCSEL arrays are potentially useful for applications in free-space optical interconnects in massively parallel computers, all optical switching, and wavelength-division-multiplexing systems. Making VCSEL arrays with precisely controlled wavelengths has been a technological problem. Emission wavelengths are usually determined during epitaxial growth by layer thicknesses, While a number of growth methods have been used to make VCSEL arrays unpredictability, nonreproducibility, noncontrolability have made wavelength control difficult. Now has developed a novel VCSEL structure that has a hybrid top mirror consisting of semiconductor distributed Bragg reflector, an oxide layer, and a metal reflector that enables arbitrary positioning of the lasing mode in individual array elements after the epitaxial growth . The method relies on anodic oxidatation of a GaAs-tuning layer to precisely set the cavity length and consequently the emitting wavelength. By this method the wavelength of each array element across an entire wafer can be arbitrarily adjusted. For understanding of device construction we need from explanation of device design.

The device contains an active region with three InGaAs QWs with GaAs barriers and AlGaAs cladding layers. The cavity is defined by a bottom DBR and hybrid top reflector that includes a four-period DBR capped by a GaAs tuning layer, SiO_2 phase-matching layer, and a gold metal reflector.

The cavity lengths of each device are precisely adjusted by anodic oxidation of the GaAs tuning layer. This process is voltage driven using a solution of ethylene glycol, tartaric acid and water to convert the surface to thin oxides - Ga_2O_3 and As_2O_3, that are then chemically removed. Forty volts creates an oxide that corresponds to a GaAs removal of about 60 nm, for example. The maximum continuous-wave optical output power reached nearly 0.5 mW, with a threshold current 6 mA and drive voltage 4 V. For optical communication systems is very convenient the other device-quantum-wire microcavity laser which can lase at room temperature with unique and exploitable polarization and modulation characteristics. The active medium in this laser is an array of quantum wires in which electron energies are confined to a narrow region. Such quantum confinement allows the laser to operate with low current thresholds, high modulation bandwidths, narrow spectral linewidths, and reduced temperature sensitivity. Photons emitted by the quantum wires resonate in a microcavity. Optical confinement in the microcavity channels reduces current thresholds. At the moment production of quantum-wire microcavity lasers is very difficult because the devices must have nanometer-sized cross section and must be uniformly sized, closely spaced, and free from contamination or defects. Also, they must be integrable into micron sized electronic components. The must convenient method for production of quantum-wire lasers is MOCVD-technique by which is possible to grow fractional-layer superlattices on tilted GaAs-substrate. This process is shown in (Figure 14). The active region, composed of GaAs/AlGaAs grown on a tilted [001] substrate, form a staircase-like patern of nanometer steps (inset of figure). Highly uniform GaAs quantum wires typically have 6x8 nm cross section with few fractional-layer superlattices layers sandwiched between dielectric mirrors set 400 nm apart. Such devices have lased at room temperature, with a 685 nm wavelength and with a lasing threshold occurring at a pulsed optical - pumping power of 7 pJ/μm^2. The output beam is linearly polarized and parallel to the quantum-wire orientation, in contrast to typical quantum-well lasers that have a random polarization direction. Output polarization properties are unusual-parallel and perpendicular polarization occurs at 710 nm and at 705 nm, respectively. Such laser can be modulated either electrically or optically via slight shifts in the cavity's effective length. Consequently, modulation rates are limited only by microcavity lifetimes, not carrier lifetimes. Modulation rates for quantum-wire lasers are expected to be more than 100 times greater then for current semiconductor lasers.

References

1. Heckt,J.(1993) Gallium arsenide laser offer an array of options,Laser Focus World 29,83-92
2. Friberg,S.(1994) Quantum-wire microcavity laser exhibits unique modulation properties, Laser Focus World 30,80-80
3. Higgins,T.(1995) The smaller,cheaper,faster world of the laser diode,Laser Focus World 31,65-76
4. Jewell,J. and lbright,G.(1992) Surface emitting lasers emerge from the laboratory, Laser Focus World 28,217-223

5.Anderson,S. and Jungluth,E. (1995) CLEO/QELS'95 celebrates the laser's 35 th anniversary, Laser Focus World 31,81-94

NEW INTEGRATED PHOTORECEIVER SYSTEMS - CHARGE COUPLED DEVICES (CCDs)

A. S. POPOV
Physics Department, University of Sofia
Bulgaria

Array detectors have been used in spectroscopic applications for more than ten years to record the spectrally dispersed light from monohromators. The earliest devices were constructed as linear photodiode arrays (typically of 1024x1 elements), which are still used in some moderate-light-level and low-resolution techniques such as absorption spectroscopy, laser-induced fluorescence and chemilumenescence. The photodiode arrays can receive information at many different wavelengths simultaneously giving them an advantage and reducing the time required to complete a spectroscopic experiment or test.

Unfortunately, the signal-to-noise ratio (S/N) of a typical diode-array pixel is very low because of the unit gain and high dark current noise of the photodiode. Consequently, these arrays cannot be used as direct-detection systems with low-light-level applications such as Raman or

M. Balkanski and N. Andreev (eds.),
Advanced Electronic Technologies and Systems Based on Low-Dimensional Quantum Devices, 175–187.

very-high resolution spectroscopy.

CCDs have been especially instrumental in the move toward solid-state imaging because they propose a solution to a fundamental problem of detector arrays: reading the image off the array. This novel approach to image presentation was first proposed in 1970 by Bell lab on the optoelectronic properties of MOS capacitor, which typically consists of an extrinsic silicon substrate and an insulating layer of SiO_2. This device is completed with two metal contacts - one is deposited on the dielectric layer and is called a gate; the other is homic contact to the silicon substrate. The MOS-structure is shown in (Figure 1 top) [1].

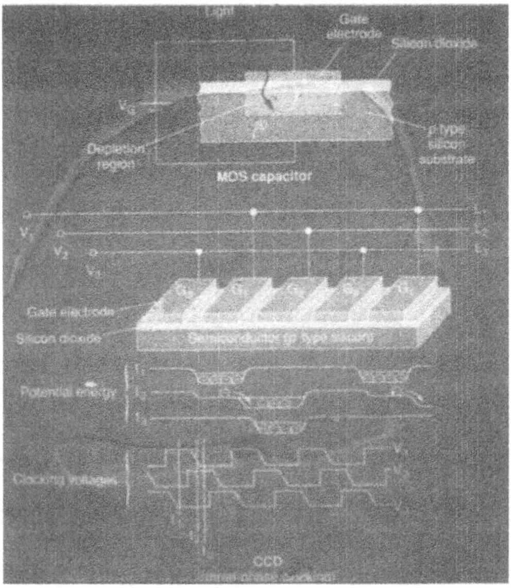

Figure 1. Absorption of light of a p-type MOS-capacitors (top) and of the three-phase clocking process

If the MOS-capacitor is voltage biased the majority charge carriers holes are pushed away from the Si-SiO_2 interface directly below the gate, leaving a region depleted of positive charge and available as a potential-energy well for any mobile minority carriers (electrons). The depth of well increases by the increasing of bias. If the device is irradiated by a light while collect in the potential energy well under the gate. The more light absorbed by the silicon, the more electrons captured at the Si-SiO_2 interface, until the well becomes saturated or the voltage is removed.

1.Architecture of CCD-matrix

A CCD array consists of an array of gates laid over a suitably doped silicon substrate as is shown in Figure 2 [2]. Under every one of gates

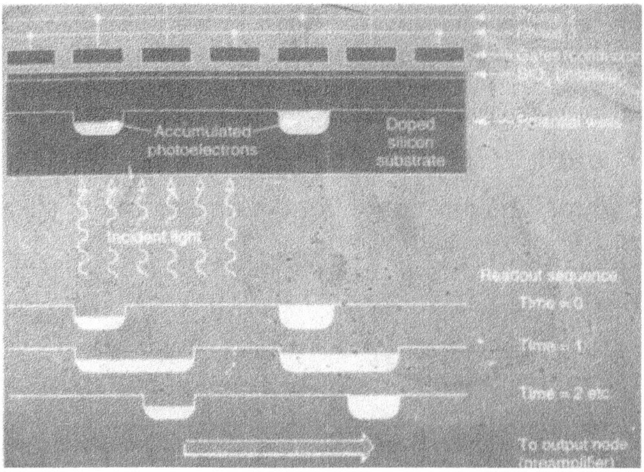

Figire 2. Scheme of the gates displaced on silicon substrate

are formed potential energy well if the voltage is applied between substrate and gates. Incident light through the substrate creates electron-hole pairs, and the electrons collect in the potential wells defined by the gate electrodes is connected by voltage line and they form three column gates as is shown in the Figure 2 [2].

A CCD array of MOS capacitors can store images in the form of trapped charge carriers under the gates. By switching the voltage on the gates of capacitors, the charge collected in the potential wells can be move off the array. This process is realized by sequence shifting of voltage on each gate so as to transfer the charges from one potential well to the next. One of the most successful voltage-shifting schemes is named three-phase clocking as is shown in Figure 1 (bottom) [1].

In a typical two-dimensional MOS array with three-phase clocking, column gates are connected to separate voltage lines L_1, L_2 and L_3 in contiguous groups of three G_1, G_2, G_3. The setup enables each gate voltage to be separately controlled, which is an essential element for moving the accumulated charges off the array.

2. Light sensitivity of CCD matrix

To make the MOS capacitor array sensitive to light, a bias voltage V_1 ia applied to the first line L_1 for a period known as the integration time. During this period all mobile electrons generated by the absorbed light will be collected under all the gates G_1 connected to the L_1-line. For the shifting of accumulated charge from this line to the L_2-line, it must be applied a bias $V_2 = V_1$ on the second line, and after that the bias V_1 becomes less than V_2, the potential wells under gates G_2 will be deeper and all collected electrons under gates G_1 will be translated under gates G_2. repeating the same voltage sequence on L_2-line and L_3-line moves the electrons from G_2-gates to G_3-gates. This process is presented in the bottom of (Figure 1), where the charges of applied voltages on the lines L_1, L_2, L_3 become at t_1, t_2, t_3 time moments. Repeating the entire V_1, V_2, V_3 voltage sequence will methodically "clock" the electrons off the array, where they are then read out as a serial signal by a linear MOS array called a readout register.

3.The principle of readout register

This principle is shown in (Figure 3) [2], with the sequential charge transfer between wells in a CCD array enables the electronic charge image to be read out in the detection process. In the picture is presented a formation and translation of image as dots combination in parallel register. 1st step - After exposure to light an electronic image accumulates as a pattern of charge in the parallel register.

2^{ond} step - Charge is shifted up the parallel register one row. The first row is shifted into the serial register.

3th step - The first pixel of the first row is serially shifted into the output node.

4th step - The charge at the output node is collected for signal processing (presented as a blue triangle).

5th step - The charge from the next pixel of the first row is shifted to the output node (the triangle is red-empty).

6th step - The charge at the output node is collected for signal processing (the triangle becomes blue again) - the same situation as 4th step.

7th step - After all pixel in the serial register are processed, the next row is shifted into the serial register and all steps from 2^{ond} to 7th will be repeat.

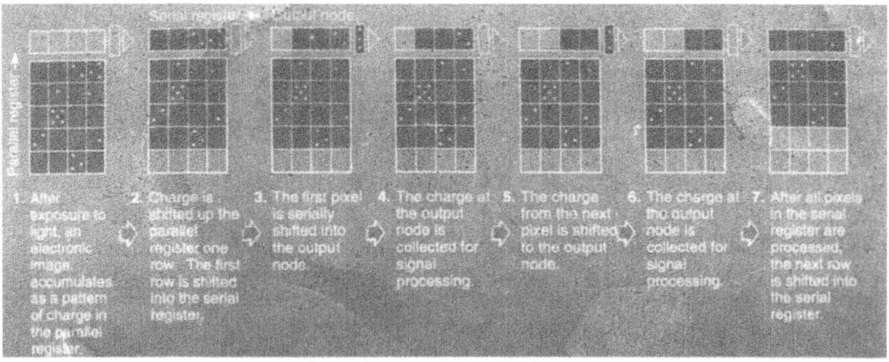

Figure 3. The principal of readout register

4.Charge transfer options

Because all of the gates in each row of the CCD matrix are connected in series, moving the accumulated charge packets in a columnar direction causes them to exit the array one row at a time as is shown in the top of (Figure 4) [1]. This method of charge transfer is called line-addressed transfer.

One problem with this method is that if light continues to shine on the array while previously accumulated charges are being clocked off, new charges can accumulate under the gates whenever voltage is applied to them. This process causes image smearing and the solution of the problem is to quickly move the charges off the array before too many new ones can gather, but this is very difficult for realization. Another solution is to mechanically cover all active area of the CCD matrix between integration times. But this is possible for a long integrations times and not very practical for a short ones. One of the most effective solution is the practical-transfer CCD, where the charges collected during integration are transferred into shielded columns of MOS capacitors, or vertical shift registers, that are interlaced between the light collecting columns of pixels. The scheme of this solution is presented in the middle of *Figure 4). From these shift-registers, the stored charges can be moved off the array while the exposed pixels integrate another image. This solution uses a big part of

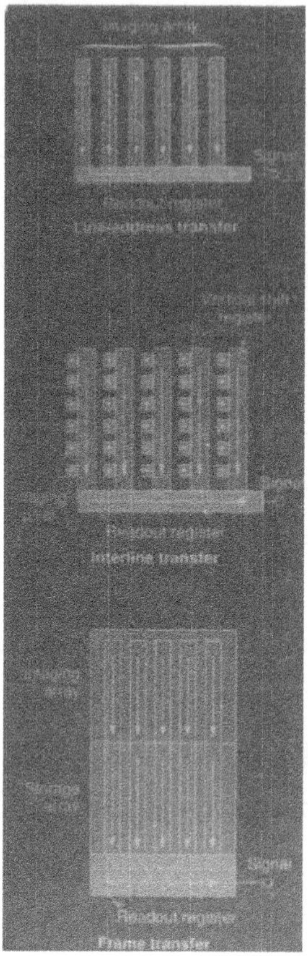

Figure 4. Three basic types of CCD charge-transfer schemes

area of CCD-matrix not for photodetection, thereby limiting the responsivity and resolution of the device.

The best solution at the moment is frame-transfer method which provides a fill factor of 100% sensitivity of CCD area. Frame-transfer CCDs quickly move all the accumulated charges into an adjacent shielded storage array the same size as the light-gathering pixel array. This structure is shown in the bottom of (Figure 4). While the pixels integrate the next image, the stored image is clocked off to the readout register. The frame-transfer method is slower than interline transfer method and also has a small image smearing. All three methods are used in CCD matrix at different schemes of application.

The charge injection device - CID is another important charge-transfer system. It operates differently from CCD because the charge is not transferred off the array. The pixels of CID consist of two MOS capacitors whose gates are separately connected to rows R_1, R_2, R_3 and columns C_1, C_2, C_3, as is shown in the top of Figure 5 [1]. Usually, the column capacitors are used to integrate charge, while row capacitors sense the charge after integration. CID readout is done by sequentially transferring the integrated charges from the column capacitors to the row capacitors, which alters the potential on the row in proportion to the amount of charge transferred. After this nondestructive signal readout, the charges can be moved back to the columns for more integration or injected back into the silicon substrate. A big advantage of the CID is that each pix el is X-Y addressable. CIDs also have higher noise, but lower blooming than CCD, because the charge spill-over between pixels.

5.CCD adaptability

Besides their low noise and good sensitivity, the greatest advantage of CCDs is their extraordinary versatility. A side from the three-phase clocking scheme described here, there is two-phase clocking, for faster image readout and four-phase clocking, named double clocking. This solution is presented at the bottom of Figure 5 [1], where on

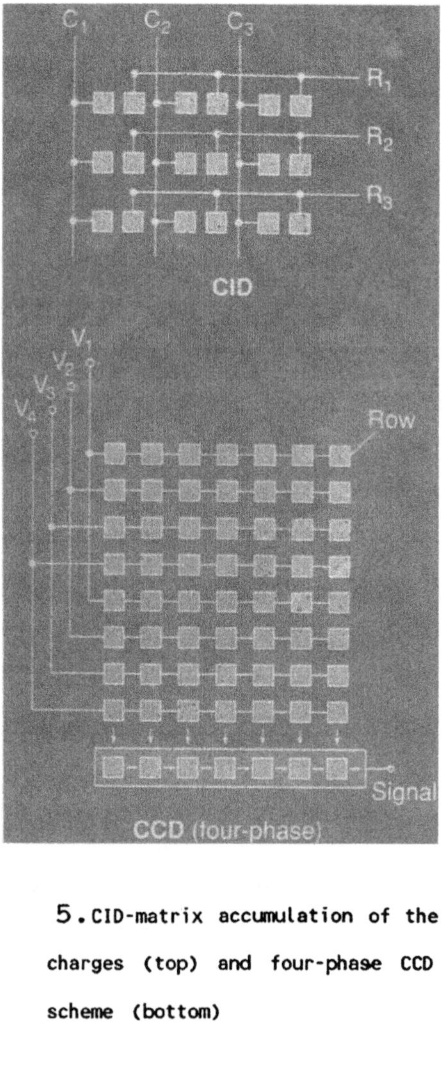

Figure 5. CID-matrix accumulation of the charges (top) and four-phase CCD scheme (bottom)

the top is shown CIDs-matrix. The CID-matrix sense charges that accumulate under the column gates - C_1, C_2, C_3 by simply transferring them under the row gates - R_1, R_2, R_3. This avoids some of the problems caused by moving charges across the entire array (such as with a four phase CCD) and makes each pixel X-Y addressable. Double clocking enables greater -charge-handling capacity, which is critical for IR imaging applications.

CCDs can be illuminated either from the front or the back. In front-illuminated CCDs, the light enters through the gates electrodes, which must therefore be made of a transparent material such as polysilicon. Back-illuminated CCDs avoid the interference and absorption effects caused by the gates but must be thinned for imaging in the visible and near IR regions.

CCDs also can be integrated with other photodetector arrays either monolithically or as a hybrid. Monolithic arrays combine the detector and CCD structures on a single chip, while hybrid arrays sandwich them together as two separate chips. Hydrides avoid some of engineering difficulties associated with growing different materials on a single chip and provide a convenient bridge between well-developed but otherwise incompatible technologies.

6.The most important parameters of CCD

1.Dark noise - arises from statistical variations

in the dark current. Dark current is defined as the number of thermally generated carriers (N) per pixel well per unit time, with the dark noise given by $N^{1/2}$. The reduction of dark current (thermally generated dark count) was made by operation of CCD with liquid nitrogen. The reduction has been made possible also by multipinned-phase device architecture, where most of the thermally generated electrons recombine with holes before they can enter a potential well.

2.Readout noise - this is the noise (in electrons) associated with each individual readout event. In the new CCD matrix a readout noise is only 4 $\bar{e}$ rms. The reduction of this parameter makes less the time which is required to accumulate spectral data with a given S/N.

With recent advances in semiconductor technologies and microlens arrays, solid state imaging arrays with lower noise, better sensitivity, higher fill factor and greater complexity have become possible. Even arrays containing 64 million pixels are within reach today. CCDs are now supplied as complete hardware/software systems for integration with a PC.

3.Binning - The process of combining charge from adjacent pixels into superpixels during the readout process. Binning is an important method of increasing the S/N of the CCD output at the expanse of spatial resolution. Because the readout noise is independent of the size of the superpixel, the S/N is improved.

Binning is useful in applications in which resolution in one or both axes is not of primary concern. For spectroscopy, this axis is usually the vertical nondispersed axis. Consequently, binning can be used to increase the S/N for a single slit image by summing each column of the CCD as a single superpixel. Even when recording multiple spectra from fibre probes or spatially varying sources, the spectra usually span several rows and can be enhanced by binning. Because of the number of charge transfer steps required, binning is much faster when the serial register is orthogonal to the binning direction: that is, if a column is to be summed or partially summed, than the serial register should be parallel to the rows.

References

1. Higgins,T. (1994) The technology of image capture, Laser Focus World 30,53-60
2. Prettyjohns,K. (1992) Improved CCDs meet special demands of spectroscopy, Laser Focus World 28,127-136

OPTICAL SWITCHES AND MODULATORS FOR INTEGRATED OPTOELECTRONIC SYSTEMS

A. S. POPOV
Physics Department, University of Sofia
Bulgaria

Optical switches and modulators are essential elements in lightwave systems for communications, spectroscopy and instrumentation. To fully exploit the potential of fibreoptic communications, these components must provide faster transmission rates and higher band widths. But one of the most important problem for these devices is integration with active optoelectronic devices into integrated circuits. Optical waveguide switches integrated with transistor drivers and photodetectors and interferometric modulators integrated with laser diodes have been demonstrated recently. The potential for lower price is monolithic integration on the base of semiconductor materials, where the added components will be defined by photolithography.

In this lecture we shell discuss the development of semiconductor optical switches and modulators in aspect of miniaturization for including into integrated optoelectronic chips.

M. Balkanski and N. Andreev (eds.),
Advanced Electronic Technologies and Systems Based on Low-Dimensional Quantum Devices, 189–200.

1. Performance of optical switches

Existing optical-waveguide switches are defined into four categories:

i - gain switches - they amplify or block the signal;

ii - interferometric switches - they switch through phase - shift control;

iii - internal reflection switches - they internally deflect the signal to an output path;

iiii - modal evolution switches - they electrically induce index asymmetry to route the signal, All four types are presented in Figure 1.

First type - gain switches passively split the signal and use optical amplifiers in the output waveguides for either amplification or absorption of the signal. In amplification waveguides, the signal is passed through. while in the absorption waveguides the signal is blocked. In the gain switches the amplifier's spontaneous emission noise is progressively added to the signal. Moreover these switches have power requirements for driving the optical amplifiers.

In the case of interferometric switches there is require a precise externally controlled phase shift to achieve a switched state with low crosstalk. The magnitude of the phase shift critically depends on the specific parameters of every one device and in addition, the needed phase shift is different at different wavelengths and polarizations. In the case of multiple-wavelength networks these switches are less useful. For large scale applications interferometric switches are generally fabricated in lithium niobate.

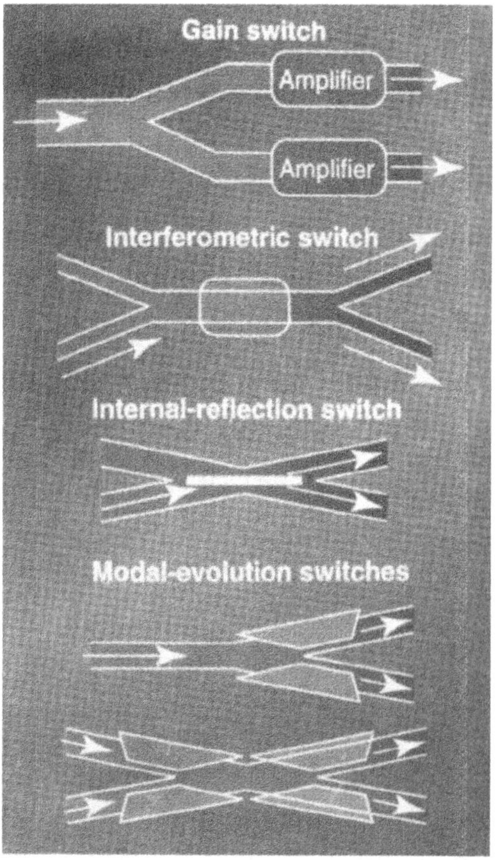

Figure 1. Four types of optical-waveguide switches [1]

Internal reflection switches consume much higher powers than gain switches to achieve total internal reflection between intersecting waveguides.

Modal-evolution switches select the propagation path of an optical signal through two intersecting waveguides by including a small index asymmetry between the waveguides. This asymmetry arises by injecting current through an electrode into the waveguide or by reverse biasing the electrode to deplete charge carriers in the waveguide. With no injected current, the output waveguides are symmetric, and the incoming light is equally divided between waveguide paths. By injecting current in one output waveguide, the symmetry is broken and light propagates down the waveguide with the large index. The physical mechanisms by which the current reduces the index in the injected waveguide are independent of wavelength and polarization, the switching process is also independent. These devices are very fast and they have possibilities for construction of digital optical switches. Digital optical switches have been demonstrated on the base of lithium niobate, Si and InP. The advantage of semiconductor optical switches is a possibility of direct including with active optoelectronic devices into integrated circuits. In Figure 2 [1] is presented the principal scheme of digital optical switch and structural details are shown in Figure 3[1]. There the vertical optical confinement is provided by an InP/InGaAsP heterojunction layer structure. The epitaxial layer structure contains a thin InGaAsP etch stop

layer that provides light control over

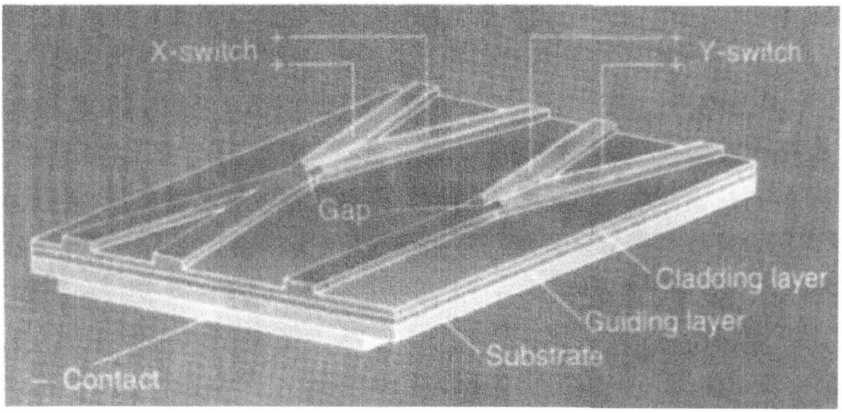

Figure 2. Principal scheme of digital optical switch with a ridge-waveguide structure[1]

ridge-waveguide dimensions to ensure single-spatial-mode operation and high extinction ratio.

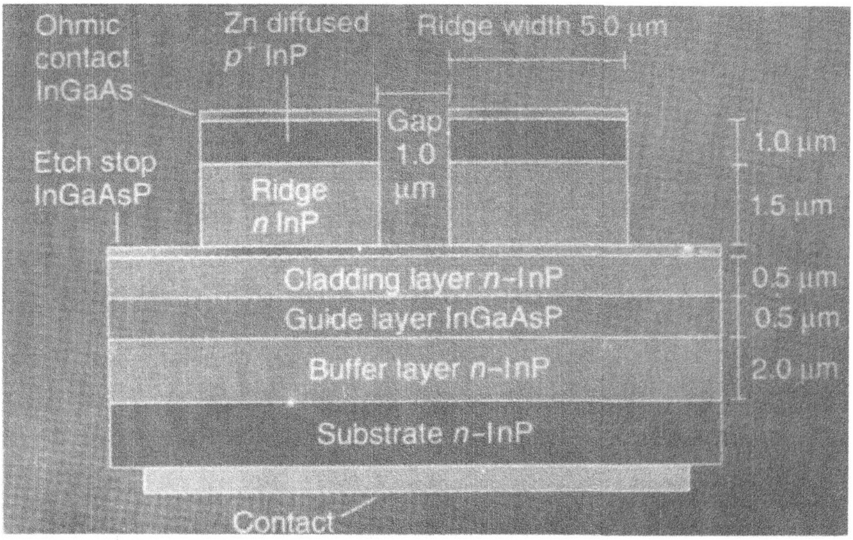

Figure 3. Structural details of digital optical switch [1]

2. Performance of optical modulators

Another very important elements in the integrated optoelectronic circuits are optical modulators which provide a simple technique to produce totally chirp-free signals in optical telecommunications. Modulators are currently the most commonly used integrated optical components. They function by controlling the amount of light transmitted into of waveguide from a continuous wave (CW) laser. The light transmitted through the modulator can be smoothly varied from zero to the maximum intensity. This transfer characteristic is a fundamental property of the modulator structure and does not vary from device with signal frequency, signal format environmental conditions. Most systems that use external modulators take advantage in relation to zero chirp. Wavelength chirp - the shifting and broadening of a laser's linewidth due to rapidly changing drive currents. Using symmetrical modulator designs and high return loss interfaces, it is possible to eliminate wavelength chirp. Separating the light generation and signaling functions produces a number of additional advantages. These include access to modulation depths greater then 20 dB, very high modulated optical powers, simplified drive electronics and excellent stability of the transmitter to the temperature and environmental fluctuations.

3.Modulator operating principles (Mach-Zehnder modulator)

Mach-Zehnder modulator is one of the most applicable structures in the optoelectronic integrated circuits. As a devices, they exist in two kind of constructions - integrated and hybrid which are implemented on the base of $LiNbO_3$ or different semiconductor materials.

The scheme of integrated Mach-Zehnder modulator is presented in Figure 4[2]. It

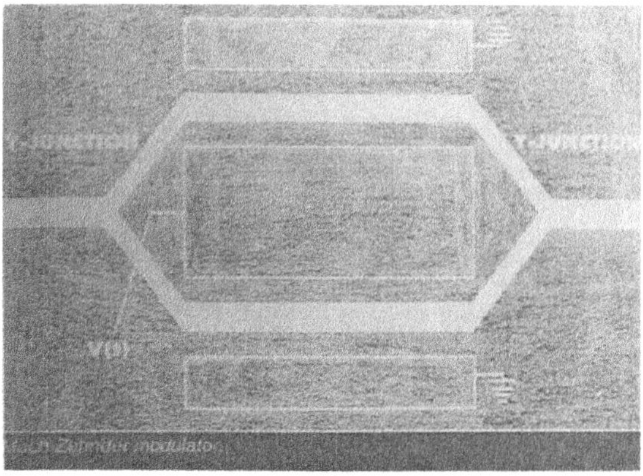

Figure 4. Integrated Mach-Zehnder modulator [2]

comprises two phase modulators. one splitter and one combiner and three operating electrodes. A voltage **V(t)** is applied to the central electrode, and the optical signal is initially split into two equal portions at the **Y** - junction. In accordance with the electro-optic effect, the induced index change causes the phase of the optical signals to be advanced in one arm and retarded in the other. When the signals are recombined in the combiner, they are coupled into the single-mode output guide if they are in phase. If they are out of phase, they are transformed into a higher-order mode and lost into the substrate. For phase differences between these two extremes, only a portion of the light is coupled into the output waveguide.

The scheme of hybrid Mach-Zehnder modulator is presented in Figure 5[2]. This is an enhanced

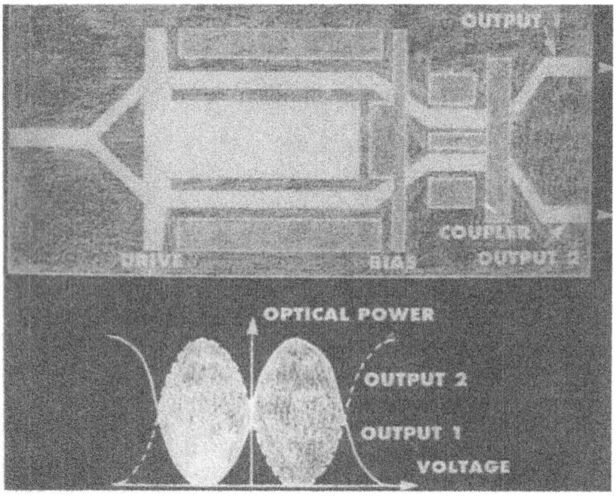

Figure 5. Hybrid Mach-Zehnder modulator [2]

modulator structure for AM applications, where the output junction is replaced by a coupler and two output waveguides. The advantage of this arrangement is that the light that is normally directed into the substrate and lost is guided by the additional output waveguide. The light in this guide is the exact complement of the light in the other. This arrangement is known as a balanced-bridge modulator and it is convenient for analog modulation applications in that the light from both output waveguides can be used effectively doubling the RF signal power available for transmission. In Figure 6[2] is presented

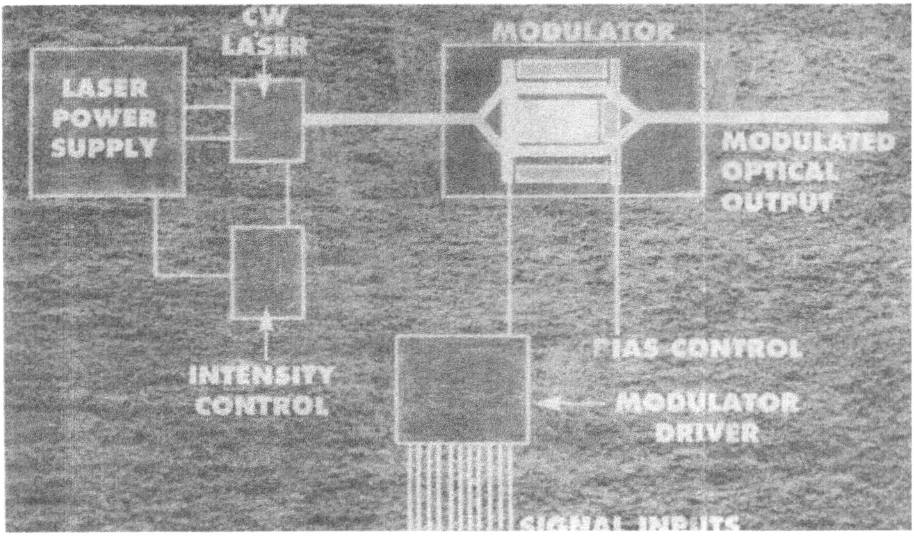

Figure 6. Integrated optical modulator in a digital transmitter [2]

optical modulator in a digital transmitter and in Figure 7[2] - integrated optical modulator in an analog transmitter.

The potential for lower price is the most-promising aspect of monolithic integration with added components defined by photolitography. This process is available only for semiconductor materials, especially A^3B^5 compounds and its solid solutions. These materials offer practical advantages as a result of well developed fabrication techniques used to make microminiature devices with increased functionality and less power consuming. A variety of active devices and

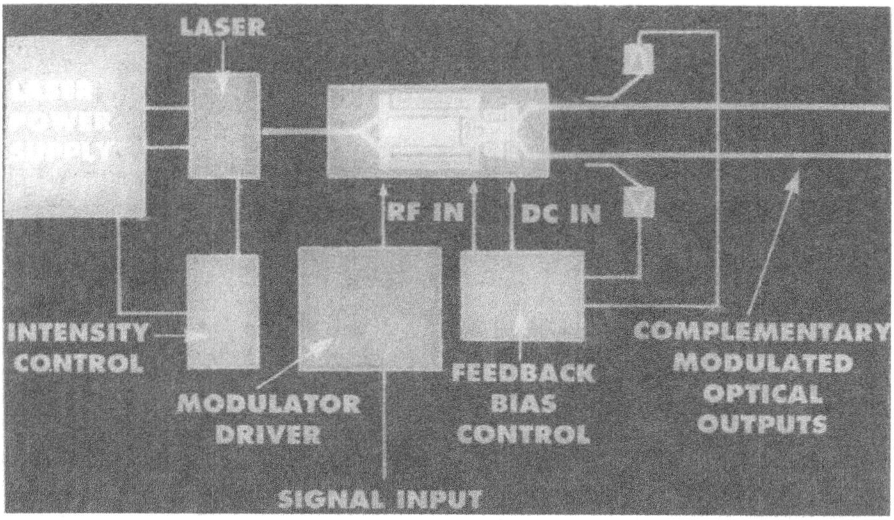

Figure 7. Integrated optical modulator in a analog transmitter [2]

passive guide structures including channel guides, directional couplers, Y- splitters and wavelength filters have been made from direct-bandgap such as GaAs, GaAlAs, InP, InAsP and GaInAsP. Some distinctive properties of semiconductors such as electroabsorption and multiple quantum wells can also be used to make advanced modulators.

In A^3B^5 compound semiconductors, the localized refractive index can be changed by varying the concentration of free carriers, by using electric-field effects, or by varying the material composition. Because of the relationship between free carrier concentration and refractive index, light is guided in layers with lower carrier concentration. Electric field induced changes in n- occur through the linear electrooptic effect, which is primarily used for phase modulation. In Figure 8[3] a Mach-Zehnder

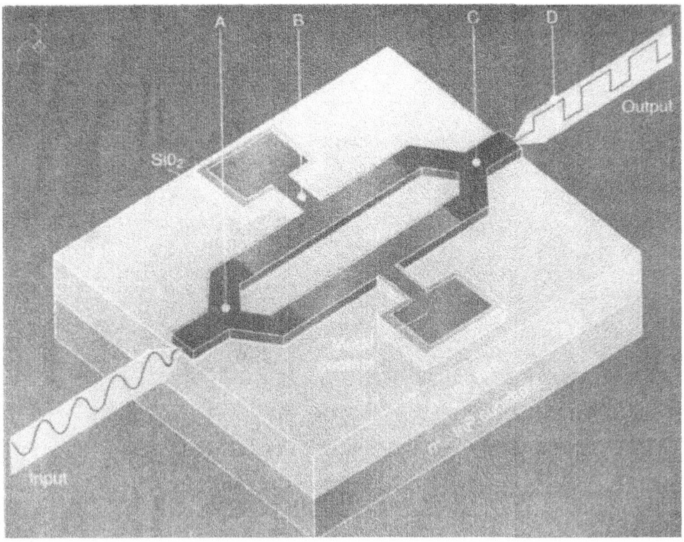

Figure 8. Scheme of semiconductor Mach-Zehnder optical modulator [3]

optical modulator made from an InGaAs is shown. The input light splits at the Y- branch (A) and recombines at the output (C). An electric field-induced 180°C phase shift applied at (B) causes optical destructive interference at (C) resulting in a 0- bit optical pulse. If the signal are phase-matched, constructive interference occurs, and the signal is 1- bit pulse. By using of conventional high technology methods in semiconductor industry is possible to obtain integrated optoelectronic schemes which include both active and passive devices. In Figure 9[3] is presented the above discussed an natural Mach -

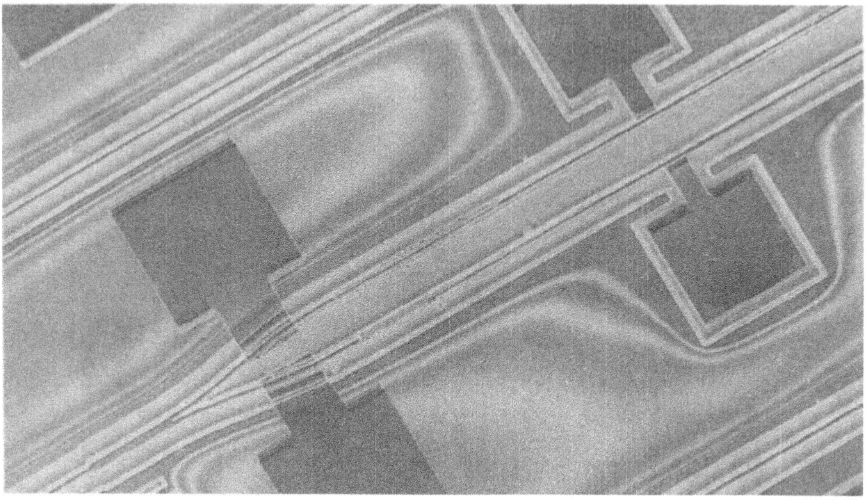

Figure 9. Natural Mach-Zehnder modulator in optoelectronic IC [3]

Zehnder modulator, which extends the signal-transmission speed to 10 Ghz/s and transmission distance to 100 km with minimal signal loss.

References

1. William, H. and Masum, A. (1994) Optical switching expands communications network capacity, Laser Focus World 30, S17-S20
2. Powell, M. (1993) An ally for high speed and the long haul, Photonics spectra 27, 102-108
3. Jungbluth, E. (1994) Optical waveguides advance to meet fiberoptic damands, Laser Focus World 30, 99-104

Scalar Off-Resonant Modulation Instabilities in General Rare-Earth Doped Fiber Amplifying Devices

T. MIRTCHEV

Quantum Electr. Dept., Sofia University, Sofia, BULGARIA
Phones: 359-2-62561887, Fax: 359-2-9625276, e-mail: todorm@phys1.phys.uni-sofia.bg

1. Introduction.

Recently, after the great progress of the rare-earth doped fiber amplifiers, the research on modulation instability (MI) has been renewed. As expected, the MI in such media demonstrates some completely new features like the expanding spectrum of the scalar and XPM-induced MI. However a complete and universal description of the MI specifics in amplifying fibers is still needed. The results published to date are based on simplified models of the amplifier gain [1]. They ignore the near-resonant refractive index dispersion (NRRID), which is allowable only if an exactly resonant propagation in amplifier with ideally symmetrical spectrum is presumed. In the same time, the position, shape and bandwidth of the gain spectrum in erbium-doped fibers, depend significantly on many parameters like the glass composition of the fiber, pump power, etc. [2]. Thus it seems difficult to assure the strictly resonant case in a real communication system. Except that, the off-resonant amplification is a prerequisite for the wavelength-division multiplexed systems.

The objective of this work is to describe the dynamics of the MI in a general nonlinear, dispersive, bandwidth-limited homogeneously broadened optical fiber amplifier. The approach is based on the traditional linear-stability analysis of a modified nonlinear Schrödinger equation (MNLSE), including rigorously both the gain dispersion and the near-resonant refractive index dispersion. Thus the derived expressions are valid for arbitrary detunings of the input wave from the resonance and arbitrary bandwidth of the amplifier. Although the results are written here for an amplifier with a simple Lorentzian spectrum, they can easily be applied to

M. Balkanski and N. Andreev (eds.),
Advanced Electronic Technologies and Systems Based on Low-Dimensional Quantum Devices, 201–205.

describe the MI in amplifiers with complex gain spectrum, as the real Er^{3+}- doped fibers.

2. SPM-induced MI in rare-earth doped fiber amplifiers.

A basic part of the present model is the starting modified nonlinear Schrödinger equation:

$$\frac{\partial A}{\partial z}+\frac{i}{2}\beta_2\frac{\partial^2 A}{\partial t^2}=i\gamma|A|^2A+\frac{1}{2}g_0\left\{\sum_{m=0}^{\infty}L^{m+1}\left(T_{21}\frac{\partial}{\partial t}\right)^m\right\}A \qquad /1/,$$

where A(z,t) is the slowly varying amplitude of the signal, β_2 is the usual nonresonant group velocity dispersion, γ is the nonlinear coefficient, T21 is the phase - relaxation time and g0 is the maximal gain coefficient. The factor $L=\left[1-iT_{21}\Delta\right]^{-1}=L^R-iL^I$ is an additional Lorentzian factor, which includes the detuning $\Delta=\omega L-\omega 21$ between the carrier frequency ωL and the resonance ω_{21} and allows to describe easily the off-resonant cases.

Fig. 1

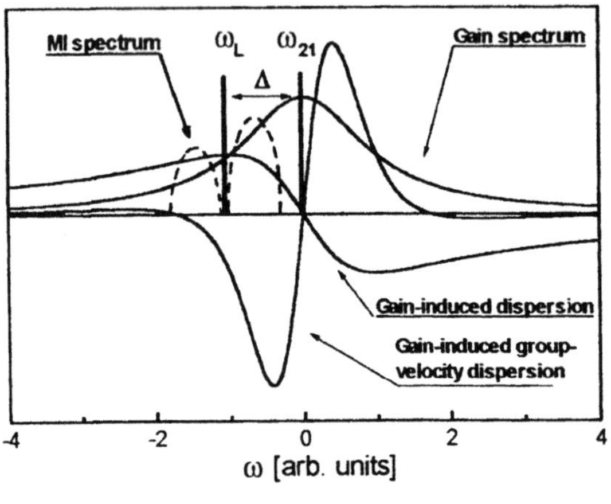

The complex infinite series in the right-hand side of /1/ result from the reverse Fourier transform of the expanded around ω_L complex gain spectrum and precisely describe the dispersive properties of the amplifier in the time domain. Fig 1 gives a general illustration of the described problem.

A standard linear stability analysis is applied to Eq. /1/. It results in the following dispersion expression

$$k(z,g_0,T_{21},\Omega,\Delta)=-\frac{i\,\mathrm{Re}(G)}{2}\pm\left[\frac{\beta_2}{2}\Omega^2-\frac{1}{2}\mathrm{Im}(G)\right]\sqrt{1+\frac{2\gamma P_0e^{g_0L^Rz}}{\frac{\beta_2}{2}\Omega^2-\frac{1}{2}\mathrm{Im}(G)}} \qquad /2/,$$

where k and Ω are the wave vector and the frequency of the perturbation a(z,t)=u+iv respectively. The quantity G is introduced as follows

$$g_0 L \sum_{m=1}^{\infty}\left[L^m\left(T_{21}\frac{\partial}{\partial t}\right)^m\right](u,v) = g_0 L \sum_{m=1}^{\infty}\left(-iT_{21}L\Omega\right)^m (u,v) = G(g_0, T_{21}, \Omega, \Delta)(u,v)$$

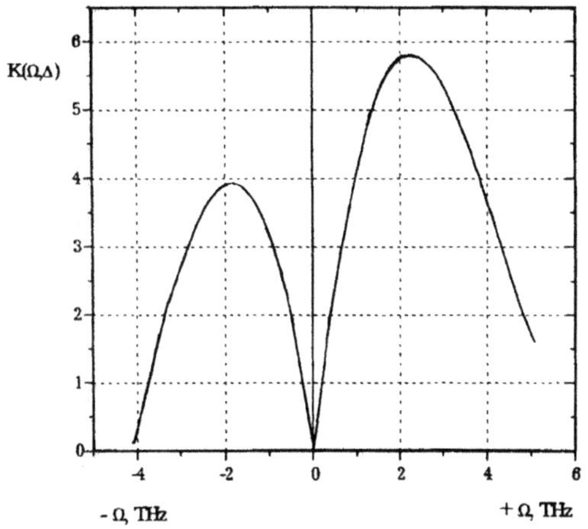

The physical meaning of the term with Re(G) is that it gives the difference between the usual stimulated emission gain coefficients of the input wave and the developing Stokes or anti- Stokes components. Its sign depend strongly on Ω and Δ, which reflects the fact that the wings of the developing spectrum can experience higher gain than a significantly detuned input wave. The rest of the right-hand side of Eq. /2/ is the contribution of the nonlinear effect of MI. Phenomenologically, only the even degree terms are used in the summation of Im(G), which is the phase-mismatch of the MI process when treated as a SFWM [3]. If the term under the square root is negative than the MI will grow exponentially. In the usual case of non-doped fibers this condition can be met only in the anomalous GVD region [3]. The influence of the NRRID of the amplifying transition (Im(G)) in fiber amplifiers alters significantly the overall phase-mismatch of the process and introduces some qualitatively new possibilities:

- ***The NRRID can suppress the MI in the anomalous GVD region.***
- ***The NRRID can induce MI in the normal GVD region.***
- ***The MI spectra in fiber amplifiers are inherently asymmetric, except in the case of exactly resonant propagation.***

The exact conditions for the onset of these effects can be easily expressed, in terms of the physical parameters of the amplifier - g_0, T_{21} and Δ. Fig 2. shows an example of a typical MI spectrum for g_0=0.15m^{-1}, T_{21} = 100fs, amplifier length l = 50m, Δ=5THz, P_0=100mW, β_2 = - 0.06ps^2/m and γ = 0.01 W^{-1}/m.

The dynamics of the process can be described also in the terms of the integral gain $K(g_0, T_{21}, \Omega, \Delta) = -2\int_0^l \mathrm{Im}\, k\, dz$. **Introducing the dimentionless normal frequency** $f^2 = \Omega^2 + \mathrm{Im}(G)/|\beta_2|/\Omega_c^2$, **K can be written as:**

$$K = \frac{8fl}{l_{NL}\ln C}\left(\sqrt{C-f^2} - f\arccos\frac{f}{\sqrt{C}} - \sqrt{C^{*}-f^2} + f\arccos\frac{f}{\sqrt{C^{*}}}\right) + l\,\mathrm{Re}(G) \qquad /3/,$$

where $l_{NL}=(\gamma P_0)^{-1}$ **is the nonlinear length,** $C = \exp(g_0 L^R l)$, $C^*=1$ **for** $f\le 1$ **and** $C^*=f^2$ **for** $f>1$.

3. XPM-induced MI in rare-earth doped fiber amplifiers.

A further generalisation of the problem of nonlinear wave propagation in rare-earth doped optical fiber amplifiers is to study the MI in the field of two co-propagating signal waves, which interact through XPM. When the two waves are allowed to be independently and arbitrarily detuned from the resonance, the problem is described by a pair of coupled equations similar to /1/ that is:

$$\frac{\partial A_1}{\partial z} + \frac{i}{2}\beta_{21}\frac{\partial^2 A_1}{\partial t^2} = i\gamma\left(|A_1|^2 + 2|A_2|^2\right)A_1 + \frac{1}{2}g_0\left\{\sum_{m=0}^{\infty} L_1^{m+1}\left(T_{21}\frac{\partial}{\partial t}\right)^m\right\}A_1$$

$$\frac{\partial A_2}{\partial z} + \frac{i}{2}\beta_{22}\frac{\partial^2 A_2}{\partial t^2} = i\gamma\left(|A_2|^2 + 2|A_1|^2\right)A_2 + \frac{1}{2}g_0\left\{\sum_{m=0}^{\infty} L_2^{m+1}\left(T_{21}\frac{\partial}{\partial t}\right)^m\right\}A_2$$

All of the variables and parameters in the set have the same meanings as in the previous section but $A, \beta_2, L, \Omega, \Delta$ **now carry an additional index j=1,2 to distinguish the two different waves. After some transformations, similar to these given above, and after introducing (to simplify the notations) the quantities**

$$g_j = \frac{\beta_j}{2}\Omega_j^2 - \frac{1}{2}\mathrm{Im}(G_j),\; h_j = 1 + \frac{2\gamma P_{j0} e^{g_0 L_j^R z}}{\frac{\beta_j}{2}\Omega_j^2 - \frac{1}{2}\mathrm{Im}(G_j)},\quad q = 16\gamma^2 P_{10} P_{20} e^{g_0\left(L_1^R + L_2^R\right)z}$$

the following expression for the complex wave vector of the signals is obtained:

$$k_j(z,g_0,T_{21},\Omega_j,\Delta_j) = -i\frac{\mathrm{Re}(G_j)}{2} \pm \left\{ \frac{\left(g_1^2h_1 + g_2^2h_2\right) \pm \left[\left(g_1^2h_1 - g_2^2h_2\right)^2 + 4g_1g_2q\right]^{\frac{1}{2}}}{2} \right\}^{\frac{1}{2}} \qquad /4/$$

It depends complexly on the amplifier parameters and multiple qualitatively different regimes of stability and instability may exist depending on the amounts of detuning, gain coefficient, input power, etc.

The expressions /2/, /3/ and /4/ are a closed form analytical solutions for the gain coefficients of the SPM- and XPM-induced MI of nonlinear waves propagating nonresonantly in a general, dispersive rare-earth doped fiber amplifier.

Acknowledgments:

This work was supported by the Bulgarian National Science Foundation with grant F-476.

References:

1. **Agrawal, G . (1992) Modulation Instability in Erbium-Doped Fiber Amplifiers, *IEEE Phot. Tech. Lett.* 4 (6), 562-564**
2. **Miniscalco, W. (1991) Erbium-Doped Glasses for Fiber Amplifiers at 1500nm., *Journal of Lightwave Tech.* 9 (2), 234-250**
3. **Agrawal, G (1989) *Nonlinear Fiber Optics*, Academic Press Inc., San Diego, CA**

QUANTUM DOT LASER WITH HIGH TEMPERATURE STABILITY OF THRESHOLD CURRENT DENSITY

A.R.KOVSH, A.E.ZHUKOV, M.A.ODNOBLYUDOV, A.YU.EGOROV, V.M.USTINOV, N.N.LEDENTSOV, M.V.MAKSIMOV, A.F.TSATSUL'NIKOV, N.YU.GORDEEV, S.V.ZAITSEV, P.S.KOP'EV
A.F.Ioffe Physico-Technical Institute,
Politekhnicheskaya 26, 194021, St.Petersburg, Russia

Quantum dot heterostructures have recently become the subject of extensive research in semiconductors physics. These structures are especially attractive for their laser application. The reduction in threshold current density (J_{th}) and increase in its temperature stability have been predicted theoretically [1]. Characteristic temperature (T_0) of laser with active region based on ideal QDs has to be infinity owing to the delta-like density of states. Laser with active region based on the In(Ga)As QD array in GaAs matrix has shown high T_0 and low J_{th} at low observation temperature. Stacking the QD sheets allowed us to reduce J_{th} and to extend the range of high T_0 (T_0~400 K from 77K to 180K). However, at higher observation temperatures J_{th} increases more rapidly due to thermal evaporation of carriers into the wetting layer and GaAs states. (Wetting layer (WL) is ultra thin two-dimensional InAs layer formed in such system). The reason for this fact is insufficient electronic confinement in In(Ga)As QDs.

In this paper we propose the way to increase the localization energy of QD relative to WL and GaAs by inserting the InAs QD array into external AlGaAs/GaAs quantum well (QW). In this case, the WL level shifts stronger than the QD one, because the former is less localized than the latter. Moreover, three-dimensional density of states of GaAs is replaced by the two-dimensional one.

To estimate theoretically the effect of external quantum well on the WL level, we used the concept of zero radius potential (ZRP) [3] to calculate binding energy in WL in the presence of external AlGaAs energy barriers. To compare theoretical data with experiment, the samples containing 1 and 1.7 monolayer (ML) of InAs inserted into GaAs or the middle of $Al_{0.22}Ga_{0.78}As$ QW were grown by MBE under the same growth conditions. 1ML of InAs is equivalent to the WL thickness (in this case no QD are formed), whereas 1.7 ML is sufficient to form quantum dots, (i.e., QDs on WL). The corresponding PL peak positions and 1ML transition energy calculated within the ZRP concept given in table 1 are in good agreement. It is seen that QD array in the external QW leads to the increase the energy separation between the WL and QD levels in agreement with our model.

M. Balkanski and N. Andreev (eds.),
Advanced Electronic Technologies and Systems Based on Low-Dimensional Quantum Devices, 207–208.

Table 1. Transition energies of InAs QD and WL (1ML of InAs)

type of structure	PL peak position		calculated 1ML transition energy, E_2^{th}, eV	E_1-E_2, eV
	1.7 ML InAs E_1, eV	1 ML InAs E_2, eV		
QD in bulk GaAs	1.266	1.443	1.440	0.177
QD in QW	1.268	1.470	1.464	0.202

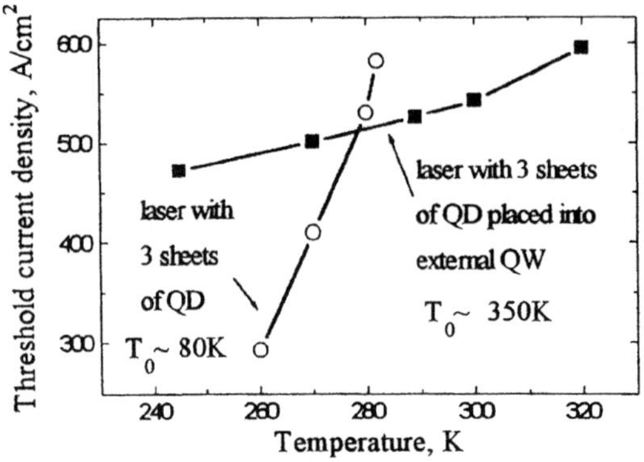

Fig.1. Temperature dependences of Jth of QD lasers

Thus, in this case the population of higher energy states should decrease.

GRIN SCH laser with active region based on three sheets of InAs vertically coupled QD (VECOD) was grown by MBE. Each sheet of QD was placed into the middle of 10 nm $Al_{0.22}Ga_{0.78}As$/GaAs QW. Temperature dependences of J_{th} for this structure and for the structure with the three QD sheets in bulk GaAs are shown in fig. 1 near room temperature (RT). One can see that laser with QD in QW has much better temperature stability of threshold current density. We believe the reason for this is the decrease of thermal evaporation of carriers from QD.

In addition our theoretical and experimental studies have shown that inserting the QDs into bulk AlGaAs should lead to the stronger increase in energy separation between the levels of AlGaAs continuum, WL and QDs. As a result laser with three sheets of InGaAs-$Al_{0.15}Ga_{0.85}As$ VECODs shows RT J_{th} as low as 63 A/cm^2 which is approximately one order of magnitude less than that of InGaAs-GaAs VECOD laser.

In conclusion modification of QD energy spectrum by external energy barriers has been studied theoretically and experimentally. Placing In(Ga)As QD in external GaAs/AlGaAs QW or in AlGaAs matrix leads to the decrease of thermal evaporation of carriers from QD states and high T_0 and low J_{th} at RT in the case of the laser with the QD active region.

This work is supported by INTAS (Grant 94-1028) and Russian Foundation for Basic Research.

[1] Arakawa, Y. and Yariv, A. (1986) QW lasers: Gain, spectra, dynamics, *IEEE J. Quantum Electron.* QE-22, 1887-1899.

[2] Ustinov, V.M., Egorov, A.Yu., Zhukov, A.E., Ledentsov, N.N., Maksimov, M.V., Tsatsul nikov, A.F., Bert, N.A., Kosogov, A.O., Kop`ev, P.S., Bimberg, D., and Alferov, Zh.I. (1995) Formation of stacked self-assembled InAs quantum dots in GaAs matrix for laser applications, *Proc. MRS*, Nov.27-Dec.01, 1995, Boston, USA, EE3.6.

[3] Bethe H.A. and Peierls P.E. 1935 *Proc R.Soc A* 148, p. 146.

SOLITON-BASED LOGIC GATES AND SOLITON TRANSMISSION SYSTEMS

A D BOARDMAN, R PUTMAN and K XIE
Photonics and Nonlinear Science Group
Joule Laboratory
Department of Physics
University of Salford
Salford, M5 4WT
United Kingdom

Abstract

This chapter provides access to basic envelope solitons concepts, the motivation for their study in the area of electromagnetic wave transmission and, finally, the elements underpinning the proposed use for solitons in telecommunications. Since temporal envelope solitons are the basic physical entities that will be used, quite a lot of attention is paid to them, with a view to equipping the reader with basic system tools, quite rapidly. Logic gates are then introduced in a simple, and well-known, way but the ideas behind soliton dragging gates are exposed through numerical studies of soliton interactions. Two major possible constraints on a soliton transmission system, namely the Gordon-Haus effect and self-frequency shifting are discussed, in detail, and remedies to overcome these problems are outlined. The chapter ends with a discussion of soliton transmission systems and the chances of practical implementation, in the near to medium future. It is emphasised, throughout, that although semiconductors, especially low-dimensional ones, are integral parts of a complete soliton system, the approach here is to concentrate on what the system actually does. Hence there is a pervasive emphasis on solitons, rather than the semiconductor materials.

1. Introduction

Since the word soliton [1-3] appears in the title it is entirely appropriate to begin by asking the question "what are they?" It is also appropriate, in a general discourse upon low-dimensional semiconductors in systems, to question the connection of solitons, in optical fibres, to semiconductors and, more importantly, to ask if solitons have any practical uses. The answer to the last question is to say that solitons are serious candidates as "information bits" in transoceanic and land-based telecommunication systems by the year 2000. Of course, the year 2000 is currently used by everyone as a

M. Balkanski and N. Andreev (eds.),
Advanced Electronic Technologies and Systems Based on Low-Dimensional Quantum Devices, 209–265.

break point, or watershed, for almost everything. Nevertheless, it is true that solitons are fascinating, robust, entities and that they could, indeed, be used for communication purposes, by this time. The answer to the question on the role of semiconductors is that they come into play because they must be integral parts of any possible communication system. This point is illustrated in figure 1, in which it can be seen that semiconductors appear in a number of important roles in a soliton transmission system. Semiconductors could be used as amplifiers, for instance, but it is more than likely now that amplification will be effected by erbium-doped fibres.

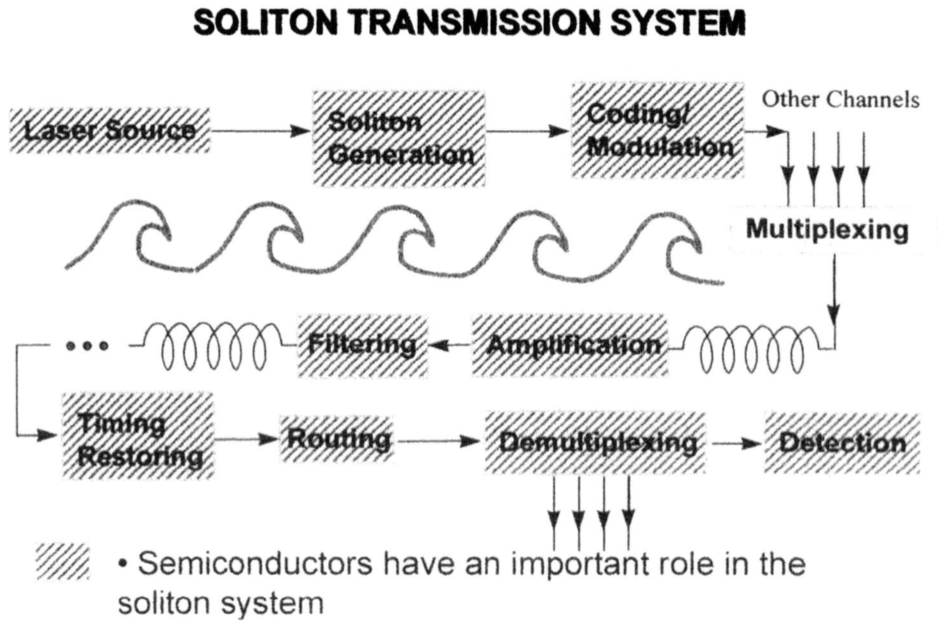

Figure 1.

The key words in any discussion of a soliton transmission system are *logic gates*, *switches*, *amplifiers*, *filters* and, above all, *solitons*. The latter will be discussed, in detail, therefore, in section 2 but, for now, they will be regarded, loosely, as very *special* [nonlinear] pulses, for which dispersion and nonlinearity act [in, as yet, an unspecified way!] against each other. This process results in very stable, potentially indestructible [robust], optical pulses, which can represent 1 in a digital stream of telecommunication information [5]. The absence of a soliton simply represents a 0. Also the soliton pulses are in the ps range, generally, and are really rather short, which means that they can be vigorously packed together to yield a high density information stream. As was said above, a soliton is a natural 'bit', so it literally is a 'bit' of information and the closeness of the 'bits' is a measure of the so-called *bit-rate*. As stated [4] in a recent publication, it is a fact of life that the whole world is to be embedded in optical fibre networks and a soliton-based system is one of the encouraging new possibilities, under active consideration by many organisations and groups. A soliton system must compete with linear systems, such as the trans-Atlantic TAT 12/13, which became operational in

October 1995. In the near future, Africa ONE is proposed as a telecommunication fibre loop around the entire coast of Africa. Behind all of these endeavours the cost/profit motive is a huge driver. To put such a comment into context, recent figures show that TAT 12/13 cost $750m to create and install. If we take into account that this system will support 300,000 *simultaneous* voice channels then, given the current cost of telephone calls, the $750m is recoverable in three days [4] i.e. just 72 hours!

Ultra-high transmission capacity is the holy grail of telecommunications and this should be possible with solitons, using multiplexing techniques. It is encouraging that an experiment using 80 Gbit/sec, over 500 km has been reported recently[5]. This is clearly an excellent step in the right direction. Returning now to the word soliton, its meaning must be now discussed, rather more precisely. The discussion will avoid the rigorous mathematical route, however, in favour of more numerical and physical illustrations.

2. Envelope solitons

Solitons are a large family but the member that "stars" here is called the *envelope soliton.* The envelope derives from the modulation of a strong carrier wave to form a nonlinear pulse. Before illustrating this concept, however, it is worth taking a little time to bring out some of the more important background ideas and to comment, briefly, on the nature of the soliton concept [1,3,6].

The *soliton* is, actually, a special form of *solitary wave,* which is defined in figure 2. The original observations of nonlinear waves by Scott-Russell, in 1834, were on shallow water and have come to be known as the Korteweg-de Vries solitons. Russell, in other words, did not see an envelope soliton but he did observe [7] that nonlinear excitations on the surface of a water channel can be *alone* or *solitary.* It took until more recent times for the additional name *soliton* to be coined, in order to say something about how solitary waves behave in a collision. As figure 2 shows, a solitary wave can assume a number of forms, including an 'edge' or a pulse. It is the latter that is interesting here because the pulse shape we need is the *envelope* modulating a carrier wave. Such a description is familiar to anyone who has even only a little knowledge of how radio or TV is broadcast [8].

Solitary Waves or Solitons?

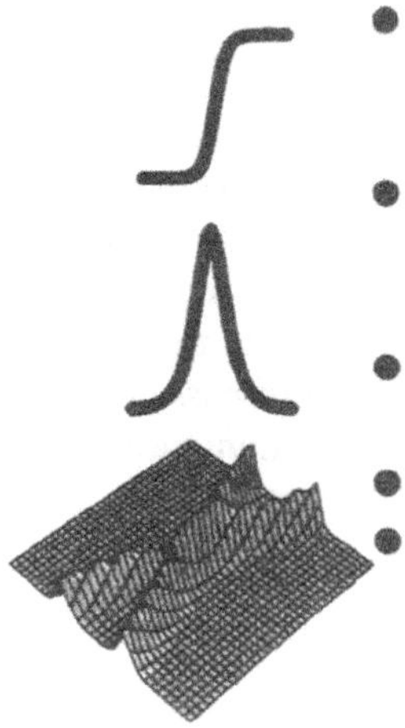

- Solitary waves can be curious entities like the kink on the left that seem to propagate without a change in shape
- A more familiar example is like the pulse on the left ... large amplitude water waves for example
- Why do we say solitary and why do we say soliton ?
- SOLITARY ... because they are alone
- SOLITON ... because they go through each other without being aware of this

Figure 2.

Figure 3 contrasts various physical conditions that can be encountered by a pulse. For example a vacuum is a linear dispersionless medium and, because of these facts, a pulse launched into it will not spread out, as would be expected in a material medium. For this reason, a pulse in a vacuum, or some [fictional] dispersion-free medium, can be viewed as a solitary wave. A pulse propagating in a glass medium, on the other hand, will suffer from dispersion and it is well-known that dispersion shows up as pulse spreading *in time.* A pulse in a linear dispersive medium is not a solitary wave, therefore. Dispersion [1] literally means that v_g, the group velocity, is a function of frequency ω. The arrival [delay!] time $\tau(\omega)$, for a monochromatic signal to travel a distance z, is therefore $\tau(\omega) = z/v_g(\omega)$. A signal [pulse] is not monochromatic but is associated with a bandwidth [frequency spread] $\Delta\omega \ll \omega_0$, where ω_0 is the carrier [centre] frequency. Hence, there is always a spread in the arrival times of frequencies inside $\Delta\omega$. The net result is a spreading in time of the pulse, as illustrated in figure 4. Dispersion is controlled, mainly, by a parameter β_2 called the GVD [group velocity dispersion] parameter, which is illustrated in figure 5. Note that very short pulses will need β_3 as part of the description.

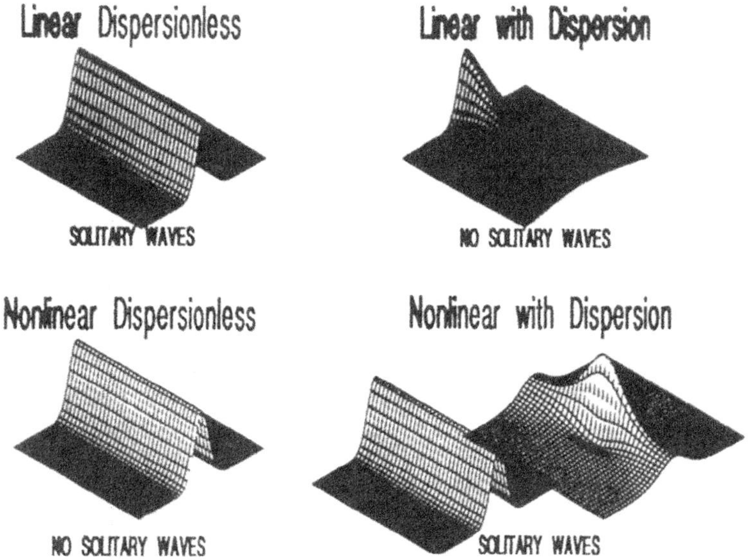

Figure 3.

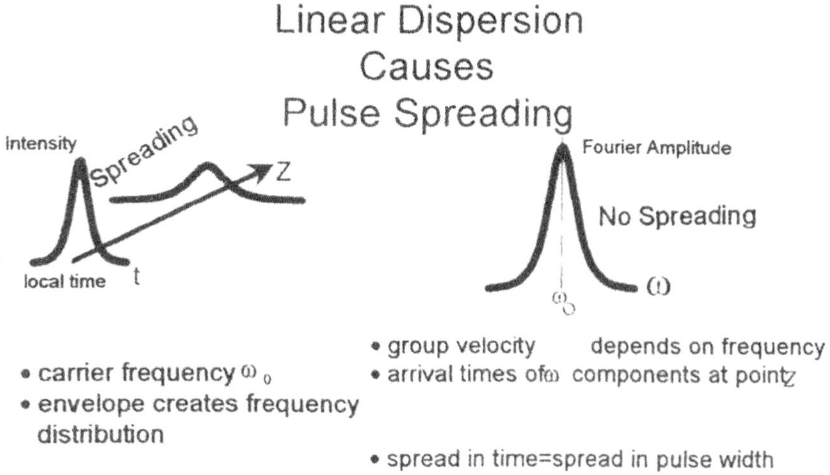

Figure 4.

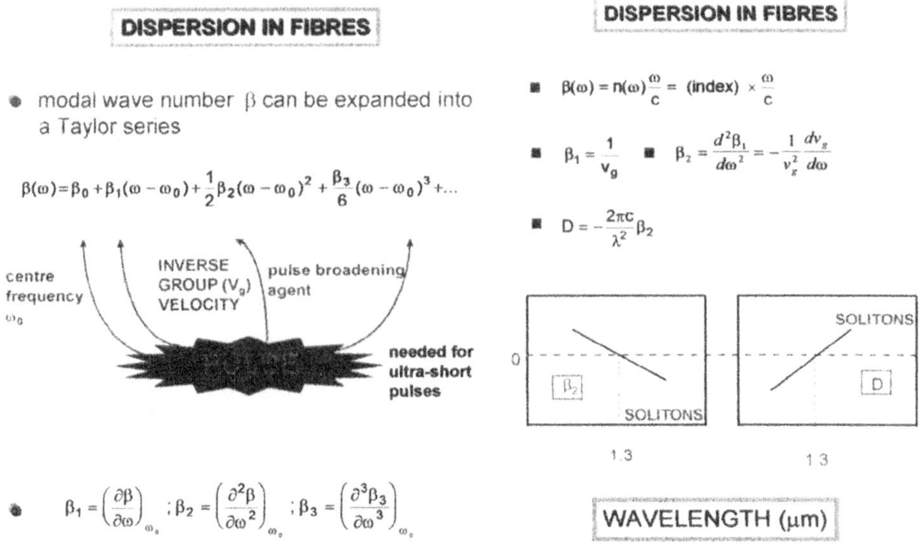

Figure 5.

The really interesting aspect of dispersion is that it introduces *chirp* [1,3]. This is explained in figures 6 and 7.

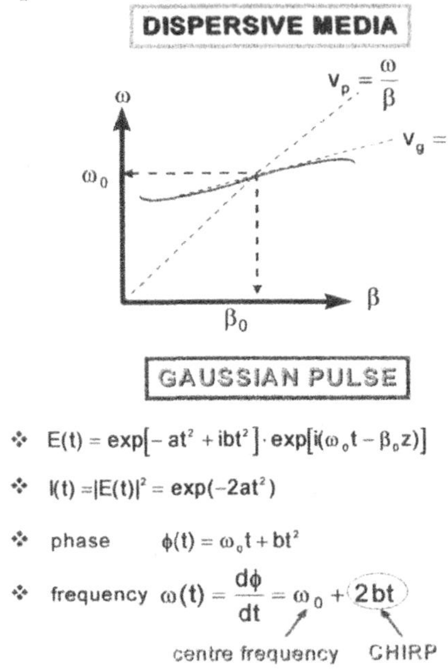

Figure 6.

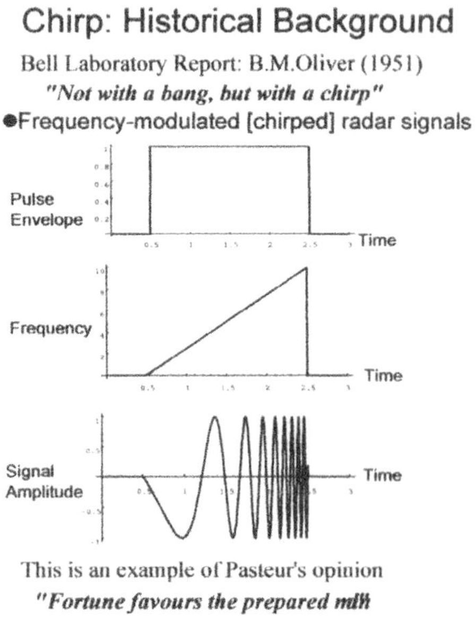

Figure 7.

In figure 6, a Gaussian pulse is used that has an *envelope* $\exp(-at^2 + ibt^2)$ modulating a *carrier wave* $\exp(i[\omega_0 t - \beta_0 z])$, where t is time, ω_0 is the centre frequency, z is length and β_0 is wavenumber. The most important step in understanding the physics of pulse spreading is to recognise that frequency is *defined* as $\frac{d\phi}{dt}$, *not* ϕ/t. Given this fact, the frequency turns out to be $\omega_0 + 2bt$ and the result shows that the carrier frequency is changed with time. This is called *chirping* and figure 7 shows how this could be done physically. In this particular example, a rectangular pulse envelope 'contains' the carrier frequency. If the frequency is 'ramped' up with time, as shown, the net result is that the oscillations inside the envelope exhibit a linearly varying frequency across the pulse. In other words, the pulse becomes a little redder at the leading edge and a little bluer at the trailing edge [positive chirp]. As a matter of fact, in a dispersive medium, this is, precisely, what happens when $\beta_2 > 0$ (normal dispersion). When $\beta_2 < 0$ (anomalous dispersion) it is the leading edge that gets bluer [negative chirp]. Historically, chirp is not really an idea associated with optics research because, as far back as 1951, radar signals were being chirped to make compressed pulses i.e. chirped radar pulses were collected by suitable delay lines that caused low frequencies to travel slower than high frequencies. This arrangement caused the pulse to 'pile up' on itself and compress. The early work of 1951 took some time to get out into the public domain but the well-known saying that "fortune favours the prepared mind" is particularly apt here. Good ideas are often simple but years of mental preparation are often a prelude to their productive reception. Birds have always known how to chirp!

How then can the concept of envelope soliton be made more quantitative? Figures 8 and 9 are simulations of what happens to a sech(t) input pulse, where t is called *local time* i.e. the variation of the pulse intensity is not plotted in real time because the origin in time is placed under the peak of the pulse. Only the *local* variation, on either side of it, is displayed. In reality, this also means that the pulse is being displayed in a frame of reference travelling with a speed equal to the group velocity. The first thing to look for, in figure 8, is pulse spreading with [propagation] distance because of material dispersion. This is accompanied by a phase [ϕ] build up [1,3] and, consequently, a chirp [$\partial\phi/\partial t$]. Anomalous dispersion, $\beta_2 < 0$, has been selected for figure 8 so a *negative* chirp builds, right across the centre of the pulse. Now consider what happens to the same pulse if dispersion is neglected and the medium is allowed to become nonlinear. This is an ideal case, because no real material is entirely free of dispersion but, if enough power is available in the pulse, this situation can be easily approximated, physically. When this happens the refractive index becomes proportional to the power [Kerr effect] and the result is displayed in figure 9. Contrary to common expectation, there is no change of pulse shape in the time domain i.e. the pulse neither spreads out, or compresses, in the time domain. Nevertheless, a phase change does occur with distance and a *positive* chirp builds up across the entire centre of the pulse. It now becomes clear, intuitively at least, from figures 8 and 9 that if nonlinearity is combined with anomalous dispersion then it will conspire to cancel the chirp development i.e. the presence of dispersion does not prevent the power level creating a Kerr effect change in the refractive index [8-13]. In other words, a soliton can exist when the chirp from

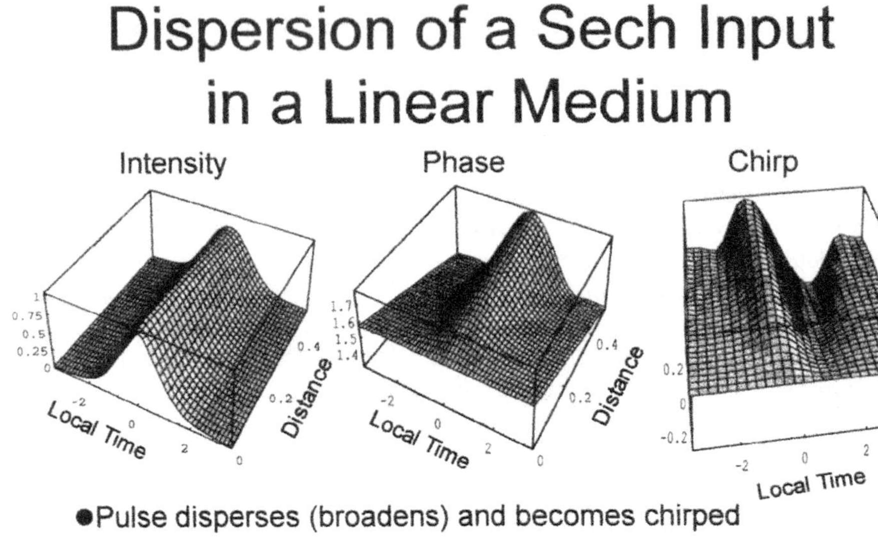

Figure 8.

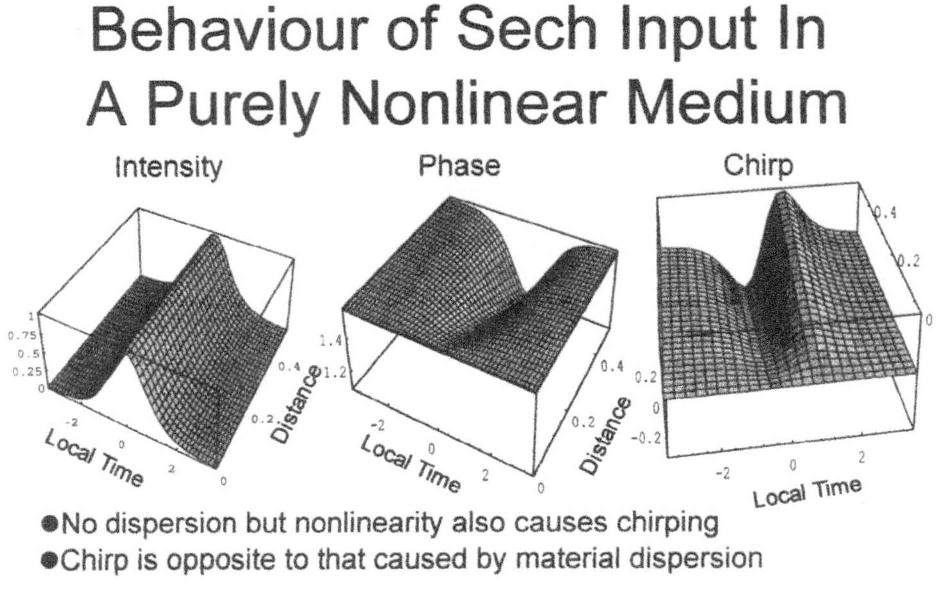

Figure 9.

dispersion cancels that due to nonlinearity and the sech(t) pulse will not spread out in the time domain even if dispersion is present. Referring back to figure 3, the situation is as if the pulse has been "tricked" into believing it is like a pulse in a vacuum and that the medium is not there. This is not the case, obviously, and, furthermore, a soliton will clearly need *enough power* to exist. Indeed, the minimum (threshold) power is needed to create a soliton is simulated in figure 10.

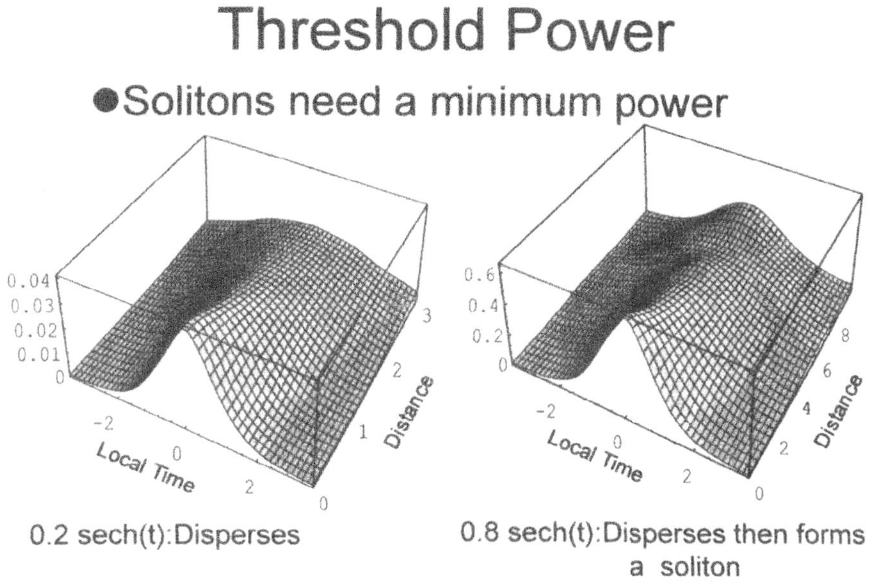

Figure 10.

As with all physical problems, the creation of solitons is associated with characteristic lengths and times but, this time a threshold power is also involved. These quantities are shown in figure 11, in which the local time, across the pulse, is $T = (t - v/v_g)$, where t is the real time, z is the propagation distance and v_g is the group velocity. A moment of reflection will reinforce the conviction that this means the pulses shown upto now are being displayed in a frame of reference moving with a speed equal to the group velocity v_g. The dotted line in figure 11 is meant to show the path of the pulse in the stationary observer, real [laboratory] frame of reference. Note that L_D is a natural combination of entities that measures the "strength" of the dispersion, while L_{NL} is proportional to P_0^{-1} and is a measure of the extent to which the nonlinear phase in figure 9 builds up. For the present purposes, γ is just some proportionality constant that does not need to be more precisely defined.

LENGTH, TIME AND POWER SCALES

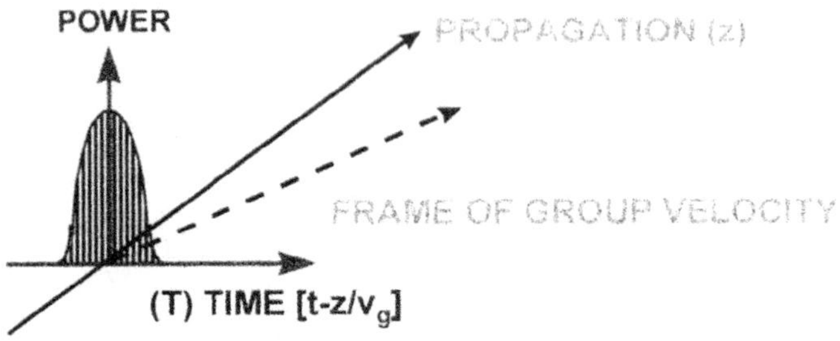

- measure T in T_0 (half-width, or full-width of pulse)
- dispersion length

$$L_D = \frac{T_0^2}{|\beta_2|} = \frac{T_0^2}{\left|\frac{\partial^2\beta}{\partial\omega^2}\right|}$$

......distanced needed to disperse to double width

- nonlinear length

$$L_{NL} = (\gamma P_0)^{-1}$$

■ P_0 is peak power ■ γ is nonlinear coefficient

- threshold for first-order soliton $L_D = L_{NL}$

$$P_0 = \frac{|\beta_2|}{\gamma T_0^2}$$

Figure 11.

Before getting to the partial differential equation governing the solitons of interest here, it is worth emphasising again that the word *soliton* has always to be qualified because it is a member of a large and extended soliton family. Some exotic family members are shown in figure 12 but the one that we want to use is the humble, *fundamental*, soliton.

This is more correctly called the lowest-order *envelope soliton* and is actually a solution of what is called the nonlinear Schrödinger equation [1], often referred to by the acronym NLS.

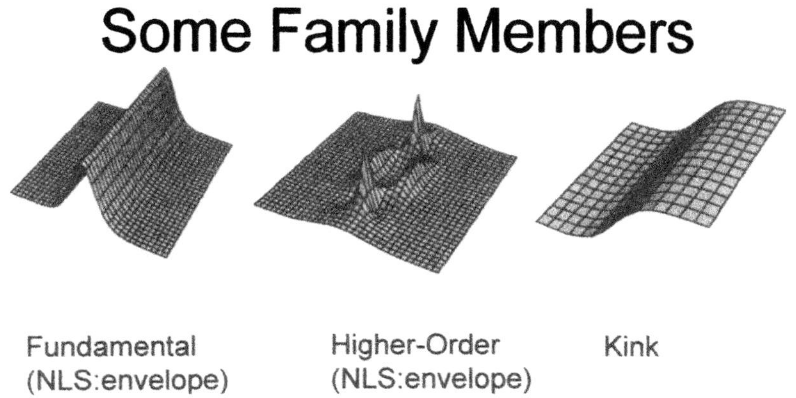

Figure 12.

If a material becomes nonlinear, due to supporting a wave with amplitude $A(z,t)$, then its nonlinear dispersion equation may be formally written as $f(\omega, \beta, |A|^2) = 0$, where ω is angular frequency and $\beta \equiv \beta(\omega)$ is modal wavenumber. If the excitation supported by the material is a pulse, then it is in the form of an envelope-modulated carrier wave, propagating along the z-axis, with an amplitude $A(z,t)$. Here t is time, and the *fast* variation, performed by the carrier wave has the explicit form $\exp[i(\omega_0 t - \beta_0 z)]$. A *small* deviation, due to the dispersion, will drive the frequency from ω_0 to $\omega_0 + \Omega$, $\Omega << \omega_0$ and a *small* nonlinearity changes the wavenumber β_0 by an amount $\frac{\partial \beta}{\partial |A|^2} \cdot |A|^2$. By a simple Taylor expansion [14], the nonlinear dispersion equation, to the lowest order that is required to account for both the pulse dispersion and the nonlinearity, is

$$\left[\beta(\omega) - \beta_0 - \left(\frac{\partial \beta}{\partial \omega}\right)_{\omega_0} \Omega - \frac{1}{2}\left(\frac{\partial^2 \beta}{\partial k^2}\right)_{\omega_0} \Omega^2 - \gamma |A|^2 \right] A = 0 \qquad (2.1)$$

where $\beta_1 = \left(\frac{\partial \beta}{\partial \omega}\right)_{\omega_0}$, $\beta_2 = \left(\frac{\partial^2 \beta}{\partial k^2}\right)_{\omega_0}$ are evaluated at ω_0, the carrier frequency and

$\gamma = \left(\frac{\partial \beta}{\partial |A|^2}\right)_{A=0}$ is called the nonlinear coefficient [14].

In operator language $\beta(\omega)-\beta_0=-i\frac{\partial}{\partial z}$ and $\Omega=i\frac{\partial}{\partial t}$, so that the nonlinear Schrödinger equation [NLS] is

$$i\frac{\partial A}{\partial z}+\beta_1\frac{\partial A}{\partial t}-\frac{\beta_2}{2}\frac{\partial^2 A}{\partial t^2}+\gamma|A|^2 A=0 \tag{2.2}$$

A(z,t) is *slowly varying* and (1) introducing a time $T = t\text{-}z/v_g$, measured in the frame of reference moving at the group velocity v_g (2) normalising T to $t' = T/T_0$, where T_0 is the pulse width (3) introducing a dimensionless amplitude u through the transformation [1] $A=\sqrt{P_0}u$ where P_0 is the peak power (4) scaling z to $z' = z/L_D$ (5) setting $N^2 = L_D/L_{NL} = \gamma P_0 T_0^2/|\beta_2| = 1$, reduces equation (2.2) to

$$i\frac{\partial u}{\partial z}+\frac{1}{2}\frac{\partial^2 u}{\partial t^2}+|u|^2 u=0 \tag{2.3}$$

in which the dashes have been dropped to make the notation easier. This derivation of the nonlinear Schrödinger equation is simple, and rather effective. A treatment of the transverse modal field structure has not been built in but this is easy to do and does not change the generic form (2.3). It will modify the quantitative form of γ but does not change its essential meaning.

Equation (2.3) is the classic, generic, form of the nonlinear Schrödinger equation, which is instantly recognised in all the literature. The quantity $N=\sqrt{L_D/L_{NL}}$ labels the envelope soliton i.e. $N = 1$ is the fundamental (first-order) soliton; $N = 2$ is the second-order soliton, and so on. The fundamental soliton has a sech shape that is preserved during propagation. It is a very stable, some say '*robust*', entity and, as pointed out in the introduction, is a natural candidate for use in optical transmission and switching systems. Higher-order solitons, as can be seen in figure 12, change shape as they propagate and this is called breathing. They are not of technical interest to systems engineers because they lack stability. For example, an $N = 2$ soliton is a bound state of two $N = 1$ solitons but the binding energy is zero. This means that an $N = 2$ soliton is easily broken up into constituent $N = 1$ solitons, by perturbations. This is an important design consideration because perturbations abound in a real system, so the $N = 1$ soliton is the best candidate for telecommunications. Even so, the $N = 1$ soliton can also be perturbed and this fact can cause problems too! The section on the Gordon-Haus effect [15] illustrates this, nicely, and shows how parameters of the soliton, like its frequency and velocity can acquire random shifts because of the necessity of the solitons having to pass through hundreds of amplifiers e.g. ~ 300 in the trans-Atlantic fibre ring loop. An $N = 2$ soliton would have no chance of survival here. But this topic is for later on. For now let us look at some gating techniques because a transmission system will be supporting a stream of solitons and gates [13] are certain to appear when processing the signals.

3. Gating

In a soliton transmission system some modulation format must be selected to code the information. Some examples are shown in figure 13a, in which the use of various components, such as a filter, is, briefly, indicated. The concept of a soliton as a 'bit' in a 'bit stream' is also illustrated in terms of Manchester coding [8,13]. Also an opportunity is taken, in figure 13b, to firm up the idea of bit rate and bit period. After deciding upon a modulation format the next major item to discuss is gating, which involves logic circuits and routing. Figure 14 shows that the main general points concern logic gates [16,17], routing, through switching and multiplexing/demultiplexing devices. As a matter of fact, logic gates are well-known in electronics and are readily implemented in a wide range of circuitry. The language, therefore, has been well-known for decades. The question for us here is the extent to which optical/soliton phenomena can be used for these purposes. This problem will be addressed in its own right, without being very concerned about how such gates will fit into a real soliton transmission system.

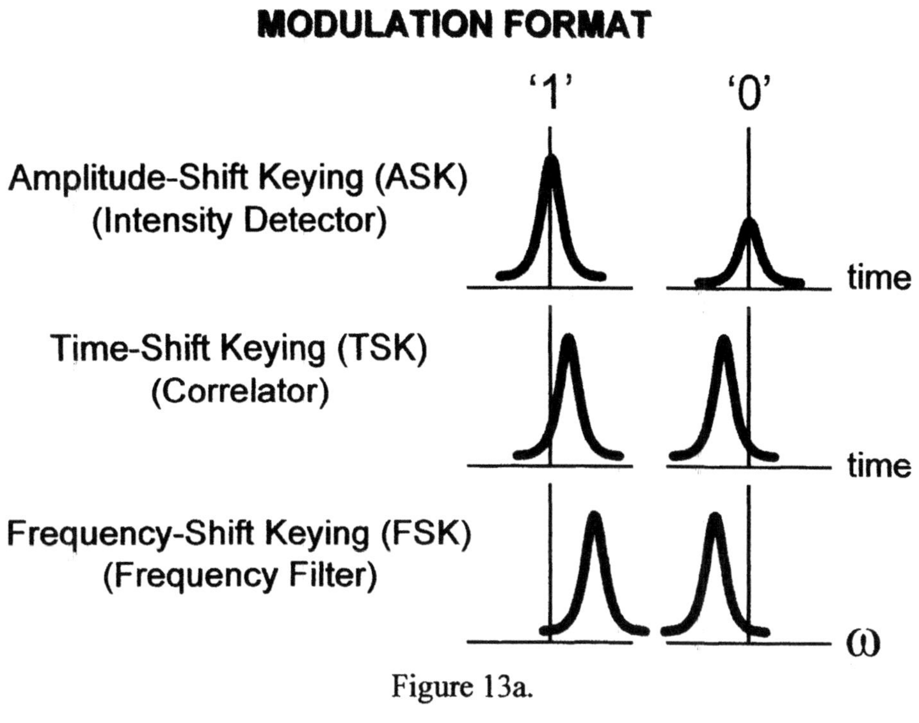

Figure 13a.

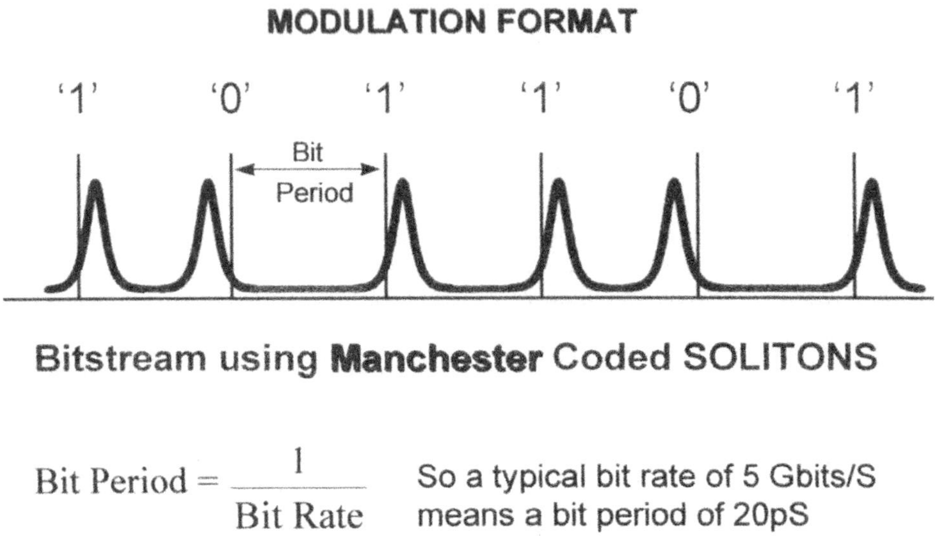

Figure 13b.

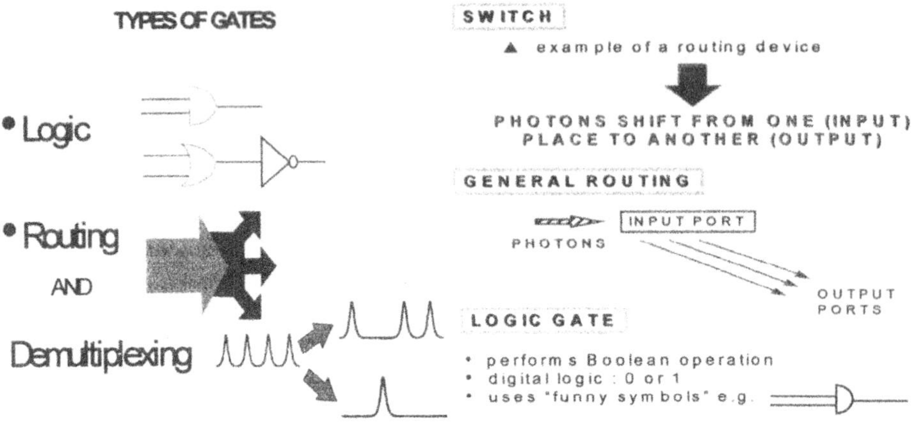

Figure 14.

Figure 15 and 16 summarise the main definitions, shows that a truth table is associated with a logic gate and shows that, in conventional electronics, at least, some 'strange' symbols are used to denote AND, *OR, NOT* and NOR logic operations. This list of logic operations is not exhaustive, of course, because *exclusive-OR* and NAND gate are also possibilities. Every possible logic gate function can be performed from the two basic logic elements NOR and NAND, however. If it is remembered that NOR=OR.NOT and NAND=AND.NOT, then a NOR gate with a single input acts as a NOT gate, for example. Similarly, NOR and NOT will generate the function OR and so on.

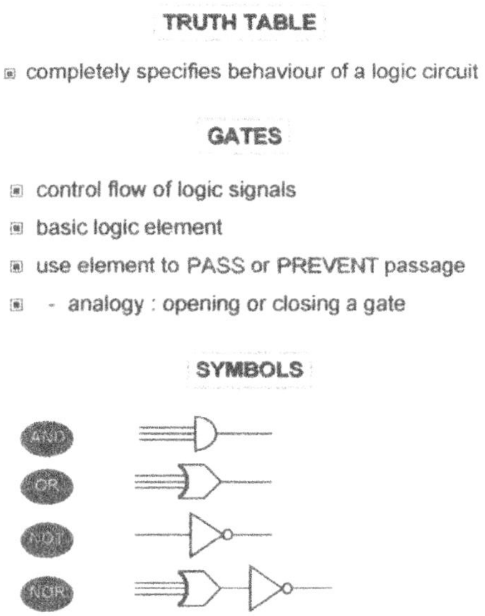

Figure 15.

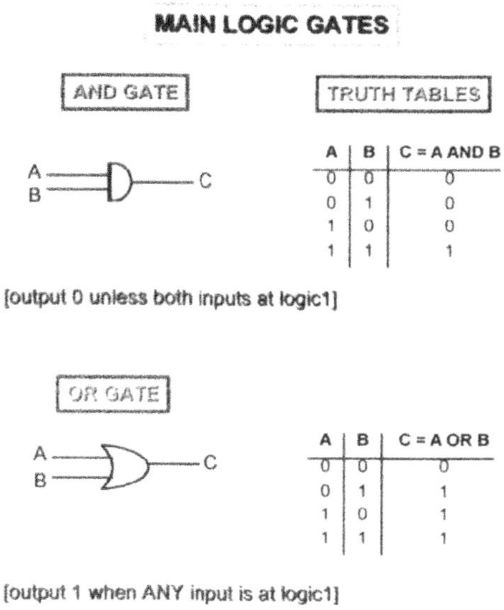

Figure 16.

Physical effects that can be exploited, using solitons, to achieve some form of switching are [1,13]

- birefringence
- four-wave mixing
- chirping
- phase repulsion

All of these phenomena involve soliton collisions which could be of the type shown in figure 17. This figure shows, in dimensionless units, two, classic, soliton interactions. Two solitons are located, initially, at time positions [in the local frame, travelling at a speed equal to the group velocity] $\tau = \pm t$. They are also assigned phases ϕ_1 and ϕ_2, respectively, so that they will behave in a manner determined by the *phase difference* $(\phi_1\text{-}\phi_2)$. For $\phi_1\text{-}\phi_2 = 0$ the solitons are drawn to each other and a 'collapse' occurs at a finite propagation distance. After the collapse, the solitons separate again but the whole process is repeated, periodically. In contrast, for $\phi_1\text{-}\phi_2 = \pi$, the solitons are kept apart and, furthermore, develop paths that diverge continually, as the propagation distance increases. Indeed, this is such a nice effect that a *soliton repulsion gate* [18] can be implemented, using a dual-core optical fibre, as shown in figure 18.

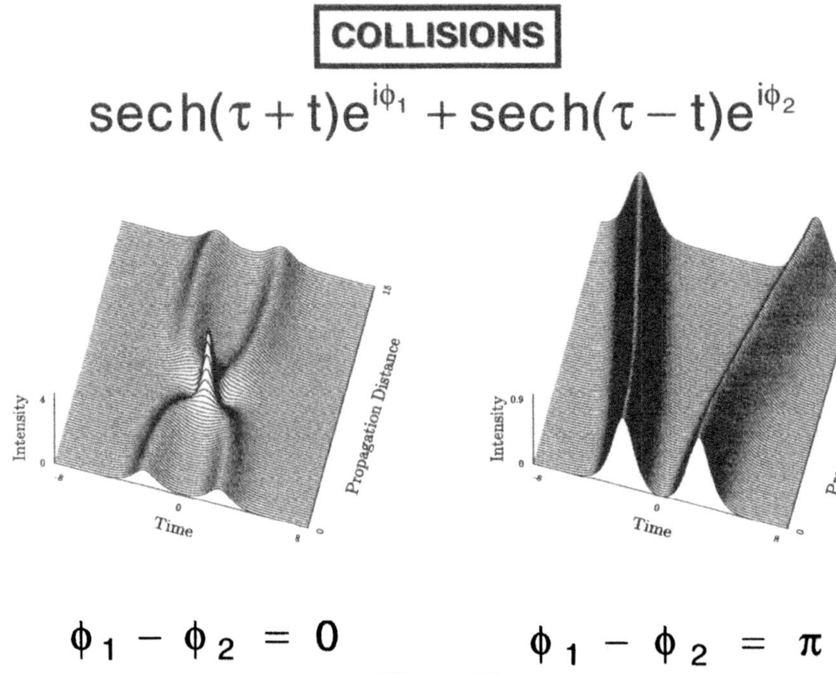

Figure 17.

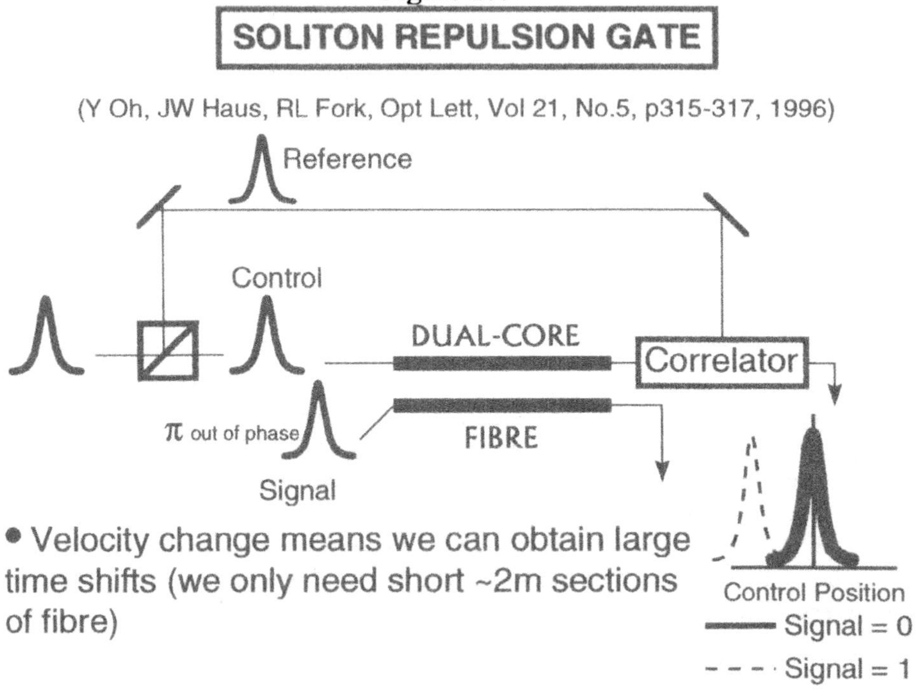

Figure 18.

Birefringence [1,9,19-23] is such a useful property to exploit in an optical fibre, that it leads to what has been termed the *soliton-dragging gate* [13]. In equation (2.3), only one amplitude component is needed because this equation models the behaviour of a soliton in an isotropic, non-birefringent, media. In other words, the polarisation does not matter because all directions, transverse to the propagation direction, are equivalent. If z is the propagation direction and (x,y) are the transverse directions, then, in a birefringent fibre, group velocities β_{1x}, β_{1y} are associated with the x and y directions [19,23]. Hence a dimensionless parameter $\delta = (\beta_{1x} - \beta_{1y})T_0/(2|\beta_2|)$, where T_0 is the width of a pulse, and β_2 is the group velocity dispersion, enter into the model of the propagation. δ is called the group velocity birefringence parameter. Since the velocity of field component polarised along the x-axis is not the same as the velocity of a field component polarised along the y-axis these axes are referred to as the *slow* and *fast* axes, respectively. The envelope pulses u and v, associated with the *slow* and *fast* linearly polarised modes satisfy the following set of *coupled* nonlinear Schrödinger equations [18]

$$i\left[\frac{\partial u}{\partial z}+\delta\frac{\partial u}{\partial t}\right]+\frac{1}{2}\frac{\partial^2 u}{\partial t^2}+\left(|u|^2+\frac{2}{3}|v|^2\right)u=0 \tag{3.1}$$

$$i\left[\frac{\partial v}{\partial z}-\delta\frac{\partial v}{\partial t}\right]+\frac{1}{2}\frac{\partial^2 v}{\partial t^2}+\left(|v|^2+\frac{2}{3}|u|^2\right)v=0 \tag{3.2}$$

Note that $\delta > 0$ makes $\beta_{1x} < \beta_{1y}$ so *u is associated with the slow axis* for data with positive δ. In equations (3.1) and (3.2) the phase velocity birefringence [22] $K = (\beta_x - \beta_y)T_0^2/(2|\beta_2|)$, where β_x and β_y are wavenumbers associated with the coordinate directions x and y, has been selected to have a negligible influence. In practice, K enters the equations quite naturally and can then be made to appear as a term with a factor e^{4iKz}, through the transformations $u \rightarrow ue^{-iKz}$, $v \rightarrow ve^{-Kz}$. A large birefringence implies a large K and the factor e^{iKz} averages to zero.

Equations (3.1) and (3.2) can easily be solved numerically and certain initial conditions for u and v can produce a fascinating evolution that suggests a type of gate known as a dragging gate. Figure 19 shows the behaviour of the u and v envelopes for the starting conditions u[z=0,t] = sech(t), v[z=0,t] = sech(t) i.e. the input distribution is a pulse v together with an *orthogonally* polarised pulse u. The two orthogonally polarised components (u,v) engage in an interaction i.e. some form of collision. Indeed, it can be seen that u *moves to the right*, demonstrating that it is slow, but it is, relatively, unscathed. v, on the other hand, is *fast* and wants to *move to the left* a little bit. The directions of propaganda in the (distance local time) plane, of figures like figure 19, can be understood, quantitatively from the transformations used to develop both the generic nonlinear Schrödinger equation and the coupled forms (3.1) and (3.2). In the arguments following (2.2) the transformation $T = t - z/v_g$ was made, which makes T the *local time*, t the real (nonlocal) time and v_g the group velocity. The generic equation (2.3) was obtained by scaling T to $t' = T/T_0$, where T_0 is the pulse

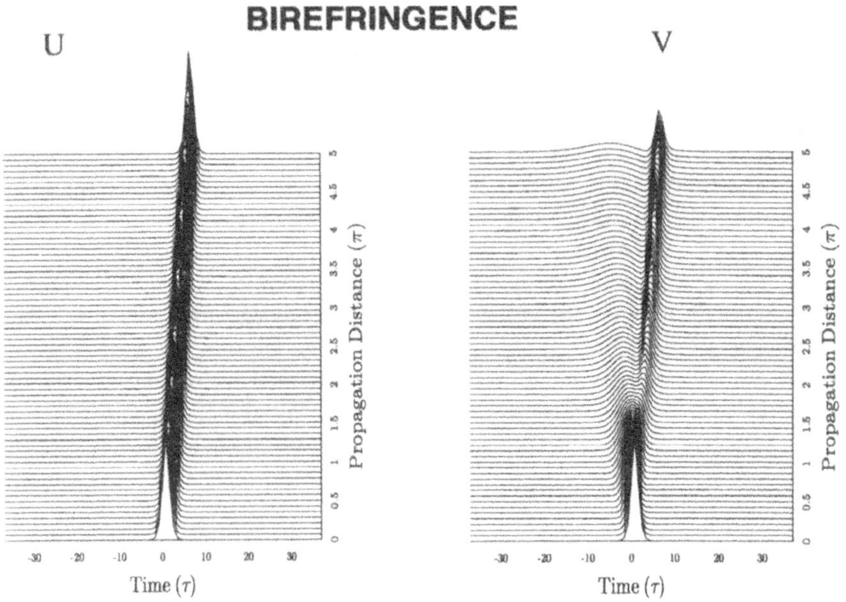

Figure 19.

half-width, and then by dropping the dash superscript, for ease of notation. Going back couple of steps, t is the real (nonlocal) time but the characteristics $[T = t - z/v_g,\ z = tv_g]$ define the plane on which intensity plots, like figure 19, are displayed. For a birefringent medium, v_g is actually the *average group velocity* of the fast and slow modes, so each of these waves has a velocity above or below this average group velocity. A velocity shift, away from the group velocity, is accompanied by shifts Δz and Δt, where

$$\Delta t = -v_g \Delta T, \quad \Delta t = \frac{\Delta z}{v_g}$$

The relationships show that the velocity of a pulse, relative to the group velocity, with which the frame of reference is moving, is

$$\Delta v = \frac{\Delta z}{\Delta t} = -v_g^2 \frac{\Delta T}{\Delta z}$$

This simple result depends upon $\Delta T/\Delta z$, which is the slope of a line in the [propagation distance, local time] frame of reference. Clearly, $\Delta v > 0$ or $\Delta v < 0$ according to whether $\Delta T/\Delta z$ is negative or positive. In other words, a line sloping to the left refers to a pulse with a velocity above that of the group velocity [or mean group velocity] so it shows the presence of a *fast wave*. Similarly, a line sloping to the right shows the existence of a *slow wave*. The dramatic thing that happens, however, is that the bulk of v actually moves *to the right*. In other words, quite a significant piece of v is *captured* or *trapped*, by u. A small piece of v, thus captured, has been called a *shadow*. The simulations in figure 20 show what is happening, rather more quantitatively. The figure shows the solutions of equations (3.1) and (3.2) for initial conditions [1,19] $u(0,\tau) = N\cos\theta\,\mathrm{sech}(\tau)$, $v(0,\tau) = N\sin\theta\,\mathrm{sech}(\tau)$, with $\theta = 30°$, $N = 1.1$ and $\delta = 0.5$. A 'snapshot' cross-section at $z \equiv \xi = 5\pi$ [dimensionless units] shows that u contributes the major part of the energy and a small piece of v [the shadow] is trapped by it. The plot of v also shows its other part, which is going to the left [dispersing] of figure 19.

BIREFRINGENCE

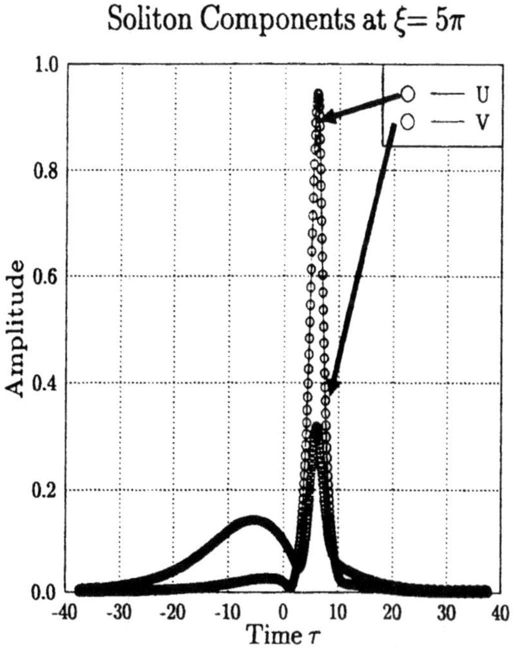

Figure 20.

Figure 21 is a sketch of the possibility of solitons *trapping* each other. The initial conditions are still $u(0,t) = N\cos\theta\,\mathrm{sech}(\tau)$, $v(0,t) = N\sin\theta\,\mathrm{sech}(\tau)$, where θ is called the *input polarisation angle* i.e. the modulus of the total electric field *intensity* is $N\mathrm{sech}(\tau)$ but the polarisation is inclined, at the input, at an angle $\theta = 30°$ to the x-axis. Figure 21 shows that, depending upon the values of δ and N, u can *trap* all of v, or only some of v. Physically the v-soliton slows down a bit, while the u-soliton speeds up a bit i.e. their carrier frequencies shift in opposite directions. The net effect is to permit travel at a common velocity and u traps v but only if the parameter N exceeds a critical value [1,19].

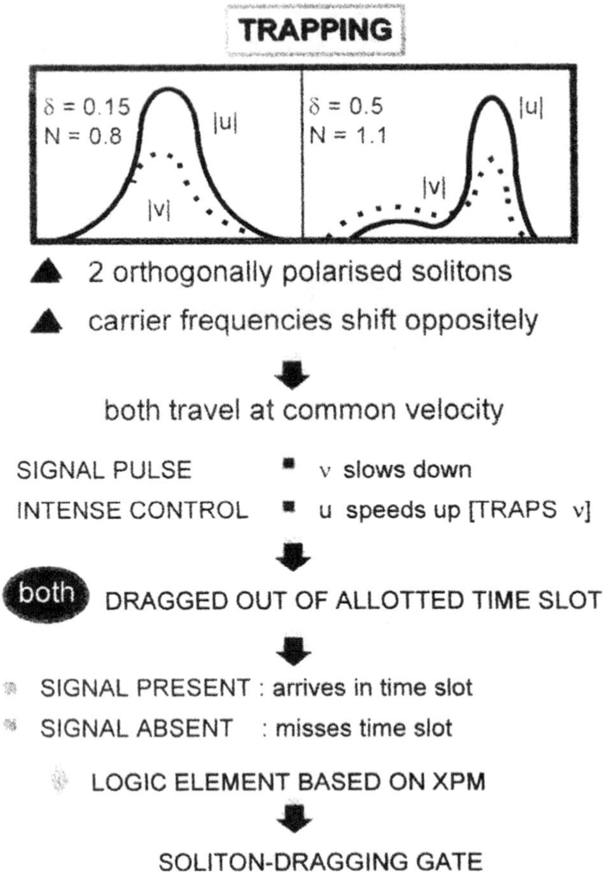

Figure 21.

This phenomenon can be exploited to make a gate [13] by assigning a signal pulse to one polarisation and a strong control pulse to the other. The strong control is there to trap the signal. If this occurs then both pulses are literally dragged out of their time slot at a given arrival point in the system. The point to emphasise here is that, in common with all digital logic, the 'bits', or solitons, have time slots assigned to them in accordance with the clock rate of the system. Anyone familiar with personal computers will readily understand the need for a 'clock'. The arrival in the time slot or the missing of the time slot is clearly controlled by the $\frac{2}{3}|u|^2, \frac{2}{3}|v|^2$ cross-phase modulation (XPM) terms in the coupled equations, so the logic element is based upon XPM. The "missing" of the slots is called dragging and so the soliton-dragging gate was born [1].

In order to emphasise the point, more quantitatively, figures 22 and 23 contrast the nature of collisions in a birefringent fibre. For envelope pulses u, v that are separated in time, at $z = 0$, figure 22 shows that a full collision occurs and, in a real sense, the solitons '*walk through*' each other.

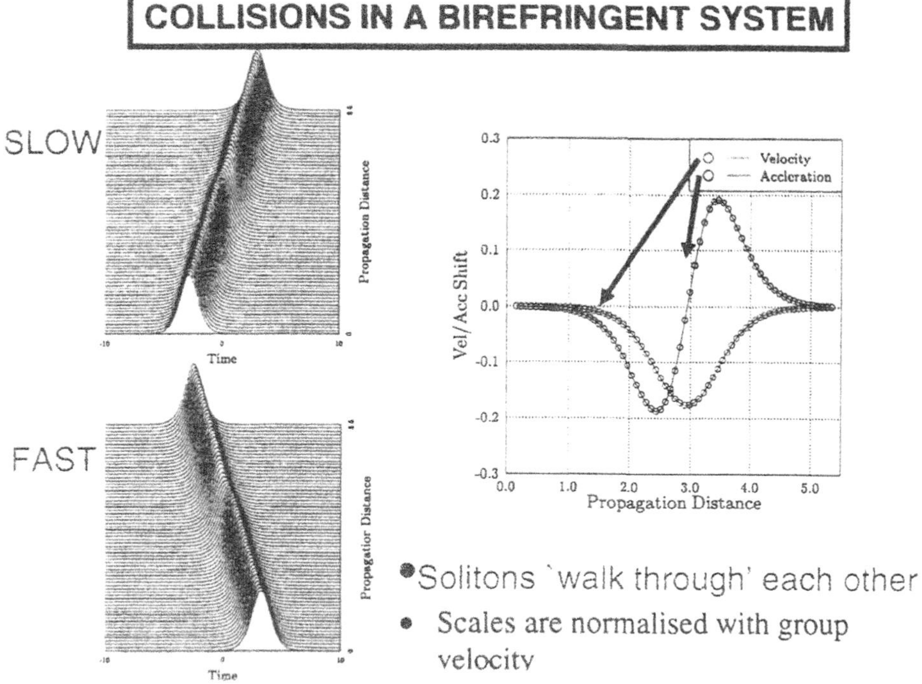

Figure 22.

When solitons pass through each other they emerged unchanged but there is a velocity shift as they go into the interaction. This collision is associated with an acceleration too. Even if the velocity shift, is symmetric and returns to zero, the area under the velocity shift- distance curve is finite and so there is a small time shift. In other words, a closer look at figure 22 shows that a slight *jump* in time has occurred, as a result of the collision. In figure 23, on the other hand, there is a ***net velocity shift*** because, in this case, the u and v are superimposed, initially, and these 'soliton envelopes' only 'walk through' half of each other. Thus for perfectly overlapped pulses the time shift goes on increasing with distance.

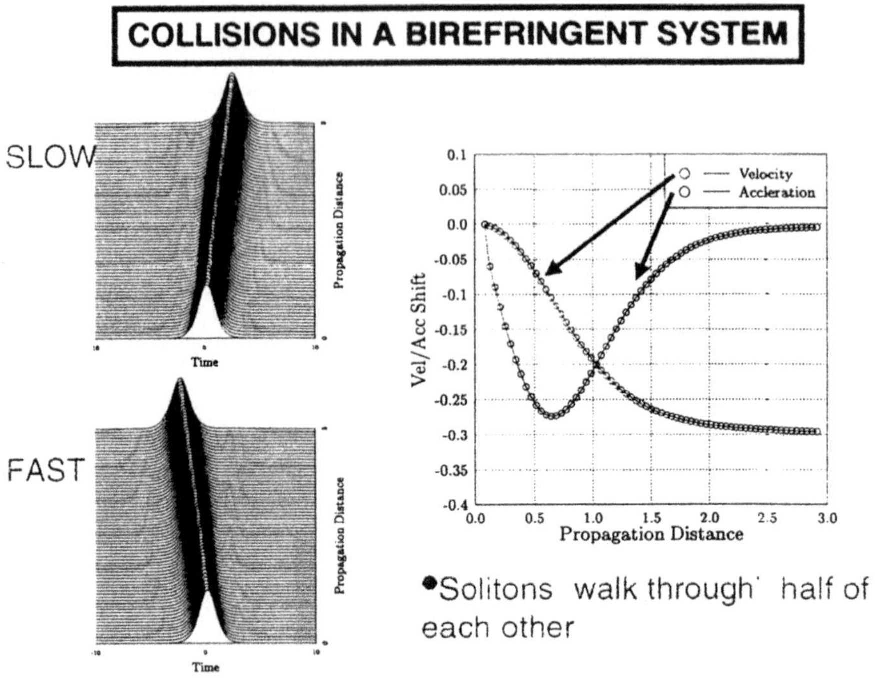

Figure 23.

From the above discussion, it is clear that a real soliton dragging logic gate can be constructed. Figure 24a shows, schematically, a proposal [24,25] for a NOR gate, in which signals A and B are used, in connection with a control signal. The control is on the fast axis, while the signals are on the slow axis. The action is that of a NOR gate and is discussed with reference to an available time window for the receipt of the signals [the information data!]. Figure 24b shows the practical layout of a very nice experiment performed by Islam but note that a clock is included, because the timing is crucial.

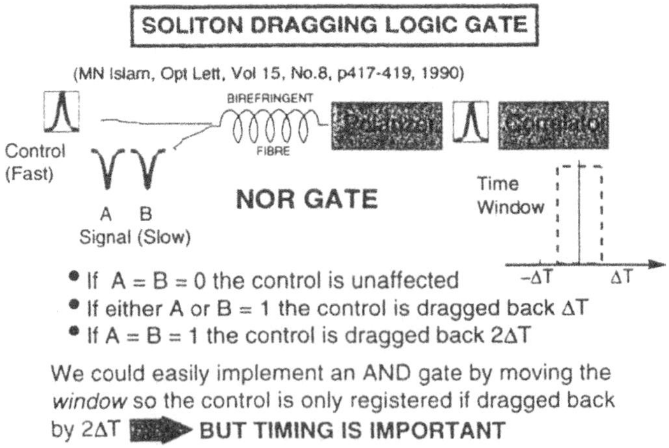

Figure 24a.

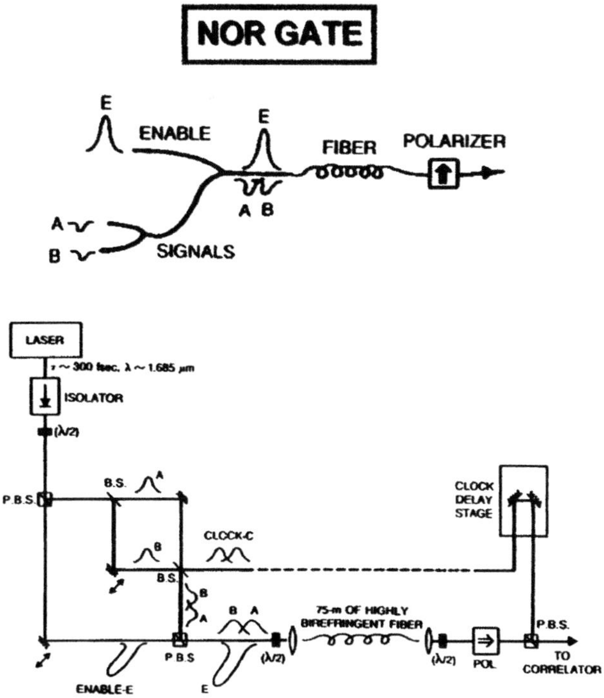

Figure 24b.

Wavelength-division-multiplexing (WDM) is very important in a transmission system. Indeed, it can be safely said that the successful implementation of WDM is going to be a necessary condition for a soliton transmission system that is competitive with other options. Without access to lots of channels the amount of information transmitted will be severely curtailed. This will soon become very important, as the demand for high capacity in FAX and even virtual reality transmissions takes yet another quantum leap. Basically, the answer to the question of how to implement a WDM system is very simple. We need to enter into the system solitons with different frequencies, and at different times, and then use a discriminator like a filter at the receiving end. Unfortunately, solitons, at different frequencies, travel at different speeds and, in any case, they interact with each other. In any real system, the design must take into account inter-channel collisions and this is an area that is still under investigation. This is especially true for WDM behaviour, in and around amplifiers, in a transmission line. These points will not be addressed here but figure 25 contains some, simple, basic, features of WDM. The figures shows a two-channel system for solitons at $\tau \pm t_0$ and at frequencies ω_1, ω_2. The collisions are shown on the basis that one soliton travels faster than the average group velocity and one travels more slowly. The filter, in this case, picks out channel 1. More interesting is the numerical analysis in figure 25b. In it, it can be seen that a net temporal shift occurs that increases with propagation distance. There is also a velocity shift that returns to its pre-collision value, after the interaction and a symmetric acceleration pattern.

COLLISIONS IN A WDM SYSTEM

$$\operatorname{sech}(\tau + t_0)e^{i\omega_1\tau} + \operatorname{sech}(\tau - t_0)e^{i\omega_2\tau}$$

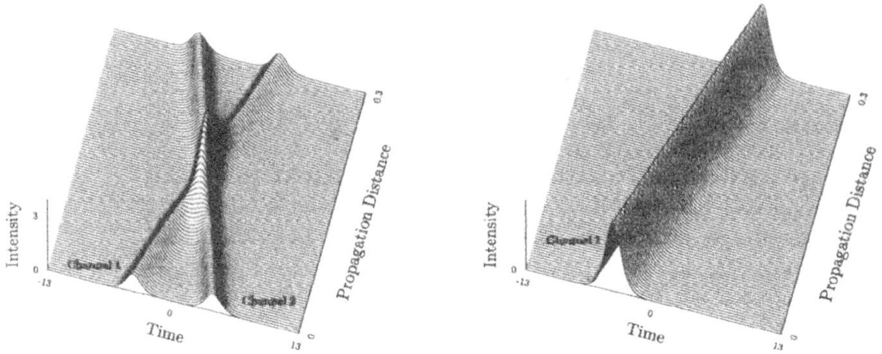

Figure 25a.

COLLISIONS IN A WDM SYSTEM

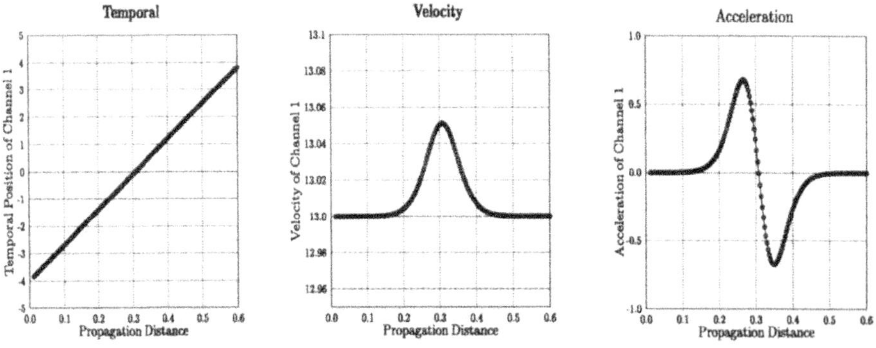

- Collision causes a time shift
- Velocity returns to its initial group velocity
- Real WDM system have many inter-channel collisions

Figure 25b.

Consider now a *semiconductor timing restorer* [13]. First of all a small (a few mm in length) piece of AlGaAs waveguide can act as a source of chirp by becoming nonlinear. In fact the AlGaAs is presented to an incoming pulse as a small, planar waveguide with a ridge on it. As shown in the earlier lectures of this series, the ridge steers the pulse in the desired direction down the waveguide. Because it is a waveguide of this type, it has birefringence i.e. it has a δ, introduced in equations (3.1) and (3.2), that is small but significant [13]. This small piece of low-dimensional semiconductor is an effective replacement for a piece of birefringent fibre, therefore.

One of the problems in a gating structure, as would be the case in arriving at an airport for a flight, is the arrival time at the gate. The pulse may be on time but more often that not it is too early or too late! In figure 26, it is a question of whether a pulse going into a gate gets there at the same time as the reference pulse is generated, or not. A timing restorer is needed to correct for this eventuality so the behaviour of this small piece of birefringent semiconductor must be assessed in the context of this kind of application. The separation stated in figure 26 refers to the *initial* separation of the input control pulse and a reference signal pulse at the semiconductor. If they are not coincident then the pulses must be shifted, relative to each other, in time. As the pulses propagate in the semiconductor a time shift will be imposed, which is a function of the initial separation ΔT, as shown. Suppose that the input [control] pulse arrives *earlier* than the reference [signal] pulse then the initial separation is *negative*, i.e. $\Delta T < 0$. To combat this state of affairs a *positive* time shift will be supplied by the nonlinear cross-phase modulation in the AlGaAs guide. This has the effect of delaying the input pulse until some *later* time [13]. Since it is also useful to compare its behaviour with that of a dragging gate, for the same value of δ, figure 26 contains two sections. The first part is just to emphasise that the AlGaAs is used [replaces] instead of a piece of birefringent fibre. This is an excellent step to take, if a compact integrated system is desired. In the second section, the resultant time and velocity shifts as a function of the initial separation of two pulses is shown and a contrast with a fibre soliton dragging gate is made. Clearly the time shift from the AlGaAs is symmetric and sharper. Thus using AlGaAs as a timing restorer looks promising.

SEMICONDUCTOR TIMING RESTORER

- The ridge waveguide has very small walkoff between mode
- Effectively a birefringent fibre with small δ
- For contrast compare with the soliton dragging case

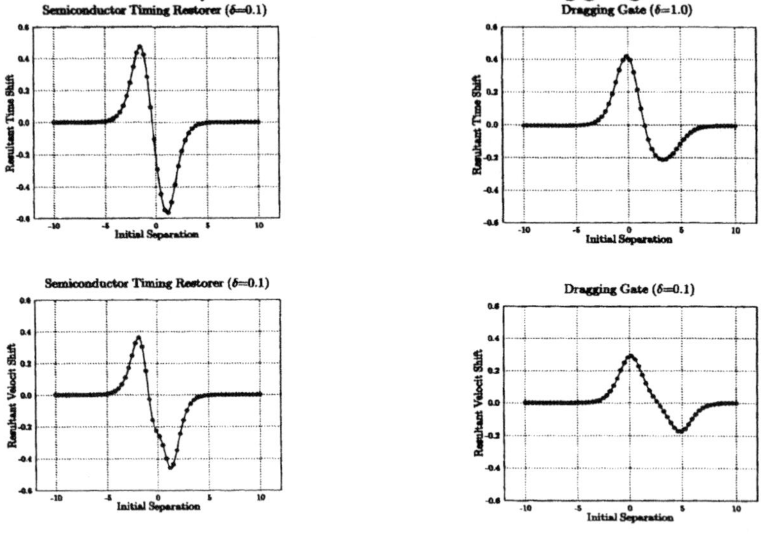

Figure 26.

Figure 27 also contains two sections, designed to illustrate the timing restoring application. In the first part, two pulses are shown, one, of which, is the reference [i.e. the benchmark pulse] pulse and the other is the signal pulse, carrying the information that needs retrieving. The reference pulse, by definition, sits completely, and properly, within the time window. The input pulse is shown as arriving too late or too early. The plot of time shift as a function of pulse separation then shows how to correct for this timing error. Figure 27b shows that it is the nonlinear chirp $\delta\omega$, developed within the AlGaAs chip that is the operative physical mechanism. With respect to the reference pulse the leading edge of the pulse that arrives too late "sees" a negative slope on the reference pulse so that the chirp is *positive*. If a pulse arrives too early, the chirp is *negative*. All that needs to be done, from a design point of view is to choose a time-shift that puts everything back into the window.

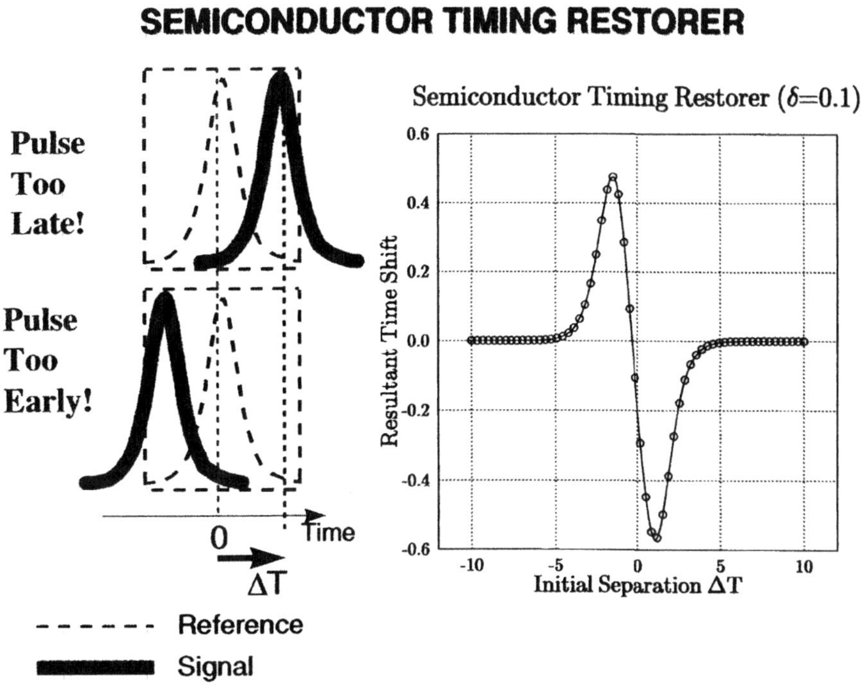

Figure 27a.

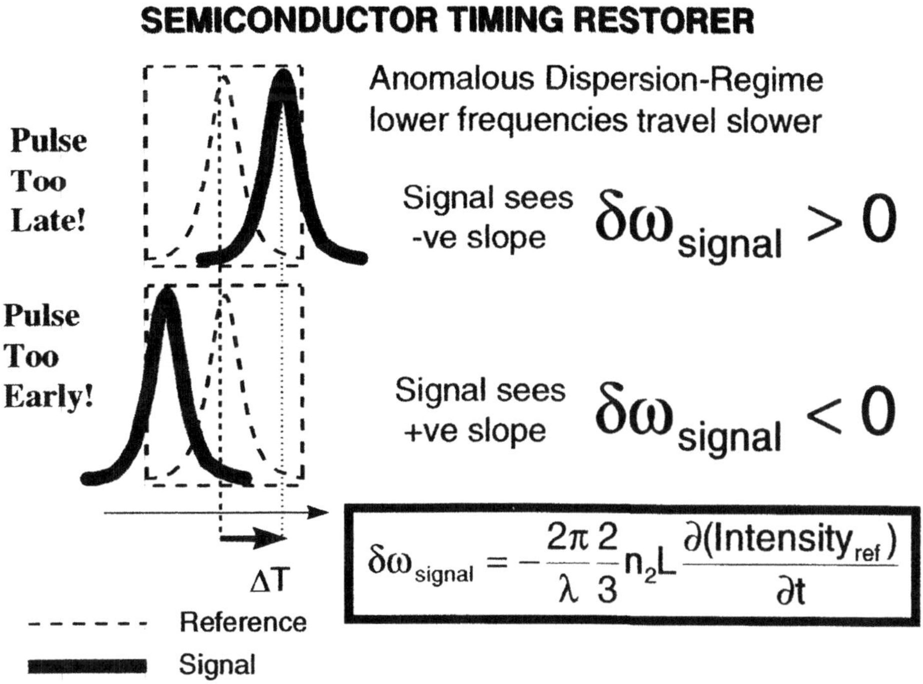

Figure 27b.

4. The Gordon-Haus effect

It is time now to address one of the possible major constraints on the implementation of a soliton-based communication system. In any real system, it is necessary to eliminate loss by some form of amplification. This can be done with semiconductor amplifiers or, as is more likely, with erbium-doped fibre amplifiers [26,27]. Either way, coherent amplification is used to overcome the degradation of the solitons due to loss. Unfortunately, there is a downside to this, because in any amplifier spontaneous emission of radiation always occurs. This is incoherent and is, therefore noise. This noise is added to the transmission, as stated in figure 28. By the time the soliton has passed through hundreds of amplifiers, or, at least, very many of them, this noise will amplify into a serious problem. Just how serious is appreciated when it is recalled that a soliton is a solution to the nonlinear Schrödinger equation and this solution is controlled by certain parameters. These parameters suffer random shifts in their values because of the presence of this noise. The greatest problem is a noise-induced shift in carrier frequency, and a consequent change in velocity, because of dispersion [15]. It might be imagined that such velocity change may well cause an arrival time problem, in the acceptance window of the system. This is a correct diagnosis and, furthermore, the timing errors build up to a significant shift, or *jitter*, in the arrival time of the soliton by the time an addition over n amplifiers is completed. In practice, this kind of effect places a limit on the system and limits the distance over which a long-distance communication system can operate without 'breaking up'. This is the Gordon-Haus effect [15], and is, potentially, devastating. To overcome this problem it will turn out that it is necessary to use filters. Before this stage is reached, however, let us derive the Gordon-Haus effect, to see just how limiting it is.

The basic equation, in dimensionless units, is equation (2.3), for which the general form of the lowest-order stationary state is

$$u_s = A\operatorname{sech}[At - q]e^{-i\Omega t + i\phi} \tag{4.1}$$

in which $q = q_0 - A\Omega z$ and $\phi = \phi_0 + (A^2 - \Omega^2)z/2$, q_0 is the soliton location in time at $z = 0$ and ϕ_0 is the initial phase at $z = 0$. This is a well-known result and it is easiest to check it by direct substitution into equation (2.3). The interesting interpretation of Ω, in (4.1) is:

$\Omega = 0$: soliton in frame of reference moving at a speed equal to the group velocity
$\Omega \neq 0$: soliton is moving slower or faster than speed of the frame of reference.

The main point to make, at this stage, is that the amplifiers cause perturbations to the stationary state, such that u_s becomes $u = u_s + \delta u$, because of the perturbations δA and $\delta\Omega$ induced by the noise.

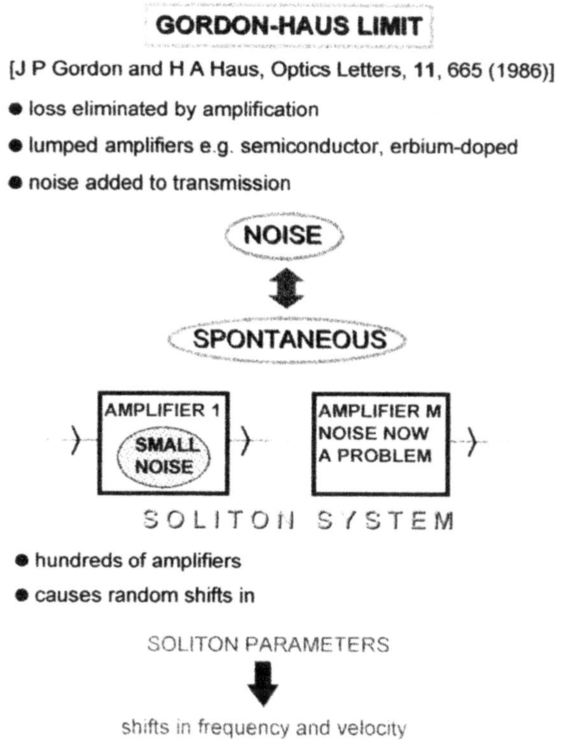

Figure 28.

Given the form (4.1) we first of all define the mean energy W where

$$W = \int |u|^2 \, dt \tag{4.2}$$

Secondly, a formula for the mean frequency can be obtained from (4.1), because

$$u\frac{\partial u^*}{\partial t} = i\Omega A^2 \mathrm{sech}^2(\tau) + A^2 \mathrm{sech}^2(\tau)\frac{\partial}{\partial t}\mathrm{sech}[At - q]$$

where $\tau = At\text{-}q$. Hence, the mean frequency is

$$W_\nu = \mathrm{Im}\int u\frac{\partial u^*}{\partial t}\,dt = \int \Omega |u|^2 dt \tag{4.3}$$

where Im denotes the imaginary part.

If u_s is perturbed to $u_s + \delta u$, then

$$W = \int\left[|u_s|^2 + \delta u u_s^* + \delta u^* u_s + |\delta u|^2\right]dt \cong 2A + 2\,\mathrm{Re}\int u_s^* \delta u dt = 2[A + \delta A] \qquad (4.4)$$

and

$$W_v \cong 2A\Omega + 2\,\mathrm{Im}\int \frac{\partial u_s^*}{\partial t}\delta u dt = 2[A\Omega + \delta(\Omega A)] \qquad (4.5)$$

where $\delta\Omega = \frac{1}{A}\left[\delta(A\Omega) - \Omega\delta A\right]$. Therefore,

$$2\delta A = \int\left[u_s \delta u^* + u_s^* \delta u\right]dt \qquad (4.6)$$

$$2\delta(A\Omega) = i\int\left[\frac{\partial u_s}{\partial t}\delta u^* - \frac{\partial u_s^*}{\partial t}\delta u\right]dt \qquad (4.7)$$

so that

$$\delta\Omega = -\frac{i}{2}\int \tanh[At - q]\left\{u_s \delta u^* - u_s^* \delta u\right\}dt = -\,\mathrm{Im}\left[\int \tanh(At - q) u_s^* \delta u dt\right] \qquad (4.8)$$

The question now is what form does a noise field, with amplitude a, take. Assuming that it is white noise that exists, exciting all modes with the same energy, then an amplifier must have a bandwidth larger than the spectral width of the soliton. Given the form (4.1) it is safe to make this assumption, and the complex conjugates appearing in (4.8) means that we first recognise that $u_s \propto$ sech [At -q] so that its differential is proportional to tanh(At-q)u_s i.e. the noise field, introducing an amplitude a, will have the form [15]

$$\delta u = i a u_s \tanh[At - q] \qquad (4.9)$$

The factor i appears because it appeared quite naturally in (4.8) and $\delta\Omega$ is a real quantity. Given equation (4.9), equation (4.8) easily integrates to give the mean value

$$\left\langle \delta\Omega^2 \right\rangle = \frac{4}{9}\left\langle a^2 \right\rangle A^2 \qquad (4.10)$$

where <> denotes a mean value.

This is a very nice and elegant little formula and further progress can be made by considering a broadband amplifier, with gain G. Actually, the calculation is not sophisticated enough to worry about the bandwidth of G but the comment had to be made. In the simplest idea of an amplifier, there a 0 photons at the *input* and G = 1 means that there is no amplification. This suggests that (G-1) must come into the explanation somewhere. Indeed, there are (G-1) photons of noise at the amplifier *output*. Crudely speaking, then, the mean number of photons in the noise field [28] is $[(G-1)+0]/2=(G-1)/2$. If the number of photons in the soliton pulse is N_0 then

$$\frac{(G-1)}{2}=N_0\left\langle\int|\delta u|^2\,dt\right\rangle \tag{4.11}$$

where the right-hand side arises from the product of N_0 and the mean energy $\left\langle\int|\delta u|^2\,dt\right\rangle$ of the noise field.

$$\frac{(G-1)}{2N_0}=\left\langle\int|\delta u|^2 dt\right\rangle=\frac{2}{3}A\left\langle a^2\right\rangle=\frac{2}{3}A\left[\frac{9}{4A^2}\left\langle\delta\Omega^2\right\rangle\right] \tag{4.12}$$

so that

$$\left\langle\delta\Omega^2\right\rangle=\frac{(G-1)}{3N_0}A \tag{4.13}$$

Now the purpose of the amplifier is to provide loss compensation and, in dimensionless terms, loss can easily be introduced through a parameter Γ. Hence, equation (2.3) becomes modified to

$$i\frac{\partial u}{\partial z}+\frac{1}{2}\frac{\partial^2 u}{\partial t^2}+|u|^2 u+i\Gamma u=0 \tag{4.14}$$

If the gain G exactly compensates for loss, sustained over a dimensionless distance $z=z_1$, then

$$(G-1)\cong\Gamma z_1 \tag{4.15}$$

since Γ is quite small in a real fibre.

Suppose that the length of the whole system is $z=L$ and that the nth amplifier is located at $z=nz_1$. Furthermore, suppose that this amplifier at nz_1 shifts the parameter Ω by an amount $-\delta\Omega_n$, where $-\delta\Omega_n$ can be interpreted now as an inverse velocity and that all these velocity shifts are *independent* of each other. The distance of the nth amplifier from the end of the system is $(L - nz_1)$, so the contribution of this amplifier to the shift in the arrival time of the soliton is $-(L - nz_1)\delta\Omega_n$. Hence the total shift in arrival time, due to the influence of all the amplifiers, is

$$\delta t=-\sum_n (L-nz_1)\delta\Omega_n \tag{4.16}$$

The number of amplifiers is $N = L/z_1$ so, because of this, the summation over n becomes

$$\sum_n \cong \int \frac{dz}{z_1}$$

and the mean of the square of the shift in arrival time is

$$\left\langle \delta t^2 \right\rangle \cong \frac{(G-1)}{3N_0} A \int_0^L z^2 \frac{dz}{z_1} = \frac{\Gamma A L^3}{9N_0} \tag{4.17}$$

Some definitions are now required. These are

- bit-rate $R \Rightarrow$ time between solitons is R^{-1}
- window of detector acceptance is $2t_w$
- full-width, at half-power of solitons is

$$t_s = \frac{1.763}{A} \Rightarrow A = 1.763 / t_s$$

- bit error rate is 10^{-9}
- for a Gaussian distribution

$$\left\langle \delta t^2 \right\rangle = \left(\frac{t_w}{6.1} \right)^2$$

Given these definitions, equation (4.17) yields the relationship

$$L^3 = \frac{9N_0}{\Gamma} \left(\frac{t_w}{6.1} \right)^2 \frac{t_s}{1.763}$$

which implies, after multiplying both sides by R^3, that

$$(RL)^3 = \frac{9}{37.21} \cdot \frac{1}{1.763} \left[\frac{t_s t_w^2 R^3 N_0}{\Gamma} \right] = 0.137 \frac{t_s t_w^2 R^3 N_0}{\Gamma} \tag{4.18}$$

The real world scales of a soliton's existence involve

- dispersion length : $L_D = \dfrac{T_0^2}{|\beta_2|}$

- nonlinear length: $L_{NL} = \frac{1}{\gamma P_0}$

where T_0 is a time unit, like the real half-width of the pulses, P_0 is the peak power, β_2 is the group-velocity dispersion and γ is the nonlinear coefficient. What has to be done is to convert (4.18) into real units but, to do that, it is necessary to know what N_0 is. Quite simply, power × time is energy and a photon, at frequency ω, has an energy $\hbar\omega$, where $\hbar = h/2\pi$ and h is Planck's constant. Hence, $N_0 = (P_0 T_0)/\hbar\omega$ is the number of photons in the pulse. This means that

$$\frac{T_0^3}{L_D^3}(RL)^3 = 0.137\frac{t_s}{T_0}\cdot\frac{t_w^2}{T_0^2}\cdot R^3 T_0^3\cdot\frac{P_0 T_0}{\hbar\omega}\cdot\frac{1}{\Gamma L_D} \tag{4.19}$$

From the literature it is well-known that $\gamma = (n_2\omega)/(cA_{eff})$ where c is the velocity of light in a vacuum, A_{eff} is the effective core area of the fibre, ω is the carrier angular frequency and n_2 is a nonlinear coefficient, measured in units m^2/W. The modulus of the dispersion parameter D is $|D| = \frac{\omega}{\lambda}|\beta_2|$, where $\lambda = 2\pi c/\omega$ is the carrier wavelength. Hence $\frac{2\pi P_0 T_0}{\hbar\omega L_D}\cdot\frac{L_D^3}{T_0^3} = \frac{A_{eff}}{hn_2 D}$ and the final result is

$$(RL)^3 = \frac{0.137 t_s t_w^2 R^2}{h\Gamma n_2 D} \tag{4.20}$$

The following data was used by Gordon and Haus [15]

- $\lambda = 1.56\mu m$
- $\Gamma = 0.0461\ km^{-1}$
- $D = 3ps/km/nm$
- $A_{eff} = 25\mu m^2$
- $n_2 = 3.18 \times 10^{-16}\ cm^2/W$
- $t_s R = 0.1$
- $t_w R = 1/3$

The substitution of these values into (4.20) gives

$$RL = 23600 GHz/km \tag{4.21}$$

This result expresses the Gordon-Haus effect, which says that there is a clear upper limit to the distance over which data can be transmitted using a soliton system. This example shows that a 10GHz data stream, converted into solitons, runs into noise trouble after 2360km. The amplifiers are the problem and when this elegant result came out, it appeared to be a disastrous conclusion for the future development of soliton transmission systems.

It is not always the case that there is a practical way out of design problems like this, but, in this case, a beautiful way forward soon presented itself. It came in the form of deploying filters in the soliton transmission line. In a very nice simulation, Marcuse [19] checked out the idea, using the following data

- $\lambda = 1.55\mu m$
- soliton width = 0.02ns
- $n_2 = 3.2 \times 10^{-20}\ m^2/W$
- loss = 0.25dB/km
- bit rate = 5 Gbit/s
- D = 1ps/nm/km
- $A_{eff} = 35 \times 10^{-12}$
- $z_1 = 33km$

The Marcuse simulation permitted a propagation distance of 9000km and it added Gaussian random variables to each component of the Fourier spectrum of the pulse, after each amplifier. In accordance with the earlier discussions, the original energy was restored by each amplifier. To make the simulation realistic Marcuse used a 64-bit word length made up from 32 zeros and 32 soliton pulses. The outcome was displayed upon an *eye diagram*. In an eye diagram, as shown in figure 29, pulses, in sequences, are placed were plotted on top of each other, in a given time window. If there is no noise in the system, then neither the pulses nor the eye diagram change, even after 9000km.

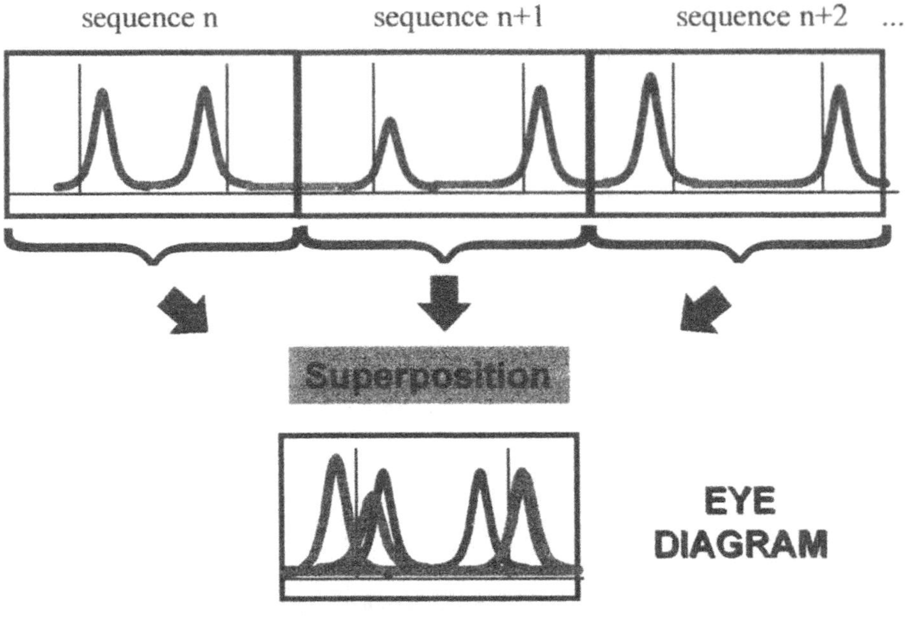

Figure 29.

The results displayed by Marcuse are quite dramatic, as shown in figure 30. First, the eye diagram for the noiseless soliton train is given in (a) and then, in (b), the *jitter* created by the noise [random time delays caused the soliton perturbations], arising from the Gordon-Haus effect is clearly seen. This is corrected in (c) by placing a [simulated] Fabry-Perot filter after each amplifier. The bandwidth of the filters is 157GHz and the noise is dramatically reduced. The reason is clear. By fixing the transmission peak of the filter, and limiting its bandwidth, the soliton, being a robust entity is not destroyed, or deterred, but is happy to pass through the filter. The soliton, in other words, will seek out the peak transmission of the filter. The filter has a transmission that is a function of frequency and will transmit the peak region of the soliton but will attenuate the frequencies on either side that arise because of noise. It is important to remember that noise is linear while the soliton is nonlinear. For this reason, solitons "follow" a filter peak but noise does not! The noise is 'clipped', therefore. The problem with an arrangement that uses the same [fixed] filter after each amplifier is that some noise always gets through, under the filter peak. It is then obvious that this noise will grow, and grow, as each successive amplifier is crossed. Also the problem is compounded by the fact that filters also cause loss, as the solitons go through them. This implies a limit as to how strongly noise can be eliminated by them. A temptation is to increase the power of the amplifiers to compensate for this but that only exacerbates the already unwanted noise problem. When this problem was identified it was felt that there had to be a way around this new difficulty. The answer came in the form of sliding-frequency filters [30-32].

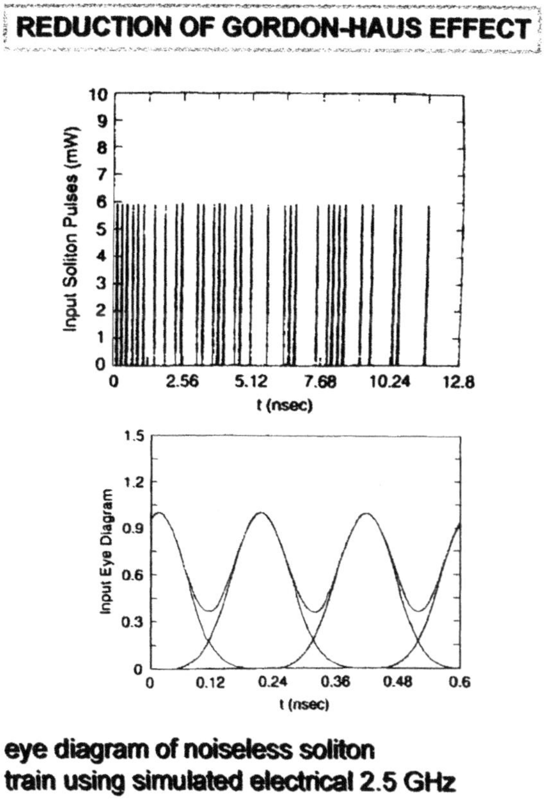

Figure 30a.

NOISE CAUSES JITTER (GORDON-HAUS EFFECT)

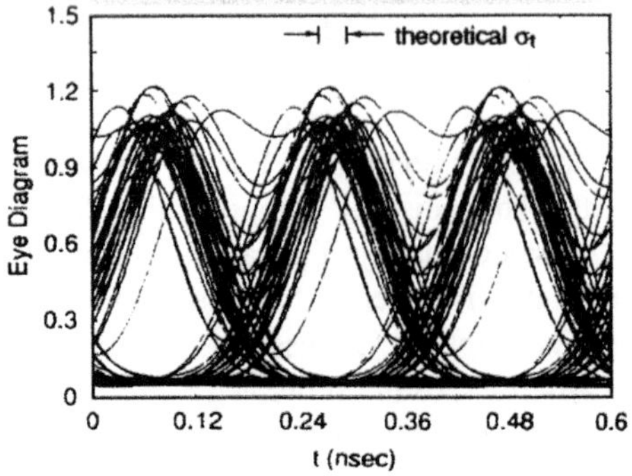

- vertical scale = mW
- fibre length = 9000km
- eye diagram of received solitons
- electrical bandwidth = 2.5GHz

  showing effect of spontaneous emission noise

Figure 30b.

REDUCTION OF GORDON-HAUS EFFECT

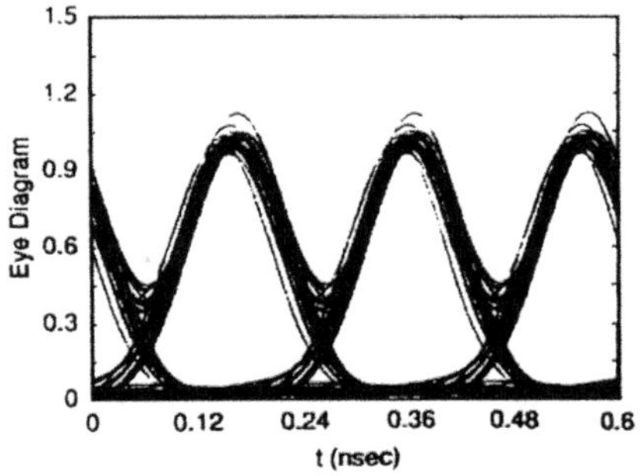

- Fabry-Perot optical filters
- bandwidth of optical filters = 157GHz
- nearly all jitter repressed
- electrical bandwidth = 2.5GHz

OPTICAL FILTERS GET RID OF NOISE (GORDON-HAUS EFFECT)

Figure 30c.

The "sliding" refers to changing the frequency of the peak transmission of the filters placed after each amplifier i.e. if a filter is located after each amplifier [30] it is not necessary for their peak transmission frequencies all to be the same. In fact, the peak frequencies can be selected to move up or down, as the soliton propagates, in any way that is desired. If the filter peak frequency changes are small enough for the soliton to be able to adjust to them then it can be exploited to keep the noise level down. The peak frequency movement is called *zigzagging* - words that nicely express the movement. The idea behind the movement is to exploit the difference between the soliton [a *nonlinear* entity] and the noise [initially, small *linear* random excitations]. The filter "captures" the soliton but keeps the noise at its, initially, low level, or virtually eliminates it, by preventing it from being amplified. It must be emphasised that the fact that the peak transmission frequency of a filter does not coincide with the envelope soliton carrier frequency when the soliton encounters a filter does cause a problem for the soliton. The soliton is a ***robust***, particle-like, entity that resists destruction. Indeed, the soliton will rapidly re-adjust to the new peak frequency dictated by the filter i.e. the soliton will *follow* the filter. The noise, on the other hand, is linear and cannot engage in this 'following'. It is left behind and falls into the 'tail' of the filter [transmission] function.

The operation of a sliding frequency filter is sketched out in figure 31. The soliton and the filter function are shown in the inset, together with the noise. The zigzagging nature of the filter, with respect to frequency-sliding, is shown, in an attempt to illustrate how the noise can be eliminated. In the first instance, only noise under the filter peak gets through. The other stages are to show the effect of sliding. For example, if the peak-frequency slides to the right the 'robust' soliton follows it but noise would find itself under the 'tail' of the filter and so would be absorbed. In a real system a filter is placed after each amplification stage. Suppose that the first filter has a peak transmission frequency equal to the carrier frequency of the emerging [soliton + noise] excitation. The frequency spectrum of the soliton will be transmitted by the filter, together with a noise spectrum lying immediately under the peak. The rest of the noise, in the wings, is eliminated. The transmitted [soliton + noise] packet then travels on to the next amplifier. At this point, the soliton is restored [amplified] once more, to compensate for loss, but the noise, lying under the peak of the filter, is also amplified. If the peak transmission frequency of every one of, perhaps, hundreds of amplifiers is absolutely the same then the noise in the system, far from being eliminated, will grow rather quickly. Moving the peak transmission of the filters backwards and forwards [i.e. zigzagging the peak transmission frequency] as the propagation proceeds keeps the noise down to acceptable levels. Sliding the other way attacks the remaining noise. Hence by setting up filters, whose peak frequency "slides" with distance, the noise from the amplifiers can be virtually eliminated and, with it, the Gordon-Haus effect.

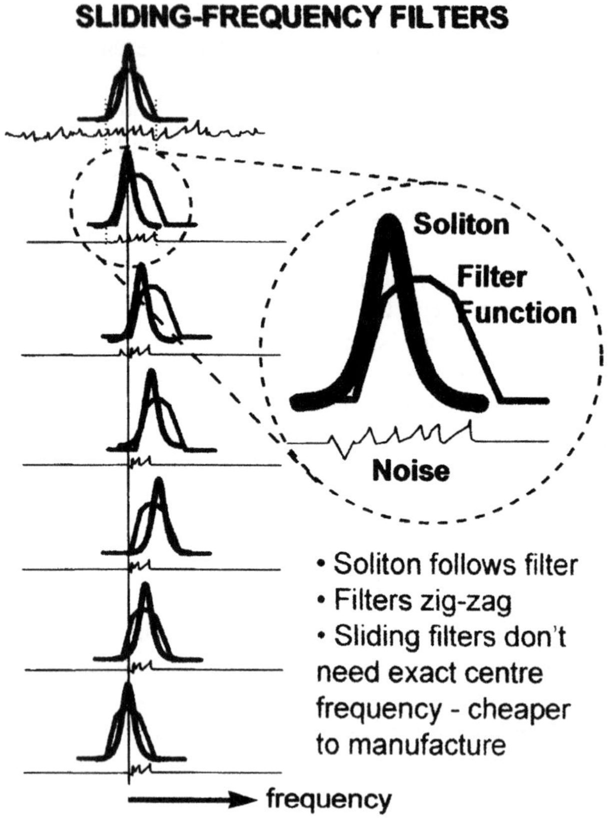

Figure 31.

5. The self-frequency shift

Now that the Gordon-Haus [15] effect has been successfully dealt with, it is time to deal with another problem. It arises through the desire to increase the bit-rate by being tempted to reduce the pulse width into the femtosecond range. The question is whether femtosecond pulses or, at least, pulses that are somewhat below a picosecond in width can be successfully deployed in a soliton transmission system. As the summary in figure 32, shows the problem with a straightforward attempt to transmit femtosecond pulses is that the carrier frequency changes because of an intrapulse Raman effect [32,33]. This means that there is energy transfer from high to low frequencies i.e. *the pulses suffer a red shift*. Because redder pulses travel slower that the original pulse, Raman-shifted pulses are *delayed*, in comparison to their expected arrival times. In the formulation adopted in this chapter pulses are displayed in a frame of reference, moving with a group velocity set by the unshifted carrier frequency. In figure 32, then, propagation down the z-axis is the 'path' of the undelayed pulse. The dotted line is the 'path' of a *delayed* pulse. Because the pulse is delayed, it will not tend to arrive in the acceptance window and is therefore, in a real sense, 'escaping'. The explanation of this self-frequency shifting and any action that can be invoked to combat it will now be addressed.

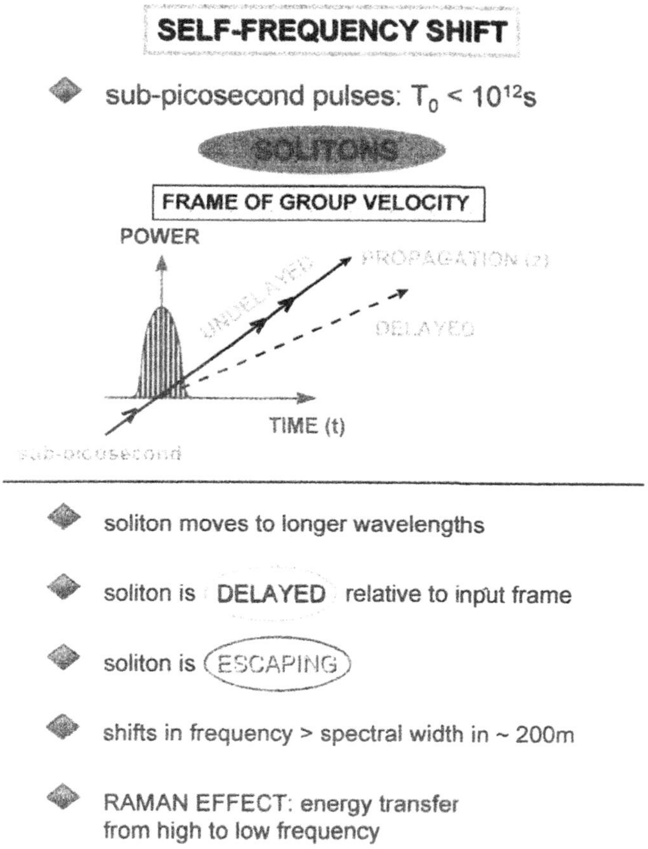

Figure 32.

Normally, soliton propagation is modelled through the nonlinear Schrödinger equation (2.3) but, for femtosecond pulses, something is clearly wrong with the description of the nonlinearity through a local instantaneous, Kerr-effect, term like $|u|^2u$. The answer is that the *local* description must be replaced by a *nonlocal* model to accommodate the Raman effect as a modification to the NLS.

In dimensionless quantities, the nonlinear Schrödinger equation to accommodate nonlocality, must be modified to the following form [32]

$$i\frac{\partial u(t)}{\partial z}+\frac{1}{2}\frac{\partial^2 u(t)}{\partial t^2}+u(t)\int dsf(s)|u(t-s)|^2=0 \tag{5.1}$$

The last term may look a little mysterious but, in actual fact, it has the classic nonlocal form and introduced through a response function f(s). Its structure is very familiar in nonlocal polarisation theory in which the polarisation $P(\tau)$ is not only related to the *local* field E(t) but also to values of field at real times $\tau' < \tau$. The response of the medium, in this case, depends only upon the time difference $(\tau-\tau')$ and *causality* demands that the effect can only be caused at times *less than* τ. In the generic equation (5.1), the function $f(\tau-\tau')$ vanishes for $\tau' > \tau$, therefore. This means that defining $s = [\tau-\tau']$ restricts s to the range $0 \le s \le \infty$ and $f(-|s|) = 0$. Since its inclusion it does not add any new physics to the model, any source of loss, other than that from the intrapulse Raman effect, will be ignored. This simplification makes f(s) real. This kind of formulation replaces the instant response of the medium with a delayed response (equivalent to including retardation in electromagnetic theory). Going to large pulse widths reduces this delay dramatically. Indeed, in the limit of negligible delay, $|u(t-s)|^2$ can be removed entirely from under the integral sign. If this is done then the nonlinear term becomes $|u(t)|^2\, u(t) \int f(s)ds$, implying that $\int f(s)ds = 1$, in these units.

The mathematical treatment of (5.1) looks set to contain some difficulties. The form of equation (5.1), however, suggests the convolution theorem, so casting (5.1) into the frequency domain appears to be a sensible approach. To do this requires the Fourier transform of (5.1) and this treatment begins with the definition

$$u(t) = \int e^{-i\Omega t}\,\tilde{u}(\Omega)d\Omega \tag{5.2}$$

where $\tilde{u}(\Omega)$ is now the Fourier transform, in Ω-space. Equation (5.1) can be transformed, therefore, to

$$i\frac{\partial\tilde{u}(\Omega)}{\partial z} - \frac{\Omega^2}{2}\tilde{u}(\Omega) + I(\Omega) = 0 \tag{5.3}$$

in which

$$\begin{aligned} I(\Omega) &= \int u(t)\left[\int ds f(s)|u(t-s)|^2\right]e^{i\Omega t}dt \\ &= \int u(t)e^{i(\Omega-\Omega'')t}\left[\int ds f(s)|u(t-s)|^2\, e^{i\Omega'' t}dt\right] \end{aligned} \tag{5.4}$$

Now use the convolution property to change $I(\Omega)$ to

$$I(\Omega) = \left[\int d\Omega''\tilde{u}(\Omega-\Omega'')\left[\int f(t)e^{i\Omega'' t}dt\right]\left[\int |u(t)|^2\, e^{i\Omega'' t}dt\right]\right] \tag{5.5}$$

The inner integration contains an interesting term that can be defined as

$$\chi(\Omega'') = \int f(t)e^{i\Omega'' t}dt \tag{5.6}$$

which is recognisable as a susceptibility function in frequency space [actually, it is a third-order susceptibility, in the language of nonlinear optics]. The integration over $|u(t)|^2$ is

$$\begin{aligned}\int u(t)u^*(t)e^{i\Omega'' t}dt &= \iiint d\nu d\gamma dt\, \tilde{u}(\nu)\tilde{u}^*(\gamma)e^{i(\gamma+\Omega''-\nu)t}\\ &= \iint d\nu d\gamma\, \tilde{u}(\nu)\tilde{u}^*(\gamma)\delta(\gamma+\Omega''-\nu)\\ &= \int d\nu\, \tilde{u}(\nu)\tilde{u}^*(\nu-\Omega'')\end{aligned} \tag{5.7}$$

so that

$$I(\Omega) = \int d\Omega''\tilde{u}(\Omega-\Omega'')\chi(\Omega'')\int d\nu\tilde{u}(\nu)\tilde{u}^*(\nu-\Omega'') \tag{5.8}$$

but, if $\nu = \Omega' + \Omega''$, then $d\Omega' = d\nu$ and

$$\int d\nu\, \tilde{u}(\nu)\tilde{u}^*(\nu-\Omega'') = \int d\Omega'\tilde{u}(\Omega''+\Omega')\tilde{u}^*(\Omega') \tag{5.9}$$

The Fourier transform of equation (5.1) then becomes

$$i\frac{\partial\tilde{u}}{\partial z}(\Omega) - \frac{\Omega^2}{2}\tilde{u}(\Omega) + \int d\Omega''\tilde{u}(\Omega-\Omega'')\chi(\Omega'')\int d\Omega'\tilde{u}(\Omega''+\Omega')\tilde{u}^*(\Omega') = 0 \tag{5.10}$$

which is now in the form first quoted by Gordon[32].

Having identified $\chi(\Omega)$ as a third-order susceptibility function it is well-known that Raman activity is associated with the imaginary part i.e. recognising that

$$\chi(\Omega) = \chi'(\Omega) + i\chi''(\Omega) \tag{5.11}$$

yields $2\chi'' = \alpha_R$, the Raman attenuation coefficient. As pointed out by Gordon, associating $\chi''(\Omega)$ with a Raman attenuation coefficient can be seen from setting up a pump-signal arrangement. This method of thinking about nonlinear optical problems is almost the paradigm of nonlinear optics. A pump [strong] signal is used and a signal [weak] is then introduced. The interaction, governed by the physics being investigated, should reveal the controlling parameters.

In frequency space a pump signal with frequency $\Omega = \Omega_p$ and amplitude u_p is written as $u_p\delta(\Omega-\Omega_p)$, where $\delta(\Omega-\Omega_p)$ is a Dirac delta function. The signal is, simply, $\tilde{u}_s(\Omega)$. The physical process is the Raman effect which transfers energy from high to low frequencies i.e. there is a continuous downshift of the mean frequency. Substituting $\tilde{u} = u_p\delta(\Omega - \Omega_p) + \tilde{u}_s(\Omega)$ into equation (5.10) gives, after some lengthy, but straightforward, manipulation

$$i\frac{\partial \tilde{u}_s}{\partial z}(\Omega) - \frac{\Omega^2}{2}\tilde{u}_s(\Omega) + \chi(\Omega - \Omega_p)|u_p|^2\,\tilde{u}_s(\Omega) + |u_p|^2\,\tilde{u}_s(\Omega) = 0 \tag{5.12}$$

where we have collected all the terms for which $\Omega \neq \Omega_p$ and we have noted that $\chi(0) = 1$, in dimensionless [soliton] units. If equation (5.12) is multiplied through by $\tilde{u}_s^*$ and the complex conjugate of equation (5.12) is multiplied by $\tilde{u}_s$ and then the resulting equations are added together,

$$i\left[\tilde{u}_s^*(\Omega)\frac{d\tilde{u}_s}{dz}(\Omega) + \tilde{u}_s(\Omega)\frac{d\tilde{u}_s^*}{dz}(\Omega)\right] + \left[\chi(\Omega - \Omega_p) - \chi^*(\Omega - \Omega_p)\right]|u_p|^2\,|\tilde{u}_s(\Omega)|^2 = 0 \tag{5.13}$$

Equation (5.13) shows, immediately, that

$$\Omega \neq \Omega_p:\; i\frac{\partial}{\partial z}|\tilde{u}_s(\Omega)|^2 + 2i\chi''(\Omega - \Omega_p)|u_p|^2\,|\tilde{u}_s(\Omega)|^2 = 0 \tag{5.14}$$

which is eq. (12) of Gordon's paper [33]. This is an interesting equation because it is well-known that, if I_s is the intensity of the frequency-shifted radiation, called the Stokes wave, I_p is the pump intensity and $G(\nu)$ is the Raman-gain coefficient, [1],

$$\frac{dI_s}{dz} = G(\nu)I_pI_s$$

This CW formula gives the early growth of the Stokes wave and $G(\nu)$, as a function of frequency shift of the signal, from the pump wave, can be obtained from experimental measurements of the spontaneous Raman-scattering cross-section. The gain curve $G(\nu)$, for a fused silica fibre has a significant, well-defined, maximum at 13.2 THz [440 cm^{-1}]. The pump wave provides the energy but the Raman effect is the name covering the scattering of higher frequency photons, by vibrational states of the material, to lower frequencies. If, for example, a signal is entered into the fibre, at frequency ω_s, then if $(\omega_p-\omega_s)$ lies under the Raman gain curve it will be amplified. The greatest amplification will be for $\omega_p-\omega_s = 13.2$ THz. If no signal is used then spontaneous Raman scattering generates enough photon noise to result in amplification right across the entire Raman gain width. $G(\nu)$ behaves like an attenuation coefficient that has dimensions [m/W]. Hence, using so-called soliton units i.e. using the dispersion length as a length scale, P_0 the peak power of the soliton and A_{eff}, the effective modal cross-sectional area, the conversion of $G(\nu)$ to α_R is $\alpha_R = \frac{(L_D P_0)}{A_{eff}} G(\nu)$. The shape of $G(\nu)$ is very convenient, too, because it is practically a linear function of ν over the range $0 \leq \nu \leq 13.2$. Finally, equation (5.14) shows that $\alpha_R = 2\chi''$.

The mean frequency of a soliton is, by definition,

$$\langle\Omega\rangle = \pi \int d\Omega\, \Omega |\tilde{u}(\Omega)|^2 \tag{5.15}$$

and although this is zero *its derivative is not*. This situation is not surprising because, even from a mathematical point of view, it is easily possible to imagine functions that behave like this. $y = x$ for instance has $y = 0$ at $x = 0$ but $\frac{dy}{dx} = 1$ at $x = 0$ i.e. $y = 0$ but $\frac{dy}{dx} \neq 0$ at $x = 0$. Having laboured this point, it is the derivative of (5.15) that is required here i.e.

$$\frac{d\langle\Omega\rangle}{dz} = \pi \int d\Omega\, \Omega \frac{\partial}{\partial z} |\tilde{u}(\Omega)|^2 \tag{5.16}$$

where $\frac{\partial \tilde{u}(\Omega)}{\partial z}$ is given by (5.10). Because for a soliton solution the phase of $\tilde{u}(\Omega)$ is not a function of Ω,

$$\tilde{u}(\Omega)\tilde{u}^*(\Omega - \Omega'') \equiv \tilde{u}^*(\Omega)\tilde{u}(\Omega - \Omega'')$$
$$\tilde{u}(\Omega')\tilde{u}^*(\Omega' + \Omega'') \equiv \tilde{u}^*(\Omega')\tilde{u}(\Omega' + \Omega'')$$

and we obtain

$$\frac{d\langle\Omega\rangle}{dz} = -\pi \int d\Omega'' \alpha_R(\Omega'') \int d\Omega\Omega \tilde{u}^*(\Omega)\tilde{u}(\Omega - \Omega'') \int d\Omega' \tilde{u}^*(\Omega')\tilde{u}(\Omega' + \Omega'') \tag{5.17}$$

as given by Gordon.

The soliton solution $\tilde{u} = \frac{1}{2}\operatorname{sech}\left(\frac{\pi}{2}\Omega\right)$ can now be used to process the right-hand side of (5.17) into a more compact form. Two of the integrals in (5.17) become

$$J_1 = \int d\Omega\Omega \frac{1}{2}\operatorname{sech}\left(\frac{\pi}{2}\Omega\right)\frac{1}{2}\operatorname{sech}\left(\frac{\pi}{2}(\Omega - \Omega'')\right) = \frac{\Omega''^2}{4}\frac{1}{\sinh\left(\frac{\pi}{2}\Omega''\right)} \tag{5.18a}$$

$$J_2 = \int d\Omega' \tilde{u}^*(\Omega')\tilde{u}(\Omega' + \Omega'') = \frac{\Omega''}{2} \frac{1}{\sinh\left(\frac{\pi}{2}\Omega''\right)} \tag{5.18b}$$

Hence, if ω_0 is the centre frequency of the soliton then

$$\frac{d\omega_0}{dz} = \frac{d\langle\Omega\rangle}{dz} = -\frac{\pi}{8}\int \frac{d\Omega''\Omega''^3\alpha_R(\Omega'')}{\sinh^2\left[\frac{\pi}{2}\Omega''\right]} \tag{5.19}$$

and it remains now to find an expression for $\alpha_R(\Omega)$.

A transformation into real world [soliton] units is required. This is done, as usual [1], through the dispersion length $L_D = \frac{T_0^2}{|\beta_2|}$ and the nonlinear length $L_{NL} = 1/(\gamma P_0)$, where T_0 is the soliton pulse width, P_0 is the peak power of the pulse, γ is a nonlinear coefficient and β_2 is called the group-velocity dispersion. In the units used by Gordon, $T_0 \equiv t_c$ and $L_D \equiv z_C$. The nonlinear coefficient is normally expressed as $\gamma = (2\pi n_2)/(\lambda A_{eff})$, where n_2 [cm^2/w] is the coefficient of the nonlinear free-space wavelength of the carrier wave and the A_{eff} is the effective modal area of the fibre. The group-velocity dispersion β_2 is often written as $\beta_2 = -(\lambda^2 D)/(2\pi c)$, where c is the velocity of light in a vacuum and D is called the *dispersion parameter*. Given these length and time scales, the following approach can be adopted.

Since the task is to convert to real units the first operation is to transform the left-hand side of (5.19) to

$$\frac{d\omega_0}{dz} \rightarrow \frac{d\nu_0}{dz}\{2\pi T_0 L_D\}$$

where z, ν_0 now have dimensions, $\omega_0 = 2\pi\,\nu_0$ and a time unit T_0 and length unit L_D are introduced. The second operation is to transform the attenuation coefficient α_R to

$$\alpha_R(\Omega) = R\left[\frac{\Omega}{2\pi T_0}\right] = R(\nu)$$

Hence,

$$\frac{dv_0}{dz} = -\frac{10^5}{2\pi T_0 L_D}\cdot\frac{\pi}{8}\cdot 2\int_0^{\infty}\frac{d\Omega\Omega^3 R\left[\dfrac{\Omega}{2\pi T_0}\right]}{\sinh^2\left(\dfrac{\pi}{2}\Omega\right)}$$

$$= -\frac{10^5\lambda^2 D}{16\pi c T_0^3}\int_0^{\infty}\frac{d\Omega\Omega^3 R[\Omega/2\tau T_0]}{\sinh^2\left(\dfrac{\pi}{2}\Omega\right)}[\mathrm{THz/km}] \tag{5.20}$$

where the length units are now in km.

The peak value of the Raman gain, $G(\nu)$ is 9.9×10^{-16} cm²/w and it occurs at -13.2 THz down from the pump frequency. The Raman loss coefficient is $(\lambda / 2\pi n_2)G(\nu)$ where n_2 is 3.2×10^{-16}, so the peak value of α_R is 0.492. Also $\lambda G(\nu)$ is independent of the wavelength. Assume, therefore, that $R[\Omega/2\pi T_0]$ is *normalised* to a peak value of 0.492. Using an exact form for $G(\nu)$ would add unnecessary complication, especially when R is, more or less, linear below the peak value at 13.2. It seems to be reasonable, therefore, to model R as

$$R(\nu) \approx 0.492\left(\frac{\nu}{13.2}\right) \tag{5.21}$$

Noting now that

$$\int_0^{\infty}\frac{d\Omega\Omega^4}{\sinh^2\left(\dfrac{\pi}{2}\Omega\right)} = \frac{16}{15\pi}$$

yields

$$\frac{dv_0}{dz} = \frac{10^5\lambda^2 D}{16\pi c T_0^3}\cdot\frac{0.492}{13.2}\cdot\frac{1}{2\pi T_0}\cdot\frac{16}{15\pi} \tag{5.22}$$

, in which $v_0 \to -v_0$.

Typical data are as follows

- $\lambda = 1.5\mu m = 1.5 \times 10^{-4}$ cm
- $c = 3 \times 10^{-2}$ cm/ps
- $D = 1500$ps/cm² [i.e. 15ps/nm/km]
- actual use of full-width at half-maximum power instead of T_0 [means using $T_0 = \tau / 1.763$].

Using the data gives [32]

$$\frac{d\nu_0}{dz} = \frac{0.0435678}{\tau^4}[\text{THz}/\text{km}] \tag{5.23}$$

Equation (5.23) shows that the frequency shift is proportional to propagation distance but it is inversely proportional to the fourth power of the pulse width. As can now be imagined, the self-frequency shift is not a problem for ps pulse widths. If, in the desire to bring in ever larger bit rates, fs pulse regimes are entered then equation (5.23) shows that an uncorrected self-frequency is going to be a problem in a real soliton communication system.

Before a way of dealing with self-frequency shifts is discussed it must be emphasised that in the foregoing analysis the attention was focussed entirely upon the delayed nonlinear response. It is, of course, legitimate to do this but there is another important consequence of using ultra-short pfs] pulses and that is *higher-order dispersion* [34,35]. The latter is easily added into the core nonlinear Schrödinger equation because it arises from the third term in the Taylor expansion [14] of the wavenumber. In other words, the third derivative $\frac{\partial^3 u}{\partial t^3}$ gains in importance as the pulse length gets smaller. The combination of third-order dispersion and self-frequency shifting is a lethal mixture that, without compensation, is a destructive force to femtosecond pulses. Figure 32, summarises this situation and indicates that the solution is to introduce band-limited optical gain. Figure 32 contains a sketch to show that, in the frame of reference moving with the group velocity, the pulses are not delayed, if compensation can be found.

In the frame of reference moving with the group velocity, v_g the time t is transformed to $t' = t - z/v_g$. Using full dimensions, therefore, the nonlinear Schrödinger is modified, for fs pulses, from equation (2.2) to

$$\frac{\partial A}{\partial z} + \frac{i}{2}\beta_2\frac{\partial^2 A}{\partial t^2} - \frac{\beta_3}{6}\frac{\partial^3 A}{\partial t^3} + \frac{\alpha}{2}A - i\gamma|A|^2 A + a_1\frac{\partial}{\partial t}(|A|^2 A) + a_2 A\frac{\partial |A|^2}{\partial t} = 0 \tag{5.24}$$

where α is ordinary linear loss, and the dash on t' has been dropped.

All of these terms are required [34,35] for pulses as short as 50 fs and the extra terms have the following interpretations. The $\beta_3 = \left(\frac{\partial^2\beta}{\partial\omega^2}\right)$ clearly comes from continuing the expansion in equation (2.1). The physical reason for its inclusion is that because the pulse is getting narrower in time, its bandwidth is getting quite large, thus forcing a possible inclusion of more terms in the Taylor expansion. The term involving a_2 is another form of the term needed to account for intrapulse Raman scattering, and is responsible for the self-frequency shift. As stated earlier in this section, the imaginary part of the Fourier transform of the third-order susceptibility is related to the Raman gain so a_2 is taken to be *pure imaginary* and defined as being proportional to the slope of the Raman gain. This fits in with our earlier model of $R(\Omega/2\pi T_0)$ being proportional to v. The form $a_2 A \frac{\partial |A|^2}{\partial t}$ is the *approximate* [1] form in the time domain for the operation of the Raman effect. As the pulses get narrower, it is quite evident that they also rise and fall much more steeply. In fact, self-steepening can occur, with the possibility of optical shock formation. a_1 is real, therefore, and, using the parameters of the material and the envelope, is given by $a_1 = -2\gamma / \omega_0$, where ω_0 is the carrier frequency. Other comments are that $a_2 = i\,T_R$, $T_R \sim 5$ fs and that, if the nonlinear response creates a refractive index change Δn, then [1]

$$T_R \frac{\partial \Delta n}{\partial t} + \Delta n = n_2 |A|^2 \tag{5.25}$$

which illustrates, rather nicely, that $\Delta n = n_2 |A|^2$ [the instantaneous value], if $T_R = 0$.

Having, discussed the viability of equation (5.24), how important are the various terms? First of all, third-order dispersion, self-steepening and self-frequency shifting are negligible for ps pulses. Clearly, when $\beta_2 = 0$ then β_3 *must* be used, but, even for $\beta_2 \neq 0$, β_3 plays a significant role when the pulse length gets smaller. We have seen that self-frequency shifting is important, as the pulse width drops into the fs range and clearly the self-steepening term is also important. Hence, although we have managed to isolate the self-frequency effect for a detailed investigation, we need to recognise that the self-frequency shift and the third-order dispersion are forces that will *destroy* femtosecond pulse propagation. The only way to combat this destruction is to introduce some form of bandwidth limited gain. In any model of femtosecond pulse behaviour, however, it is worth including the shock term, even though the shock formation distance is likely to be very large in a typical soliton transmission system. What really concerns us is whether the third-order dispersion, taken together with the self-frequency shifting, can be *completely* compensated. It appears that this is possible and what is needed to make this happen is the addition of an optical gain term $-G(\omega)A$ to equation (5.24). $G(\omega)$ is the net optical gain coefficient $G(\omega) = g - \alpha/2$ and g is the actual optical gain.

Since a femtosecond pulse has a bandwidth approaching the optical gain bandwidth, it is obvious that $G(\omega)$ must be expanded around the carrier frequency ω_0, just as was done for the wavenumber in equation (2.1), and for the same reasons. This being the case then [34]

$$\begin{aligned} G(\omega) = G(\omega_0) &+ \left(\frac{\partial G}{\partial \omega}\right)_{\omega_0} (\omega - \omega_0) \\ &+ \frac{1}{2}\left(\frac{\partial^2 G}{\partial \omega^2}\right)_{\omega_0} (\omega - \omega_0)^2 + \frac{1}{6}\left(\frac{\partial^3 G}{\partial \omega^3}\right)_{\omega_0} (\omega - \omega_0)^3 + \ldots \end{aligned} \tag{5.26}$$

By identifying $(\omega-\omega_0)$ with the time operator $i\frac{\partial}{\partial t}$, $G(\omega)$ can be put into the time domain and equation (5.24) modifies to

$$i\left[\frac{\partial A}{\partial z}+(\beta_1-iG')\frac{\partial A}{\partial t}-G_0 q\right]-\frac{\left[\beta_2-iG''\right]}{2}\frac{\partial^2 A}{\partial t^2}$$
$$+\gamma|A|^2 A-i\frac{\left[\beta_3-iG'''\right]}{6}\frac{\partial^3 A}{\partial t^3}+ia_1\frac{\partial}{\partial t}(|A|^2 A)+ia_2\frac{\partial|A|^2}{\partial t}A=0 \qquad (5.27)$$

where $G_0=G_0(\omega)$, $G'=\left(\frac{\partial G}{\partial \omega}\right)_{\omega_0}$, $G''=\left(\frac{\partial^2 G}{\partial \omega^2}\right)_{\omega_0}$, $G'''=\left(\frac{\partial^3 G}{\partial \omega^3}\right)_{\omega_0}$ and the equation is, once again, expressed in the laboratory frame coordinates.

An elegant analytical solution to (5.24) is

$$A=A_0\,\mathrm{sech}\left[\frac{(t-z/v)}{\tau}\right]\exp(i\Delta kz) \qquad (5.28)$$

where

- $|A_0|^2=-\frac{\beta_2}{(\gamma\tau^2)}$ or $\frac{-\beta_2}{\left[3a_1+2a_2^r\right]}\tau^{-2}$

[r denotes real part]

- $v^{-1}=\beta_1+\frac{\beta_3}{6}\tau^{-2}$
- $\Delta k=-\frac{\beta_2}{2}\tau^{-2}$

Δk is the shift to the carrier wavelength, v is the velocity of the pulse, $G_0=G''=0$, $G'=\frac{a_2^i}{3}|A_0|^2$ [i denotes the imaginary part]. Hence [34]

$$G(\omega)=\frac{a_2^i}{3}|A_0|^2\left[(\omega-\omega_0)+(\omega-\omega_0)^3\tau^2\right] \qquad (5.29)$$

Equation (5.29) operates in the following way.

- $G(\omega)$ compensates for the self-frequency shift by increasing energy of blue components $[\omega-\omega_0>0]$ and decreasing energy of red components $[\omega-\omega_0<0]$
- $G(\omega)$ compensates for third-order dispersion.

In conclusion, it can be asserted that, just as the Gordon-Haus effect was overcome, the problems concerning the joint attack of third-order dispersion and self-frequency shifting on fs pulses can also be solved. The way is open, therefore, to extremely high bit rate optical communication systems. For the remainder of this chapter, however, we will summarise what is currently being achieved with the first generation, ps soliton-pulse systems.

6. Soliton transmission systems

In the preceding sections, the basic ingredients of a possible soliton transmission system have been reviewed, together with a detailed discussion of some major hurdles. Basically, as sketched in figure 33 an electrical bit stream ends up as a modulated optical signal, which is in the form of a stream of solitons. To describe this there is a technical language and it includes terms like IM/DD, which are almost self-explanatory. For *linear* current commercial systems, dispersion-limiting [spreading of the pulses] is the major concern but using solitons as the 'natural bits' of information, at first sight, seems limited by the Gordon-Haus effect. Although this can be sorted out with filters, there is going to be a basic difference between modulation formats. As figure 34 shows that soliton systems will use RZ but it is NRZ that is in the marketplace.

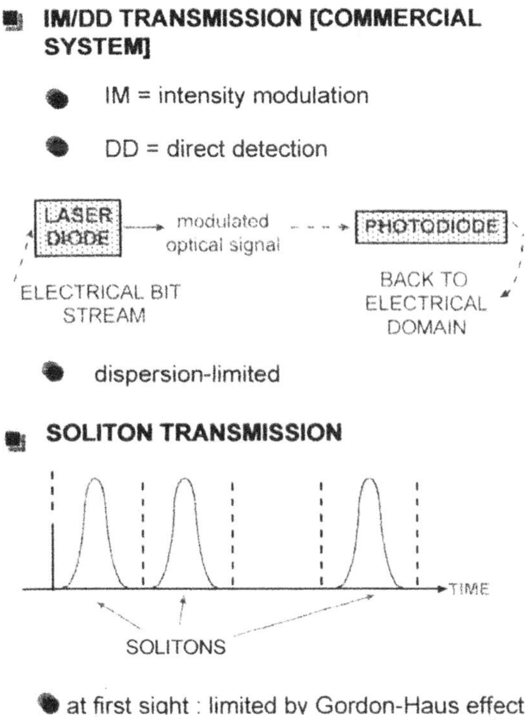

Figure 33.

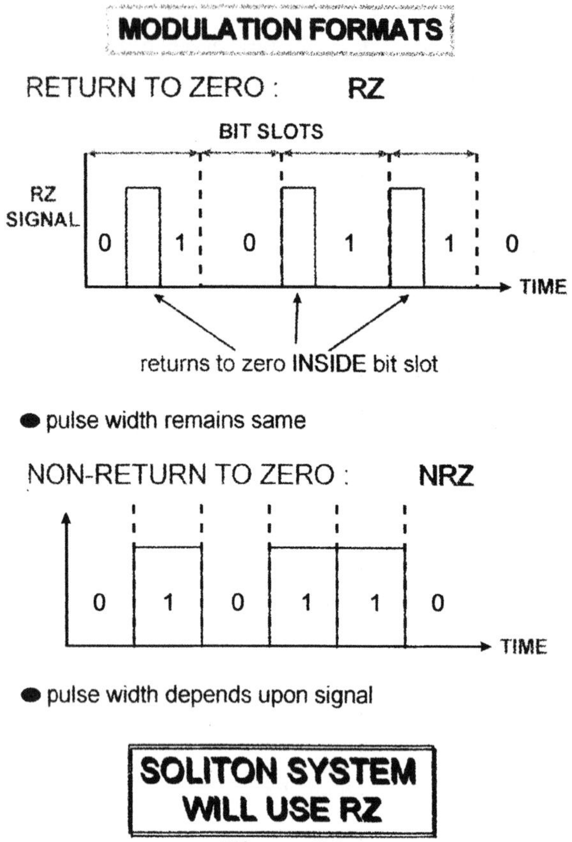

Figure 34.

Even in a *supposedly linear* telecommunication system, nonlinear effects [36] cannot be avoided, especially when multichannel operation is attempted. To emphasises this, figure 35 summarises the nonlinear crosstalk situation encountered by the *linear system* engineer. Four nonlinear effects are listed and the power per channel, in mW, as a function of the number of channels requested is also given. Raman scattering is not very limiting but Brillouin scattering is, potentially, a very real problem. The good news is that the latter has a gain bandwidth that is very narrow, so the channel spacing must match this bandwidth pretty well exactly, if stimulated Brillouin scattering is to be a problem. Also, any two channels must sustain *counterpropagating* waves to get Brillouin crosstalk [36]. All we have to do to avoid this is to make sure everything in each channel is going the same way. Brillouin never goes away, however, and 10mW seems to be an upper channel power limit to avoid Brillouin effects. Other limitations like four-wave mixing (FWM) and cross-phase modulation (XPM) are also shown in figure 35.

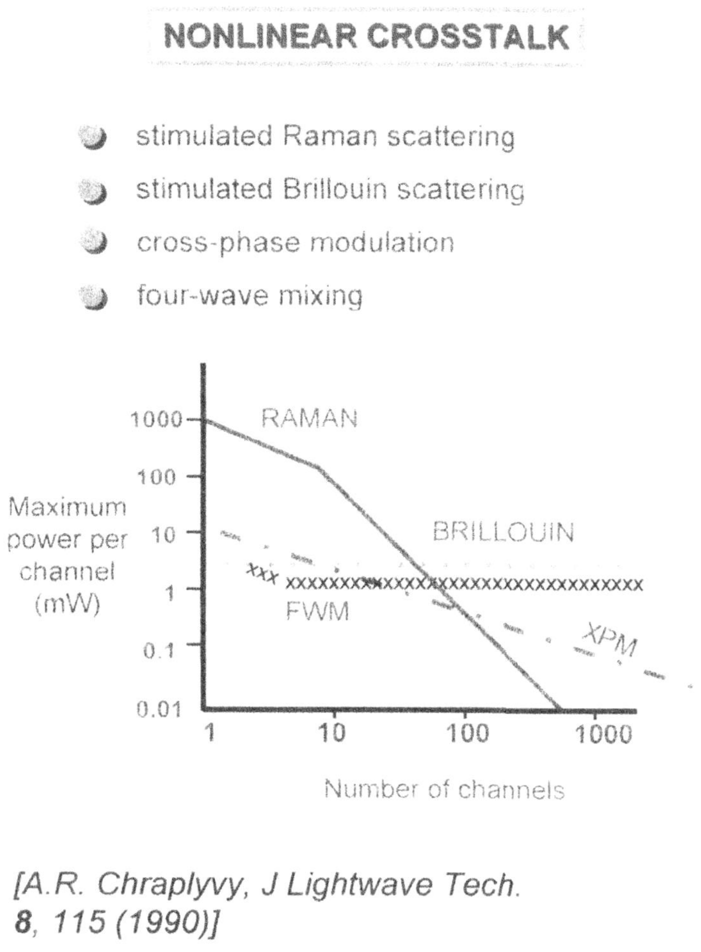

Figure 35.

The transmission distances of both conventional (linear) and soliton transmission systems against bit rates are shown in figure 36. In a linear system, the pulses will spread out due to group-velocity dispersion, so the value of $|\beta_2|$ clearly places an upper limit to the total transmission distance, for a given bit rate. Although the figure shows where the limit lies for solitons, if they are controlled by the Gordon-Haus effect, the breaching of this limit is addressed in detail, in this chapter. Hence this line, in figure 36, is there to show what could will if filters are not used. Even so solitons still offer a superior system if the bit rate gets high enough.

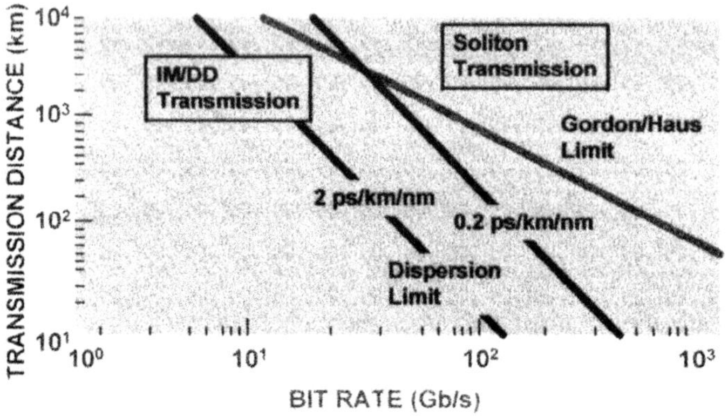

[M. Nakazawa, Photonics Spectra, February, 1996]

- 10-20 Gb/s systems ready for commercial use
- solitons → higher speeds
- lower bit rates < 10 Gb/s and few thousand of km and GVD = 0.2 ps/km/nm

⇓

CONVENTIONAL SUPERIOR TO SOLITON

- high bit rates → **SOLITON WINS**

Figure 36.

In any real system there will be linear loss and, obviously, that is the need, in the first place, for in-line amplifiers. In dimensionless quantities, ps pulse propagation is controlled by the usual nonlinear Schrödinger equation, which has an added linear loss term. If this loss is characterised by a dimensionless parameter Γ, then the, modified, nonlinear Schrödinger equation is

$$i\frac{\partial u}{\partial z}+\frac{1}{2}\frac{\partial^2 u}{\partial t^2}+N^2|u|^2 u=-i\Gamma u \tag{6.1}$$

where N is the order of the soliton. Note that it is common to define a new quantity u' = Nu so that N disappears from view. For this discussion, however, the parameter N is needed so that if (6.1) is solved numerically, N = 1 specifies the lowest-order soliton, provided that the initial condition is $u(0,t) = \text{sech}(t)$. If a lumped amplifier scheme is used then each amplifier, with gain $G_0 = G(\omega_0)$, ω_0 being the centre frequency, completely restores the reduction in the soliton amplitude arising from linear loss encountered between the amplifier stages.

Before amplifiers are introduced, however, a very neat transformation can be made to equation (6.1). If $V = u(z,t)\, e^{-\Gamma z}$ is introduced to equation (6.1) it becomes

$$i\frac{\partial v}{\partial z}+\frac{1}{2}\frac{\partial^2 v}{\partial t^2}+N^2 \exp[-2\Gamma z]|v|^2 v=0 \tag{6.2}$$

Equation (6.2) is now like the unmodified nonlinear Schrődinger equation, except for the fact that N *reduces* as the soliton propagates. Of course, equation (6.2) is more difficult to handle in its present form, than (6.1), because of the $\exp[-2\Gamma z]$ factor, so it would be useful to be able to average it over a distance equal to the amplifier spacing and then try to put the Γ in terms of the amplifier gain G_0. If a soliton drops significantly in amplitude between amplifier stages its width will increase, accordingly, in order to conserve energy. A real system is designed, however, to make sure that the soliton does not actually suffer much of a change in its peak amplitude, between amplifiers. It is in any case, *restored* at each amplifier. Since the peak amplitude does not change much an adiabatic approximation holds and the $\exp(-\Gamma z)$ term in equation (6.2) can be averaged over an amplifier spacing i.e. an average value N_{av} can be introduced, in a similar sense to the *average soliton* or *guiding-centre* soliton, [2,37]

$$N_{av}^2=\frac{N^2}{L}\int_0^{L_A} e^{-2\Gamma z}dz=N^2\left[\frac{1-\exp(-\alpha L_A)}{\alpha L_A}\right] \tag{6.3}$$

In equation (6.3), the *real* quantity α is used, together with the *real* distance L_A, as opposed to dimensionless quantities. On *average* N_{av} is required to be 1 if the system is to behave as if, on average, a fundamental soliton is propagating, so this restraint gives

$$N=\left[\frac{\alpha L_A}{1-\exp(-\alpha L_A)}\right]^{\frac{1}{2}} \tag{6.4}$$

In other words, this must be the selected value of N to compute the Nsech(t) input pulse to the system, if it is, on average, to propagate as an $N = 1$ soliton. This is called the *pre-emphasis* method. The argument does not stop there, however, because the amplifier gain can now be brought into the argument.

The fact that each amplifier stage has a gain G_0 means that, given a distance L_A, between amplifier stages,

$$G_0 e^{-\Gamma L_A} = 1 \qquad (6.5)$$

which implies that $\Gamma L_A = \ln(G_0)$ i.e.

$$N = \sqrt{\frac{G_0 \ln(G_0)}{G_0 - 1}} \approx \sqrt{\ln(G_0)} \quad [G_0 >> 1] \qquad (6.6)$$

From equation (6.6), it can be seen that, $N_{av} > 1$, since, typically, $G_0 = 10$, or more. In fact, for $G_0 = 10$, $N = 1.6$.

The concept of average soliton, or guiding centre soliton, is another saving grace for the possibility of realising a soliton transmission system. All that needs to be done is make sure that the *initial* pulse value at the beginning of a 'run' between amplifiers, has a non-integer value $N > 1$, as specified by (6.6). As propagation proceeds, the soliton drops in amplitude but is *pumped up* again at each amplifier position. Simulations show, convincingly, that, *on average*, an $N = 1$ soliton is transmitted.

If the soliton in a transmission system sits in a time slot, of width T_B, and its half-width is T_0, then B, the bit rate, is $1/T_B$. If two solitons, in dimensionless units, appear at $z = 0$ and they are represented as $u(0,t) = \text{sech}[t-q_0] + r\,\text{sech}[r(t+q_0)]$, then their time separation is $2q_0T_0$ and $B = 1/(2q_0T_0)$. Dispersive build up could occur in a system containing a soliton stream, because after solitons lose out to damping and they then *regain their shape* through amplification. They regain shape by shaking loose "unwanted" energy as a dispersive wave, so the amplifier spacing must be less than the dispersion length to avoid this nuisance. This implies that L_A, the dimensionless distance between amplifiers, satisfies the relationship [36]

$$B^2 L_A < \left[4q_0^2|\beta_2|\right]^{-1} \qquad (6.7)$$

For $\beta_2 = -20\text{ps}^2/\text{km}$, $B = 5\text{Gb/s}$ @ $1.55\mu\text{m}$, equation (6.7) implies an amplifier spacing of 5km.

Manipulating the system to overcome loss is not the only thing engineers [5,38,39] can do to promote soliton transmission. In this context, Nakazawa has presented a very nice review of what is called *dispersion allocation*, which is summarised in figure 37. The idea is to use already-installed optical fibre for soliton transmission. As we saw at the very beginning of these lectures, however, β_2 must be negative to permit soliton transmission i.e. we must have anomalous dispersion. It would be very nice, therefore, if an examination of already-installed fibre showed that it possessed a negative β_2 value at $1.55\mu\text{m}$. This is unlikely to be the case because they were not designed for use at $1.55\mu\text{m}$. The best hope is that the already-installed fibre has anomalous dispersion, *on average*, during the propagation of an initial soliton pulse. In other words, perhaps the concept of *average soliton* applies in this case as well. Figure 37 shows some work of

Nakazawa, who emphasises that dispersion-allocated solitons [38,39] do rather well, over vast propagation distances. In figure 37 the amplifier spacing is 90km and the 'run' between the amplifiers has anomalous dispersion for 60km and normal dispersion for 30km. The average is -0.5ps/km/nm and because this number is negative, it is precisely what was being hoped for. Nakazawa went further and checked out the fibres in the Tokyo metropolitan area [38], shown in figure 38. He discovered that two-thirds of the installed fibre did, indeed, exhibit anomalous dispersion and the experiment with solitons in the metro area of Tokyo went very well.

SOLITON TRANSMISSION SYSTEM

Dispersion sequence and eye-diagram of 20ps DISPERSION ALLOCATED soliton.

10 Gb/s over 1170km

[M. Nakazawa, Photonics Spectra, February, 1996]

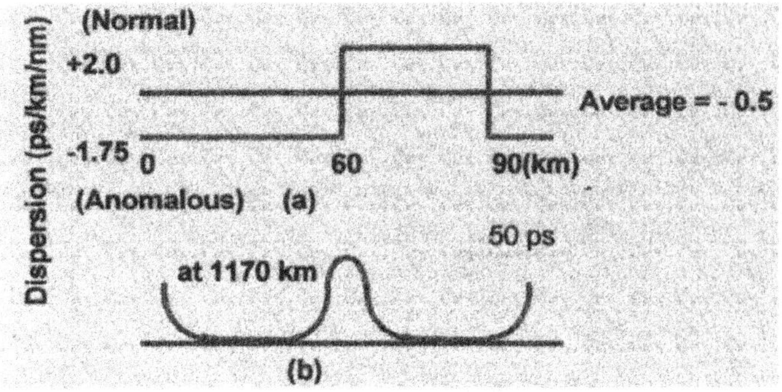

- technique for using already installed fibre

↓

average GVD is anomalous

- amplifier spacing = 90km
- 30km in positive GVD region
  60km in negative GVD region
- soliton period = 317 km >> amplifier spacing

Figure 37.

FIELD DEMONSTRATIONS

1995 : M. Nakazawa et al, Elec. Lett. ***31****, 992*

- soliton transmission at 10 Gb/s over 2000km in Tokyo metropolitan optical loop network

1995 : M. Nakazawa et al, Elec. Lett. ***31****, 1478*

- soliton transmission at 20 Gb/s over 2000km in Tokyo metropolitan

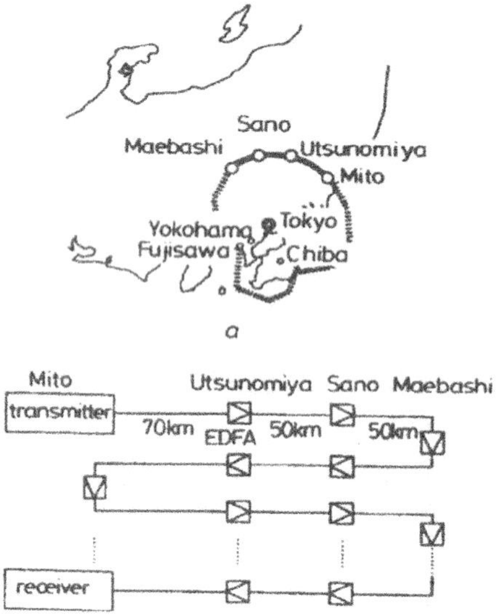

Figure 38.

Over the years a lot of experimental progress has been made that owes much to the pioneering work of Mollenauer and Nakazawa. Major achievements have been made and many more are still to come. The soliton transmission system is not here yet but the green shoots of success are clearly visible. Maybe NRZ systems will win in the end but this article will finish on a note of optimism by first of all noting that very short pulses (fs) have still to be investigated properly, especially in the area of soliton interactions. It is also the case that the highly desirable wavelength division multiplexing (WDM) schemes remain to be analysed satisfactorily. There are other problems involving modulation, switching and so on, which no doubt will keep the subject alive for some time to come.

References

1. Agrawal, G.P. (1995) *Nonlinear Fiber Optics*, Academic Press, San Diego.
2. Hasegawa, A. and Kodama, Y. (1995) *Solitons in Optical Communications*, Clarendon Press, Oxford.
3. Remoissenet, M. (1995) *Waves Called Solitons*, Springer-Verlag, Berlin.
4. Doran, N. and Bennion, I. (1996) Optical networks to span the globe, *Physics World* **9**, 35-39.
5. Nakazawa, M. (1996) Telecommunications rides a new wave, *Photonics Spectra* (February) 98-104.
6. Boardman, A.D. and Xie, K. (1996) Waveguide-based devices : linear and nonlinear coupling in *Devices Based on Low-Dimensional Semiconductor Structures*, M Balkanski (Ed.), Kluwer Academic Publishers, London, 231-292.
7. Russell, J.S. (1844) *Report on Waves*, 14th Meeting of British Association of Advances in Science, York, 311-390.
8. Boardman, A.D. and Kie, X. (1993) Theory of spatial solitons, *Radio Science* **28**, 891-899.
9. Boardman, A.D. and Cooper, G.S. (1988) Power-dependent polarization of optical pulses, *J. Opt. Soc. Am* **5**, 403-418.
10. Boardman, A.D., Brown, G. and Cooper, G.S. (1982) Solitons in single-mode optical fibres in *Tenth International Optical Computing Conference*, S Horovitz (ed.) *SPIE* **422**, 369-377.
11. Boardman, A.D. and Cooper, G.S. (1988) Evolution of circularly polarised pulses in nonlinear optical fibres, *Journal of Mod. Opt.* **35**, 407-417.
12. Boardman, A.D., Mehta, H.M., Sangarpaul,. A.K., Putman, R. and Xie, K. (1994) Interactions in bright N-soliton trains propagating in birefringent optical fibres, *Opt. Comm.* **116**, 208-218.
13. Islam, M.N. (1992) *Ultrafast Fiber Switching Devices and Systems*, Cambridge University Press, Cambridge, U.K.
14. Boardman, A.D., Nikitov, S.A., Xie, K. and Mehta, H.M. (1995) Bright magnetostatic spin-wave envelope solitons in ferromagnetic films, *Journal of Magnetism and Magnetic Materials* **145**, 357-378.
15. Gordon, J.P. and Haus, H.A. (1986) Random walk of coherently amplified solitons in optical fiber transmission, *Optics Letters* **11**, 665-667.
16. Tocci, R.J. (1991) *Digital Systems : Principles and Applications*, Prentice-Hall, Englewood Cliffs, New Jersey.
17. Morris, N.M. (1983) *Logic Circuits*, McGraw-Hill, New York
18. Oh, Y., Haus, J.W. and Fork, R.L. (1996) Soliton-repulsion logic gate, *Optics Letters* **21**, 315-317.
19. Menyuk, C.R. (1988) Stability of solitons in birefringent fibres II. Arbitrary amplitudes, *J. Opt. Soc. Am. B.* **5**, 392-402.
20. Kivshar, Y.S. (1990) Soliton stability in birefringent optical fibres: analytical approach, *J. Opt. Soc. Am. B.* **7**, 2204-2209.

100GHz*100km Optical Soliton Data Transmission System, Based on Gradient Distributed Er^{3+}-Doped Fiber Amplifiers

T. MIRTCHEV

Quantum Electr. Dept., Sofia University, Sofia, BULGARIA
Phone: 359-2-62561887, Fax: 359-2-9625276

e-mail: todorm@phys1.phys.uni-sofia.bg

1. Introduction.

Recently it became clear that the direct implementation of the reported so far soliton stabilising techniques [1,2] in practical communication systems with standard distributed Er^{3+}-doped fiber amplifiers (DEDFA) is not possible. A significant excursion of the pump power and hence of the gain coefficient inevitably exists even in bidirectionally pumped long DEDFA's because of pump attenuation [3]. Thus the stabilising values [1] of the gain coefficient and the detuning from the resonance frequency can not be maintained along the whole length.

In this work a new kind of device is proposed and analysed - a distributed Er-doped fiber amplifier with longitudinally gradient dopant concentration (GFA). The performance of GFA's is numerically studied and it is shown that an almost perfectly constant gain coefficient can be maintained. Further, the GFA gain distribution data are used to model the dynamics of solitary pulse propagation in such devices. The results indicate that an optimal amplification configuration, which simultaneously minimises the influence of the background attenuation, the soliton self-frequency shift and the third order dispersion on the solitary signal pulses can be chosen and maintained in GFA's. Although in the general case the perturbations cannot be completely cancelled [4] in practical devices, the results of the computer model clearly show that a gradient fiber amplifier can be designed as

M. Balkanski and N. Andreev (eds.),
Advanced Electronic Technologies and Systems Based on Low-Dimensional Quantum Devices, 267–275.

a transparent transport medium for high-speed, long distance optical fiber communications.

The proposed devices and stabilisation technique may prove to be a viable alternative to the other reported methods for stabilising soliton pulses against attenuation, SSFS and/or third order dispersion [1,2].

2. Numerical model of the gain distribution.

In the case of resonant pumping in the 1.48µm band, the erbium doped fibers act as quasi-two level amplifiers, i.e. the emission cross sections from the metastable level 2 and the pump level 3 are equal [5]. Hence, in the present notations the rate equations for the population densities have the form:

$$\frac{dN_1(z,t)}{dt}=\frac{\Gamma_P}{h\nu_P A_{eff}}\left[\sigma_{pe}N_3+\sigma_{pe}N_2-\sigma_{pa}N_1\right]\left[P_p^{+}+P_p^{-}\right]+A_{21}N_2=0 \qquad /1/$$

$$\frac{dN_2(z,t)}{dt}=-\frac{\sigma_{pe}\Gamma_P}{h\nu_P A_{eff}}\left[P_p^{+}+P_p^{-}\right]N_2-A_{21}N_2+A_{32}N_3=0 \qquad /2/$$

$$N_3=N_t\left(z\right)-N_1-N_2 \qquad /3/,$$

where $N_{1,2,3}$ are the densities of the ground, metastable and pump levels respectively, N_t - the total Er^{3+} concentration, P_p^+ and P_p^- are the co- and counterpropagating (with respect to the signal) pump powers, ν_p is the frequency of the pump wave and σ_{pa} and σ_{pe} are respectively the absorption and emission cross. The other parameters in the system are the effective area of the fiber core A_{eff}, the spontaneous emission rate A_{21}, the rate A_{32} of the nonradiative relaxation from level 3 to level 2 and the pump mode-to-erbium transverse profile overlap factor Γ_p.

The coupled equations describing the longitudinal evolution of the pump waves are of the form [3,5]:

$$\frac{dP_p^{+}(z,t)}{dz}=-P_p^{+}\,\Gamma_p\left[\sigma_{pa}N_1-\sigma_{pe}N_3-\sigma_{pe}N_2\right]-\alpha_p P_p^{+} \qquad /4/$$

$$\frac{dP_p^{-}(z,t)}{dz}=+P_p^{-}\,\Gamma_p\left[\sigma_{pa}N_1-\sigma_{pe}N_3-\sigma_{pe}N_2\right]+\alpha_p P_p^{-} \qquad /5/,$$

where α_p is the background loss at the pump wavelength $\lambda_p = 1480$ nm.

Note in eq. /3/ the explicitly pointed out dependence of N_t on the space coordinate z. In the following section the aim will be to choose the analytical form of $N_t(z)$ in a way that renders as nearly independent of z the population densities N_1, N_2 and the resonant signal gain coefficient

$$g(z) = \sigma_{se} N_2 - \sigma_{sa} N_1 \qquad /6/,$$

where σ_{se} and σ_{sa} are the emission and absorption cross sections at the signal wavelength λ_s .

3. Gain distribution in GFA's.

Using the fact that in the standard DEDFA's the pump power decreases almost exponentially over the length of the fiber, it was reasonable to expect that gradient amplifiers with a similar dependence $N_t(z)$ - EGFA's - would give acceptable results. It is *a priori* clear, that longer distances can be achieved by bi-directional pumping. We propose a bi-directionally pumped EGFA, in it's simplest symmetrical form, as a device, in which the dopant concentration increases bilaterally toward the midpoint of the amplifier. The z dependence of the dopant concentration is described by eq. /7b/ for $0<z<(L/2)$ and by /7a/ for $(L/2)<z<L$, where L is the full length of the EGFA .

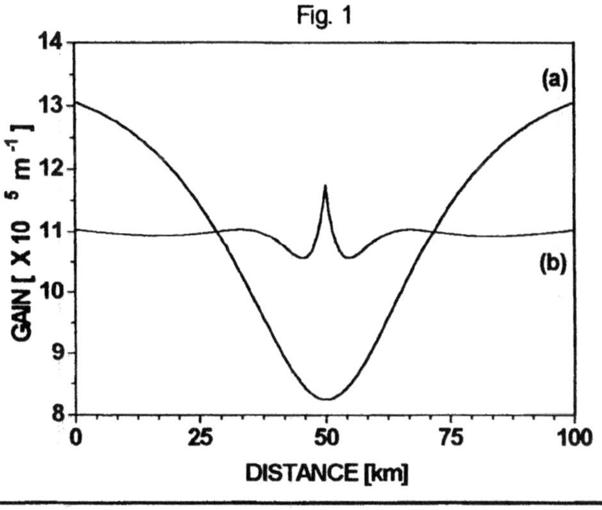

Figure 1.: Gain distributions in 100km long DEDFA (curve (a)) and EGFA (b) with N_t=0.95x10^{15}[cm^{-3}]. The parameters of the a, c_1, c_2, N_t^{max} and the pump powers are given in the text.

$$N_t = N_t^{\max}\left[c_1 + c_2 \exp\left(-a\frac{z}{L}\right)\right]$$ **(backward pumping) /7a/**

$$N_t = N_t^{\max}\left[c_1 + c_2 \exp\left(-a\frac{(L-z)}{L}\right)\right]$$ **(forward pumping) /7b/**

Fig. (1) illustrates the variations of the resonant signal gain coefficient g(z) along two 100 km long, bidirectionally pumped fiber amplifiers EGFA. Curve (a) corresponds to a usual DEDFA with constant concentration and curve (b) - to a EGFA with parameters a=4.35, c_1=0.33 and c_2=0.67 and N_t^{max}=1.97x10^{-15}[cm^{-3}]. The average dopant concentration of the two amplifiers was N_t=0.95x10^{15}[cm^{-3}]. The amplifier from cases (a) is pumped with P_p^+=P_p^-=85mW. The results for the EGFA corresponding to case (b) were most uniform when the pump powers were P_p^+=P_p^- =65mW. Obviously such an amplifier is almost perfectly transparent for the signal in every point of its length. The computer optimisations showed that the EGFA's are relatively insensitive to the amount of pump power and to the values of c_1 and c_2. The factors that influence significantly their behaviour are the average dopant concentration, which changes strongly the average gain coefficient, and the value of the parameter *a* (see eq. /7/), which is decisive for the gain coefficient uniformity.

4. A model of short solitary pulse propagation in GFA's.

In a number of papers the DEDFA's have recently been successfully modelled as unsaturable, homogeneously broadened amplifiers [5,6]. It is well known that, starting from the Maxwell-Bloch equations, the complex gain coefficient spectrum of such amplifier can be written as [7]

$$g(\xi,\omega) = \frac{Lg_0(\xi)}{1 - iLT_{21}(\omega-\omega_L)} \qquad /8/,$$

where T_{21} is the phase-relaxation time and $g_0(\xi)$ contains the dependence of the scaled gain coefficient on the normalised space coordinate ξ. Eq /10/ describes properly the off-resonant cases by introducing the additional Lorentzian factor $L = \left[1 - iT_{21}\Omega\right]^{-1}$, which includes the detuning $\Omega=\omega_L-\omega_{21}$ between the carrier frequency of the signal pulses ω_L and the resonant frequency ω_{21}. Eq. /8/ can now be expanded in the vicinity of ω_L to obtain

$$g(\xi,\omega) = g_0(\xi)\left\{ L\sum_{m=0}^{\infty}\left[iLT_{21}(\omega-\omega_L)\right]^m \right\} \qquad /9/$$

Defining $D_R^{(m)}(\xi) = g_0(\xi)(-T_{21})^m \operatorname{Re}\left(L^{m+1}\right)$, $D_I^{(m)}(\xi) = g_0(\xi)(-T_{21})^m \operatorname{Im}\left(L^{m+1}\right)$

and Fourier transforming eq. /9/, the expression for the gain operator in the time domain becomes

$$g(\xi,t) = \sum_{m=0}^{\infty}\left[\left[[D_R^{(m)}(\xi) + i\,D_I^{(m)}(\xi)\right]\frac{d^m}{dt^m}\right] \qquad /10/$$

Included in a propagation equation, eq. /10/ properly accounts for the gain dispersion and the near-resonant dispersion of the refractive index for arbitrary detunings Ω. The modified nonlinear Schrödinger equation [8], extended with the truncated after m=3 series /10/ has the form:

$$i\frac{\partial A}{\partial \xi} = -\frac{1}{2}\frac{\partial^2 A}{\partial \tau^2} - |A|^2 A - i\gamma A + \tau_R \frac{\partial |A|^2}{\partial \tau} A + i\beta\frac{\partial A^3}{\partial \tau^3} + iD_R^{(0)}(\xi)A$$
$$+ D_I^{(1)}(\xi)\frac{\partial A}{\partial \tau} + iD_R^{(2)}(\xi)\frac{\partial^2 A}{\partial \tau^2} + D_I^{(3)}(\xi)\frac{\partial^3 A}{\partial \tau^3} + iD_R^{(3)}(\xi)\frac{\partial^3 A}{\partial \tau^3} \qquad /11/,$$

where ξ and τ are respectively the space and time coordinates and A is the slowly varying amplitude of the solitary pulses. Except the complex gain dispersion (up to third order) this variant of the equation includes also the perturbative effects of the background attenuation γ, the soliton self-frequency shift (the term with τ_R) and the nonresonant third order material dispersion (the term with β). All the quantities in /11/ are expressed in the well known "soliton units" [8].

The next step is to apply the perturbative approach [9] to eq. /11/. The following set of equations, describing the evolution of the soliton amplitude η and speed k is easily obtained from /11/ :

$$\frac{\partial \eta}{\partial \xi} = -2\gamma + 2\eta D_R^{(0)}(\xi) - 2k\eta\, D_I^{(1)}(\xi) - 2D_R^{(2)}(\xi)\left(\frac{\eta^3}{3} + \eta k^3\right) + 2D_I^{(3)}(\xi)\left(\eta^3 k - \eta k^3\right)$$

$$/12/,$$

$$\frac{\partial k}{\partial \xi} = -\frac{8}{15}\eta^4\tau_R - \frac{2}{3}D_I^{(1)}(\xi)\eta^2 - \frac{4}{3}D_R^{(2)}(\xi)k\eta^2 + D_I^3(\xi)\left(\frac{14}{15}\eta^4 - 4k^2\eta^2 + \frac{4}{3}k^4\right)$$

If for a moment the fact that $D_{R,I}$ depends on ξ is ignored, eq. /12/ directly give two quite simple conditions for the stable propagation of a soliton with constant amplitude η=1 and constant speed k=0

$$A\left(g_0,\Omega\right) = -2\gamma + 2D_R^{(0)} - \frac{2}{3}D_R^{(2)} = 0 \qquad /13a/$$

$$B\left(g_0,\Omega\right) = -\frac{8}{15}\tau_R - \frac{2}{3}D_I^{(1)} + \frac{14}{15}D_I^{(3)} = 0 \qquad /13b/$$

This set of conditions can be extended with the obvious condition for the mutual cancellation of the nonresonant 3rd-order material dispersion and the resonant 3rd-order dispersion terms in /11/ :

$$C\left(g_0,\Omega\right) = \beta + D_R^{(3)} = 0 \qquad /13c/$$

The physical meaning of the three expressions /13a,b,c/ is that the perturbations must be counterbalanced by the corresponding gain-induced factor. Mathematically they form a set of nonlinear algebraic equations, from which the stabilising values of the amplifier parameters g_0, T_{21} (or equivalently the input pulse duration T=0.568T_{fwhm} , which scales the phase-relaxation time) and Ω must be determined. Hence, the problem for the stabilisation of short solitary pulses by off-resonant bandwidth-limited distributed amplification must be treated as the nonlinear optimisation problem :

$$M\left(g_0,\Omega\right) = \min\ (w_A A^2 + w_B B^2 + w_C C^2) \qquad /14/,$$

where w_A, w_B and w_C are weight coefficients. The optimal values g_{opt}, T_{opt} and Ω_{opt}, which simultaneously minimise the influence of the perturbations, can be obtained after the iterative numerical minimisation of /14/. The procedure must be iterated, because the values of g_0, Ω and T enter the "soliton units", thus influencing the normalised values of the perturbations and the right hand sides of /13a,b,c/. Normally 3 or 4 steps are sufficient to get constant optimal values.

5. Stabilised short solitary pulse propagation in GFA's.

In this section the rate equations /1/-/5/ for the graded fiber amplifiers and the set of perturbative equations /12/ are combined and simultaneously solved to study the propagation of short solitary pulses in bilaterally pumped EGFA's. Here the values $\Delta\lambda$=10nm and λ_1=1330nm are used. Thus the phase-relaxation time, entering eq /8/-/14/ is T_{21}=350fsec. The cross-sections and fiber parameters, characterising the

amplifier have typical values [3] corresponding to λ_p =1480nm. The initial signal pulse duration is T_{fwhm} =1psec. The input pulses are assumed to be exact fundamental solitons. The group velocity dispersion at the signal wavelength is $k_s^{(2)}$=3ps^2/km (D=2.4ps/km/nm), the third order dispersion coefficient is $k_s^{(3)}$=0.1ps^3/km, the characteristic Raman self-scattering time is T_R=5fsec and the background attenuaton is α_s=0.25dB/km .

For the given set of parameters the procedure /14/ were applied with w_A=1, w_B=1 and w_C=0.3 . In real units the optimal values are g_{opt} =1.096x10^{-4}[m^{-1}] and Ω_{opt}=-1THz, (signal wavelength λ_s≈1533.5nm).

Fig. (2) and fig. (3) show respectively the calculated evolutions of η and k. Curves (a), on both figures, illustrate the case of standard DEDFA and curves (b)- the case of EGFA. The dissimilarities of these propagation regimes, are rather prominent. As it is seen from curves (a) in both figures, in a standard DEDFA the pulses totally slip off the resonance and are absorbed. Much better results are obtained for a EGFA (curve (b)). Although perturbed by the local gain peak in the centre of the fiber (see fig. 1b) the signal pulses accumulate minor speed and amplitude changes. Thus the EGFA, although pumped with less power than the standard DEDFA, can assure almost completely stable propagation of signal with repetition rate up to 100 GHz over 100 km long communication links.

Fig. 2

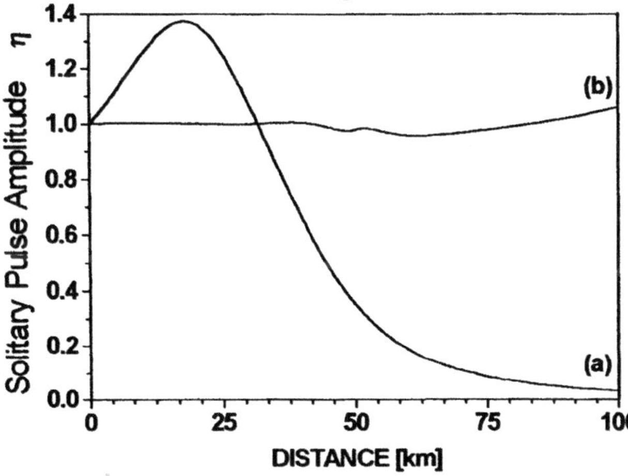

Figure 2.: The evolution of the amplitude η of a solitary pulse with T_{fwhm} =1psec. along a 100km long DEDFA (curve (a)) and EGFA (curve (b)) . The values of the perturbation constants, the optimal gain coefficient, detuning, pump power, etc. are given in the text

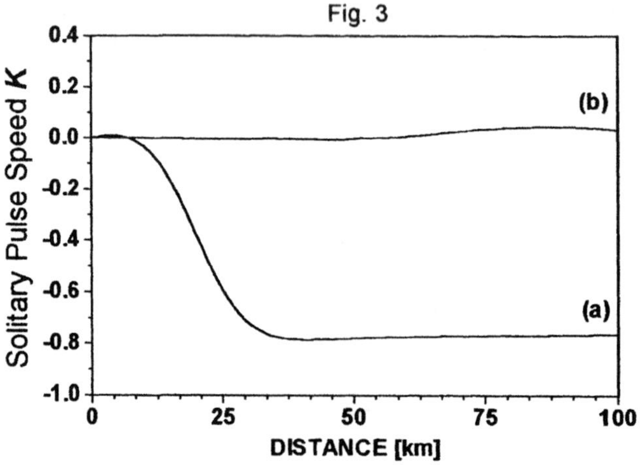

Figure 3.: **The evolution of the solitary pulse speed *k* for the same cases as in fig. 2 .**

6. Conclusion.

The present paper studies the problem of signal stabilisation in high-speed, long-distance repeaterless soliton based optical fiber communication lines. A new device is proposed - an exponentially longitudinally gradient distributed fiber amplifier, in which the concentration of the rare - earth dopant gradually increases in the direction of the pump power decrement along the nearly constant gain coefficient and hence optimal stabilising conditions can be realised in every point of the fiber length.

The computer models of EGFA's clearly show that 100 km long devices can be designed as a truly transparent transport media for data streams with repetition rate up to 100 GHz.

The values of the wavelengths, cross-section and other constants used in this paper correspond to erbium-doped graded amplifiers, but the same stabilisation principle can be applied to distributed fiber amplifier with other dopants (Nd^{3+},Pr^{3+}) as well.

Acknowledgements.

This work was supported by The Bulgarian Science Fund, grant F-476 and the "Young Scientists" Fund, grant MU-IS-3/94.

References.

1. Blow, K., Doran, N. , Wood, D. (1988), Suppression of the soliton self-frequency shift by bandwidth-limited amplification, *J. Opt. Soc. Am. B* 5 (6), 1301-1304
2. Nakazawa, M., Kubota, H., Kurokawa, K., Yamada, E. (1991) Femtosecond optical transmission over long distances using adiabatic trapping and soliton standardisation, *J. Opt. Soc. Am. B* 8 (9), 1811-1817
3. Desurvire, E. (1994), *Erbium-doped Fibre Amplifiers, principles and Applications*, John Wiley & Sons, New York
4. Liu, S., Wang, W. (1993) Complete compensation for the soliton self-frequency shift and third-order dispersion of a fiber, *Opt. Lett.* 18 (22), 1911-1912
5. Giles, C., Desurvire, E. (1991), Propagation of signal and noise in concatenated erbium-doped fiber optical amplifiers, *J. of Ligthwave Tech.* 9 (2), 147-154
6. Romagnoli, M., Locati, F., Matera, F., Settembre, M., Tamburrini, M., Wabnitz, S. (1992) Role of pump-induced dispersion on femtosecond soliton amplification in erbium-doped fibers, *Opt. Lett.* 17 (13), 923-925
7. Rudolph, W., (1984) Calculation of pulse shaping in saturable media with consideration of phase memory, *Opt. and Quant. Electr. 16* (5), 541-549
8. Agrawal, G. (1989), *Nonlinear Fiber Optics*, Academic Press Inc., San Diego, CA
9. Kodama, Y., Hasegawa, A. (1992), Generation of asymptotically stable optical solitons and suppression of the Gordon-Haus effect, *Opt. Lett.* 17 (1), 31-33

CONTROL OF LIGHTGUIDING IN H:LiTaO$_3$ AND H:LiNbO$_3$ THIN FILMS

C.C. ZILING[1], V.V. ATUCHIN[1], I. SAVATINOVA[2], S. TONCHEV[2], M.N. ARMENISE[3] and V.N. PASSARO[3]
[1]Institute of Semiconductor Physics, Novosibirsk, Russia
[2]Institute of Solid State Physics, Sofia, Bulgaria
[3]Politecnico di Bari, Italy

1. Introduction

Lithium niobate (LiNbO$_3$) and lithium tantalate (LiTaO$_3$) are important ferroelectrics for modern optoelectronic device applications. Optical waveguides formed in these materials by using proton exchange (PE) technology (replacement of Li with H at 200-250°C), have the advantage of an easy and fast low-temperature process and decreased susceptibility to photorefractive effects (optical damage). Despite of the successful development of the PE technology [1-3], these waveguides require further investigations.

In this work some new properties of PE LiTaO$_3$ (H:LiTaO$_3$) and PE LiNbO$_3$ (H:LiNbO$_3$) thin layers are described. Experiments have been performed with samples quenched directly to room ambient from their annealing temperatures as well as with samples subjected to slow cooling to ensure the attainment of equilibrium. The effective mode indices measured after the two regimes of cooling were applied may differ up to 0.0010 for red light and up to 0.0050 for green light. Moreover, the mode indices of H:LiNbO$_3$ show an additional photoinduced splitting at high-intensity lightguiding. The results may have important implications concerning the conditions under which annealing of these materials can be conducted.

M. Balkanski and N. Andreev (eds.),
Advanced Electronic Technologies and Systems Based on Low-Dimensional Quantum Devices, 277–280.

2. Experimental results

Planar optical waveguides were fabricated in a melt of benzoic or pyrophosphoric acid at 240°C for 8hrs. After the initial PE reaction, the samples were annealed in air at the desired temperature T_a for 1h (APE samples). At each T_a first a slow than a quick cooling was applied. Slow cooling was realized in an programmable oven at a rate of 1 deg/min which is supposed to lead to an steady (equilibrium) state ("e-sample"). At quick cooling the sample was taken directly out of the oven in air (quenched). Within 5-7 minutes it accepts the room temperature, remaining in the high-temperature phase ("q-sample"). After each annealing step the values of effective mode indices $n_{eff,m}$ were measured with an accuracy of about ±0.0002 using the two prisms-coupling technique.The profiles were reconstructed applying a standard inverse WKB method. The as-exchanged samples have a step-like index shape with an increase of the extraordinary index Δn_e about 0.02 for $H:LiTaO_3$ and 0.13 for $H:LiNbO_3$ (λ=633 nm).

An interesting effect has been observed in both types of waveguides: the values of the mode indices in q- samples, $\Delta n_{eff,m}(q)$, are higher than those of e-samples, $\Delta n_{eff,m}(e)$. Depending on the conditions of the initial PE reaction and the wavelength of the guided light, the difference $\delta n=\Delta n(q)-\Delta n(e)$ varies between 0.0010 (λ=633 and 647nm) and 0.0050 (λ=514.5 nm) estimated with an accuracy of ±0004. Altering quick and slow coolings the changes of n_{eff} are reversible. The variations in n_{eff} are accompanied by a change of the profile area. Typical (q)-(e) profiles of $H:LiTaO_3$ and $H:LiNbO_3$ are shown in Figs.1 and 2. The mode spectrum $n_{eff,m}(q)$ is not stable - in a period of 3 months the values of $n_{eff,m}(q)$ were reduced to those of $n_{eff,m}(e)$.

In the case of $H:LiNbO_e$ waveguides, the (q) and (e) phase states were found to be rather sensitive to the power of the guided light P_g. The dependence of the out-coupled intensity measured as a function of the in-coupled intensity demonstrated a deviation from linearity at a level of P_{in}=50mW or P_g≈7 $\mu W/\mu m^2$ (λ=647nm). The loss of linearity is usually considered as a threshold of the photorefractive effect (optical damage).

Photoinduced (q)-(e) experiments were performed with a homogeneous z-cut $H:LiNbO_3$ waveguide made for 8h at 240°C and using a suitable APE procedure. Fig.2 shows the behaviour of the (q) and (e) waveguide profiles obtained below and above the photorefractive threshold for red light. At P_{in}=90μW, (≈0.1$\mu W/\mu m^2$) the distance between the plateaux of the corresponding (q)-(e) profiles determines an evarage value of $\delta n=\Delta n(q)-\Delta n(e)\approx 0.0010$. However at P_{in}=360mW (≈50 $\mu W/\mu m^2$) the (q)-(e) splitting is much higher. The plateaux of the corresponding (q)-(e) profiles are shifted from their low-power positions almost symmetrically in opposite directions by a value of about ±0.0010. So the total (q)-(e) splitting being $\delta n\approx 0.0030$.

The preliminary experiments with the green line of an Ar laser (λ=514.5nm) demonstrated an increased photorefractive sensitivity: at P_{in}=320mW $\delta n\approx 0.0110$.

3. Conclusions

At present, we have difficulties to explain the (q)-(e) effect in $H:LiTaO_3$ and $H:LiNbO_3$ waveguides. In principle, such changes in the refractive index may occur from bild-up or relaxation of stresses (strains), phase transitions, or small changes in atomic positions leading to change in the macroscopic polarizability. The low temperature at which the (q)-(e) effect is observed and the magnitude of the induced changes are an evidence that the changes are connected with the hydrogen network in the exchanged layers.

Acknowledgments
This work was supported by the Bulgarian Foundation of Science (Contract no. F-419) and the Commission of the European Communities (Contract no. COP959).

References

1. Ziling C.C., Atuchin V.V., Savatinova I., and Kuneva M., (1992) Intern J. Optoelectr. **7**, 519-532.
2. Yuhara T., Tada K., and Li Yu.S., (1992) J. Appl. Phys., **71(8)**, 3966-3979.
3. Ahlfeldt H., Webjorn J., F. Laurell F., and G. Arvidsson G., (1994) J. Appl. Phys. **75(2)**, 717-727.

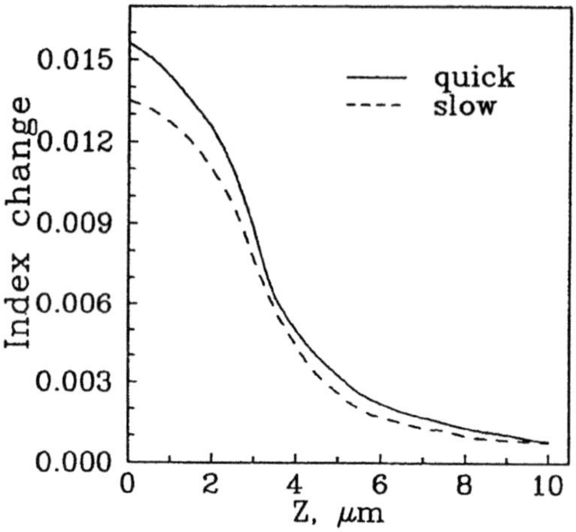

Figure 1. Profiles of a z-cut H:$LiTaO_3$ sample (λ=647nm) formed at 240°C for 8h; APE procedure: 400°C, 2h.

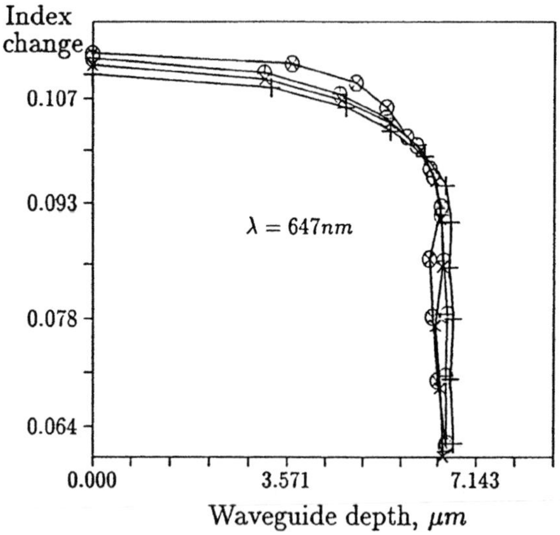

Figure 2. Profiles of a (q) and (e) type H:$LiNbO_3$ sample at different power of red guided light: below (⊕⊕⊕,xxx) and above (⊗⊗⊗, +++) the photorefractive threshold

SELF-PHASE MODULATION DUE TO THIRD-ORDER CASCADING: APPLICATION TO ALL OPTICAL SWITCHING DEVICES

S. TANEV, K. KOYNOV*, S. SALTIEL*, K. XIE**, A. BOARDMAN**

Technical University of Sofia, Institute of Applied Physics,
8 Climent Ohridski Blvd., 1756 Sofia, Bulgaria
Fax:359 2 987 78 70 E-mail: sit@vmei.acad.bg

* *University of Sofia, Faculty of Physics, Quantum Electronics Department, 5 J. Bourchier Blvd., 1164 Sofia, Bulgaria*
Fax:359 2 9625 276 E-mail: ssv@phys.uni-sofia.bg

** *University of Salford, Department of Physics, Joule Laboratory, M5 4WT, UK*

It is considered theoretically the possibility of all optical switching on the base of third order cascading in centrosymmetric media. Due to cascaded third order processes the fundamental beams participating in the four wave mixing processes obtain an additional high order nonlinear phase shift. Our calculation shows that the contribution of this cascade type phase modulation is comparable with the direct $\chi^{(3)}$ contribution and that such "cascadable" nonlinear media can be used for construction of all-optical switching devices.

Our calculations were applied to the case of third harmonic generation (THG) in media with center of inversion. Reduced stationary amplitude equations have been solved by two approaches:

a) analytically with fixed intensity approximation valid for low power input beams. In contrast to our first preliminary results published in [1] here we consider also the possibility of non zero third harmonic signal at the input of the nonlinear media;

b) numerically - approach that allowed us to evaluate the characteristics of the cascade type phase shift at arbitrary power for the input beams.

The process of third harmonic generation in a nonlinear media with length L and transparent at both the fundamental and the generated wavelength is described by the following slowly varying envelope equations valid for plane, linearly polarized waves:

M. Balkanski and N. Andreev (eds.),
Advanced Electronic Technologies and Systems Based on Low-Dimensional Quantum Devices, 281–287.

$$\frac{dA_1}{dz} = i\left(\gamma_1|A_1|^2 + \gamma_2|A_3|^2 + \delta|A_1|^4\right)A_1 + i\gamma_{13}A_1^{*2}A_3\exp(i\Delta kz) \tag{1a}$$

$$\frac{dA_3}{dz} = i\gamma_4|A_1|^2 A_3 + i\gamma_{31}A_1^3\exp(-i\Delta kz), \tag{1b}$$

where the γ and δ are the nonlinear coupling coefficients as defined in [1], the wavevector mismatch is $\Delta k = k_3 - 3k_1$ and A_j (j=1;3) incorporate both real amplitudes a_j and the phases φ_j:

$$A_j(z) = a_j(z)\exp[i\varphi_j(z)]$$

The input values for the phases and the amplitudes of the two waves will be denoted by φ_{j0} and a_{j0}. The system (1) was solved analytically by fixed intensity approximation [2] that was previously used by us to study analytically the effect of self-phase modulation due to second order cascading in non-centrosymmetric media [3,4]. This approximation suggests no depletion for the intensity of the fundamental wave and possible changes for their phases [2]. In fact this approximation is valid for low conversion coefficients (η<0.1) into third harmonic wave.

Following the same approach as in [1,3,4] we obtained for the third harmonic amplitude and phase:

$$a_3(z) = \left[(a_{30}\cos\Lambda z - p_2\sin\Lambda z)^2 + p_1^2\sin^2\Lambda z\right]^{1/2}, \tag{2}$$

$$\varphi_3(z) = \varphi_{30} - \frac{\Delta}{2}z + \operatorname{arctg}\frac{p_1\sin\Lambda z}{a_{30}\cos\Lambda z - p_2\sin\Lambda z}, \tag{3}$$

where

$$\Delta = \Delta k - (\gamma_4 + 3\gamma_1)a_1^2 - 3\gamma_2 a_{30}^2,$$

$$S = \left[\Delta k\gamma_4 + 3(\gamma_{13}\gamma_{31} - \gamma_1\gamma_4)a_1^2 - 3\gamma_2\gamma_4 a_{30}^2\right]a_1^2,$$

$$\Lambda^2 = \Delta^2/4 + S,$$

$$p_1 = \frac{\left(\Delta/2 + \gamma_4 a_1^2\right)a_{30} + \gamma_{31}a_1^3\cos\Delta\varphi_0}{\Lambda}, \quad p_2 = \frac{\gamma_{31}a_1^3\sin\Delta\varphi_0}{\Lambda}$$

The value of $\Delta\varphi_0 = 3\varphi_{10} - \varphi_{30}$ determines the role of the input phase of the cascaded wave. Using (2) and (3) we obtain following result for the nonlinear phase shift $\Delta\varphi_1^{NL} = \varphi_1 - \varphi_{10}$ of the fundamental beam at the output of the media:

$$\Delta\varphi_1^{NL} = \Delta\varphi_{1,dir}^{NL} + \Delta\varphi_{1,casc}^{NL} + \frac{\Delta k}{2}\left(\frac{a_{30}}{a_{10}}\right)^2 L \tag{4}$$

$$\Delta\varphi_{1,dir}^{NL} = \left[\gamma_1 a_{10}^2 + \delta a_{10}^4 + \frac{1}{2}(\gamma_4 - 3\gamma_1)a_{30}^2 + \gamma_{31}a_{10}a_{30}\cos\Delta\varphi_0\right]L \tag{5}$$

$$\Delta\varphi_{1,casc}^{NL} = \left[W_1\Lambda L + \frac{1}{2}W_2\sin(2\Lambda L) - a_{30}p_2(1-\cos(2\Lambda L))\right]\frac{\left[(\gamma_2-\gamma_4)a_{10}^2 - \Delta/2\right]}{2\Lambda a_{10}^2} \quad (6)$$

where: $W_1 = a_{30}^2 + p_1^2 + p_2^2$ and $W_1 = a_{30}^2 - p_1^2 - p_2^2$.

For the case of zero input third harmonic signal [1] the expressions (5) and (6) reduce to:

$$\Delta\varphi_{1,dir}^{NL} = \left[\gamma_1 a_{10}^2 + \delta a_{10}^4\right]L \quad (7)$$

$$\Delta\varphi_{1,casc}^{NL} = -\gamma_{13}^2\left[\frac{\Delta}{2} - (\gamma_2-\gamma_4)a_{10}^2\right]\frac{a_{10}^4 L}{2\Lambda^2}\left[1 - \mathrm{sinc}(2\Lambda L\right] \quad (8)$$

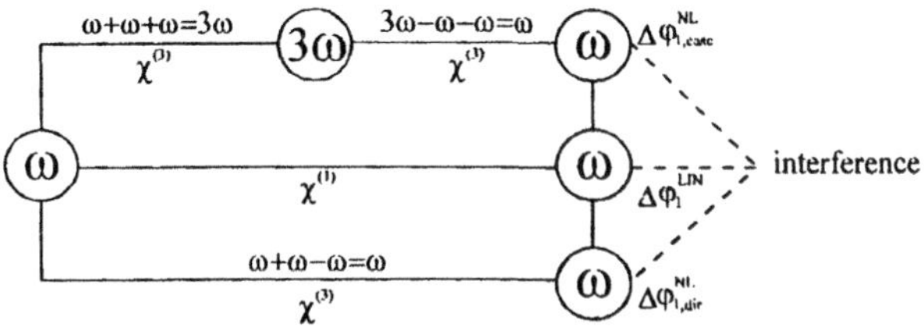

Figure 1. Schematic drawing for illustration of the physical explanation of the process of formation of nonlinear phase shift in nearly phase matched third harmonic media in the case of zero input third harmonic signal. The upper channel represents the generation of the wave by two step third order processes. The lower channel - the generation of the wave by degenerate four wave mixing. The middle channel represents the linear shift of the input wave.

The first term $\Delta\varphi_{1,dir}^{NL}$ in (4) is a result of direct, one-step nonlinear processes as self- and cross- phase modulation described by the natural $\chi^{(3)}$ and $\chi^{(5)}$ susceptibilities of the media. The second term $\Delta\varphi_{1,casc}^{NL}$ in (4) is a result of two step third order process ($\chi^{(3)}:\chi^{(3)}$ cascading). The physical explanation of the process of the nonlinear phase shift due to the cascaded third order processes is based on the interference of the three waves: (i) the linearly shifted fundamental wave, (ii) the wave generated as a result of a single step third order processes - degenerate and non degenerate four wave mixing processes and (iii) the wave generated as a result of the two step non-degenerate four wave mixing interactions $\omega = 3\omega - \omega - \omega$. The resulted from this interference wave is phase shifted with respect to the free running fundamental wave. On figure 1 is shown a schematic drawing of the interference process in the case of zero input third harmonic signal ($a_{30} = 0$). In fact the phase shift due to cascaded third order processes (the upper channel in fig. 1) is a result of effective fifth-order process with $\chi_{eff}^{(5)} \propto \left(\chi^{(3)}\right)^2$. That is why the nonlinear phase shift with cascade origin should be called high order nonlinear phase shift. As it is estimated in [1] the contribution of $\chi_{eff}^{(5)}$ can exceed the contribution of the intrinsic $\chi_{int}^{(5)}$ of the media.

Figure 2 illustrates the dependence of the cascaded part $\Delta\varphi_{1,casc}^{NL}$ of the nonlinear phase shift as a function of the normalized phase mismatch ΔkL for different values of

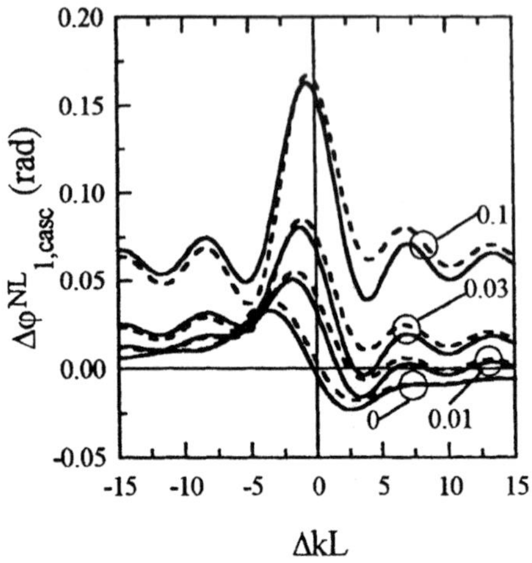

Figure 2. High order nonlinear phase shift (due to cascading) as a function of the normalized phase mismatch ΔkL. Normalized input intensity at fundamental wavelength is $\gamma a_{10}^2 L = 0.3$, input phase difference is $\Delta\varphi_0 = 0$. The parameter is the normalized value of the seeded wave intensity $(a_{30} / a_{10})^2$.

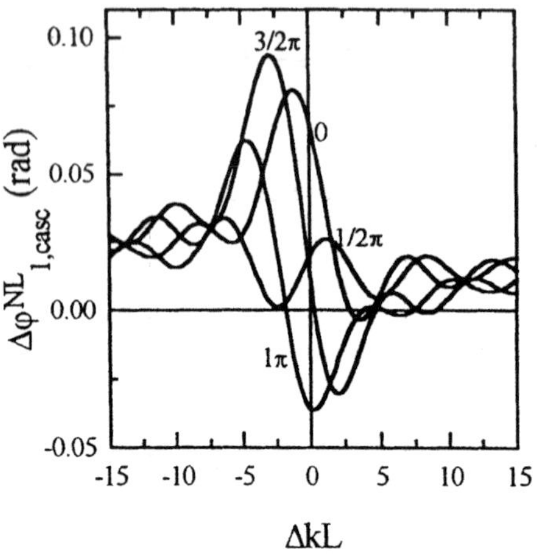

Figure 3. High order nonlinear phase shift (due to cascading) as a function of the normalized phase mismatch ΔkL. Normalized input intensities are $\gamma a_{10}^2 L = 0.3$ and $(a_{30} / a_{10})^2 = 0.03$. The parameter is the input phase difference $\Delta\varphi_0 = 3\varphi_{10} - \varphi_{30}$.

the seeded third harmonic signal $(a_{30} / a_{10})^2$ and normalized input intensity $\gamma a_{10}^2 L = 0.3$. We accept here the following equalities [5]: $\gamma_{31} = \gamma_{13} = \gamma$ and $\gamma_4 \approx 3\gamma_2$. In our calculations we took that $\gamma_1 = \gamma_{13}$ and $\gamma_4 = 6\gamma_{13}$. The phase shifts shown on this curves are calculated with two approaches: exact numerical (solid line) and the fixed intensity approximation (dashed line). It is seen that fixed intensity approximation can be used for description of this process at this levels of input intensities. For lower levels of input intensities the agreement between the numerical approach and the analytical formula is improving; for higher levels of input intensities - the numerical approach have to be used.

On figure 3 is shown the role of the relative input phase of the seeded third harmonic signal on the cascaded part of the nonlinear phase shift. The normalized input intensities used in this calculation were $\gamma a_{10}^2 L = 0.3$ and $(a_{30} / a_{10})^2 = 0.03$. We see that the phase of the seeding can be used for control of the output phase of the fundamental wave. In contrast to the case of zero seeding where the nonlinear phase shift is very small at exact phase-matching condition [1], here by adjusting of the seeding phase one can obtain the maximal (positive or negative) value of the nonlinear phase shift at the exact phase-matching condition $\Delta kL = 0$.

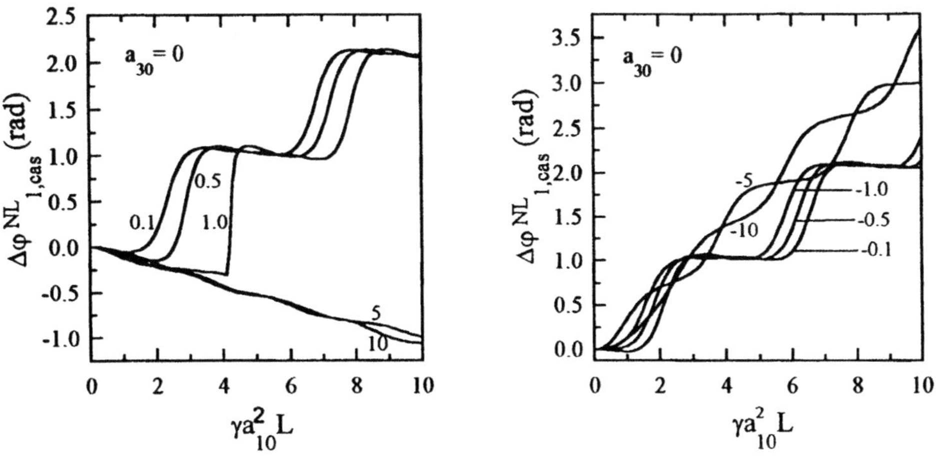

Figure 4. Calculated by the numerical approach cascaded part of the nonlinear phase shift as a function of the normalized pump intensity $\gamma a_{10}^2 L$ for positive (left figure) and negative (right figure) values of the normalized phase mismatch ΔkL. No seeding wave is assumed at the input of the nonlinear media.

The dependence of the cascaded part of the nonlinear phase shift on the normalized input intensity of the input fundamental wave is shown on figure 4 for positive and negative values of the normalized phase mismatch ΔkL and zero seeding. It is seen from figure 4 that the cascaded part of the nonlinear phase shift is 2 rad for normalized intensity $\gamma a_{10}^2 L = 6$, what means that the $\Delta\varphi_{1,casc}^{NL}$ is 30% of the value of $\Delta\varphi_{1,dir}^{NL} = \gamma a_{10}^2 L$.

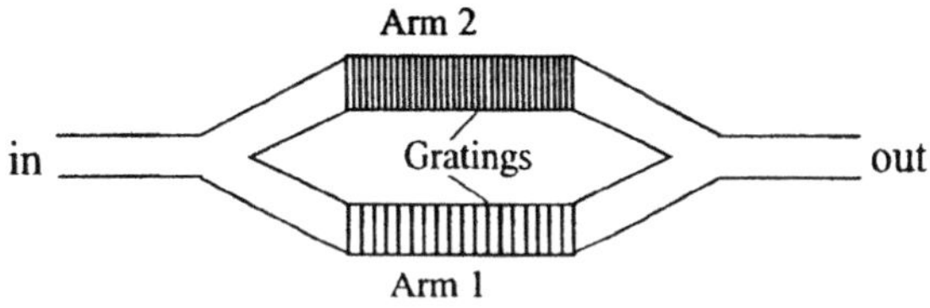

Figure 5. Schematic drawing of the proposed integrated Mach Zehnder interferometer for observation of all optical switching as a result of high order, cascade origin nonlinear phase shift.

As an application example we suggest construction of integrated Mach Zehnder interferometer similar to that proposed in [6] but with two third harmonic generation media in both arms figure 5. Each arm of the interferometer have to be designed in such a way so the third harmonic generation process to have phase mismatch that will results in a different sign for the phase shift arising from the cascaded third order nonlinear effects in both arms. Appropriate technique to achieve phase matching condition is so called Quasi Phase Matching (QPM) method [7,8].

The interferometer can be constructed in two ways. In the first one both arms are with one and the same nonlinear media. In this case the switching of the interferometer will be due to the nonlinear phase shift with cascaded origin. The interferometer will switch when the difference $\Delta\varphi^{NL}_{arm1} - \Delta\varphi^{NL}_{arm2}$ between the collected in both arms high order nonlinear phase shifts with cascade origin reach the value of π. On figure 6 is shown the dependence of the phase difference $\Delta\varphi^{NL}_{arm1} - \Delta\varphi^{NL}_{arm2}$ as a function of the normalized intensity at the input of the interferometer. The calculations are done by the numerical approach with assumption of zero seeding and 50/50 splitting of the input intensity.

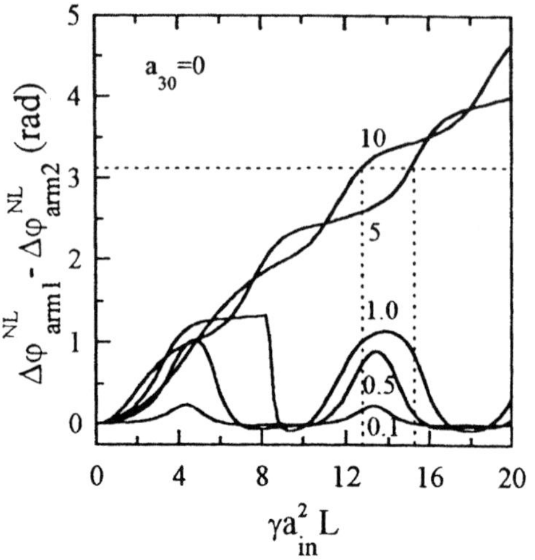

Figure 6. Two arms phase difference as a function of the normalized intensity at the input of the interferometer. The phase mismatches ΔkL in both arms are equal in amplitude and opposite in sign. The parameter shown on the figure is the absolute value of the normalized phase mismatch ΔkL.

In the second option in both arms can be used media with different signs of their n_2 coefficient. In this case the sum of both the direct and the cascade type phase shifts contributes to the switching, reducing in this way the switching power.

The estimations show that if in the both arms is used 1 cm long AlGaAs and the phase-mismatch is respectively -5 and $+5$ rad then the switching power that correspond to $\gamma a^2_{in}L = 13$ will be less than 100 MW/cm^2. With the recently discovered new nonlinear media polydiacetylene *para*-toluene sulfonate (PTS) [9], and again assuming L = 1 cm, the switching power will be 120 KW/cm^2!

In conclusion, we report for the first theoretical investigation of the third-order non-linear phase shift due to two step third order processes in centrosymmetric media at high pump power. The nonlinear phase shift due to cascading can be 30-40% of the value of the nonlinear phase shift due to intrinsic $\chi^{(3)}$ of the media. Its sign can be controlled by the sign of the phase mismatch of the third harmonic generation process. In the presence of seeding the high-order phase shift can be achieved at exact phase matching condition.

Acknowledgments

We would like to express our acknowledgements for the support by the Bulgarian Science Foundation (contract MU F-01/96).

References

1. Saltiel, S., Tanev, S. and Boardman, A. D. (accepted) High order nonlinear phase shift due to cascaded third-order processes, *Optics Letters*

2. Tagiev, Z., Chirkin, A. (1977) Fixed intensity approximation in nonlinear wave theory, *ZETPh* **73**, 1271-1281,
3. Saltiel, S., Koynov, K., Buchvarov, I. (1995) Analytical and numerical investigation of opto-optical phase modulation based on coupled second order nonlinear processes, *Bulg. J. Phys.* **22**, 39 - 47.
4. Saltiel, S., Koynov, K., Buchvarov, I. (1996) Analytical formulae for optimization of the process of low power phase modulation in a quadratic nonlinear medium, *Appl. Phys. B* **62**, 39 - 42.
5. Midwinter, J. E. and Warner J. (1965) The effect of phase matching method and crystal symmetry on the polar dependance of third-order non-linear polarization, *Brit. J. Appl. Phys.* **16**, 1667 - 1674.
6. Ironside, C. N., Aitchison, J. S., Arnold, J. M. (1993) An all-optical switch employing the cascaded second order nonlinear effects, *IEEE JQE* **25**, 2650-2654.
7. Fejer, M. M., Magel, G. A., Jundt, D. H. and Byer, R. L. (1992) Quasi-Phase-Matched Second Harmonic Generation: Tuning and Tolerances, *IEEE J. QE..* **28**, 2631 -2636.
8. Williams, D., West, D. and King, T. (1996) Quasi-phasematched third harmonic generation in doped sol-gel derived multilayer stacks, *Technical Digest of the European Quantum Electronics Conference (8-13 sept. 1996, Hamburg, Germany),* paper QWK4.
9. Wright, E. M., Lawrence, B. L., Torruellas W. and Stegeman, G. (1995) Stable self-trapping and ring formation in polydiacetylene *para*-toluene sulfonate, *Opt. Lett.* **20**, 2481-2483.

INDEX

GPSR Compliance

The European Union's (EU) General Product Safety Regulation (GPSR) is a set of rules that requires consumer products to be safe and our obligations to ensure this.

If you have any concerns about our products, you can contact us on ProductSafety@springernature.com

In case Publisher is established outside the EU, the EU authorized representative is:

Springer Nature Customer Service Center GmbH
Europaplatz 3
69115 Heidelberg, Germany

Batch number: 10755284

Printed by Printforce, the Netherlands